Lecture Notes in Economics and Mathematical Systems 455

Springer
Berlin
Heidelberg
New York
Barcelona
Budapest
Hong Kong
London
Milan
Paris
Santa Clara
Singapore
Tokyo

Rafael Caballero   Francisco Ruiz
Ralph E. Steuer (Eds.)

Advances in Multiple Objective and Goal Programming

Proceedings of the Second International Conference on Multi-Objective Programming and Goal Programming, Torremolinos, Spain, May 16–18, 1996

Springer

Editors

Prof. Rafael Caballero
Prof. Francisco Ruiz
University of Málaga
Departamento de Economía
Aplicada (Matemáticas)
Facultad de Ciencias Económicas
y Empresariales
Campus El Ejido s/n
29071 Málaga, Spain

Prof. Ralph Steuer
University of Georgia
Faculty of Management Science
206 Brooks Hall
Athens, Georgia 30602-6255, USA

Cataloging-in-Publication Data applied for

Die Deutsche Bibliothek - CIP-Einheitsaufnahme

Advances in multiple objektive and goal programming : proceedings of the Second International Conference on Multi-Objective Programming and Goal Programming, Torremolinos, Spain, May 16 - 18, 1996 / Rafael Caballero ... (ed.). - Berlin ; Heidelberg ; New York ; Barcelona ; Budapest ; Hong Kong ; London ; Milan ; Paris ; Santa Clara ; Singapore ; Tokyo : Springer, 1997
(Lecture notes in economics and mathematical systems ; 455)
ISBN-13: 978-3-540-63599-4 e-ISBN-13: 978-3-642-46854-4
DOI: 10.1007/978-3-642-46854-4

ISSN 0075-8442
ISBN-13: 978-3-540-63599-4

Typesetting: Camera ready by author

SPIN: 10548050 42/3142-543210 - Printed on acid-free paper

Contents

Introduction

Introduction 1
R. Caballero, F. Ruiz, R.E. Steuer

Abraham Charnes Distinguished Lecture

On the Education of a Multi-Criteria Researcher: A Personal View 5
S. I. Gass

Part 1: Goal Programming

An Example of Good Modelling Practice in Goal Programming: Means for Overcoming Incommensurability 29
M. Tamiz, D. Jones

Pattern Classification by Linear Goal Programming and its Applications 38
H. Nakayama, N. Kagaku

Goal Programming and Heuristic Search 48
L. Mandow, E. Millán

Goal Programming Model for Evaluating Hospital Service Performance 57
M.M. Arenas, E. Lafuente, M.V. Rodríguez Uría

An Application of Interactive Multiple Goal Programming on the Warsaw Stock Exchange 66
C. Dominiak

Methodology for an Analysis of Decision Criteria: An Application to Irrigated Farms in Southern Spain 75
A. Rodríguez-Ocaña, J. Berbel, P. Ruiz-Avilés

Socioeconomic Impact Evaluation of the Drought in Irrigated Lands in Southern Spain: A Multicriteria Decision Making Approach 84
J.A. Gómez-Limón, P. Lara, A. Rodríguez, F.J. Sánchez

Goal Programming in Distribution System Design 93
E. Konarzewska-Gubala, A. Zajaczkowski

Distance Games and Goal Programming Models of Voting Behaviour 102
F. Turnovec

Dynamic Goal Programming Models 111
T. Trzaskalik

Estimation Through the Imprecise Goal Programming Model 120
B. Aouni, O. Kettani, J.-M. Martel

Part 2: Multiple Objective Theory

Characteristics of the Efficient Solutions of Bicriteria and Tricriteria Network Flow Problems 131
A. Mustafa, M. Goh

Non-connected Efficiency Graphs in Multiple Criteria Combinatorial Optimization 140
M. Ehrgott, K. Klamroth

An Augmented Lagrangian Scalarization for Multiple Objective Programming 151
M.L. TenHuisen, M.M. Wiecek

Dominance and Efficiency in Multiobjective Stochastic Linear Programming 160
F. Ben Abdelaziz, P. Lang, R. Nadeau

A Multi-Criteria Decision Making Problem Associated to Preference Modelling 170
J. González-Pachón, M.I. Rodríguez-Galiano

On Optimality and Duality in Multiobjective Nonsmooth Programming 178
V. Preda, I.M. Stancu-Minasian

Sensitivity in Multiobjective Programming by Differential Equations Methods. The Case of Homogeneous Functions 188
A. Balbás, P. Jiménez Guerra

On Pseudo-Boolean Multicriteria Optimization Problems with Incomplete Information 197
V. Donskoy, A. Perekhod

Properties of Efficient Points Sets and Related Topics 201
V. Postolică

Existence Theorems for Cone Saddle Points and Vector-Valued Minimax Theorems 210
T. Tanaka

A New Approach to Second Order Optimality Conditions in Vector Optimization 219
A. Cambini, L. Martein, R. Cambini

Invex and Pseudoinvex Functions in Multiobjective Programming.. 228
R. Osuna Gómez, A. Beato-Moreno, P. Luque Calvo, A. Rufián-Lizana

Vector-Values Risk in Multicriteria Problems 235
M. Salukvadze, A. E. Bardin

Utopian Efficient Strategies in Multicriteria Matrix Games 245
F.R. Fernández, A. Mármol, L. Monroy, J. Puerto

About the Existence of Nash-Slater Equilibrium for a Non-Cooperative Game under Uncertainty 255
M. Larbani

Part 3: Implementation and Practice of Multiple Objective Programming

An Algorithmic Package for the Resolution of Dynamic Multiobjective Problems 265
R. Caballero, T. Gómez, M. González, L. Rey, F. Ruiz

Closest Solutions in Ideal-Point Methods 274
E. Carrizosa, E. Conde, A. Pascual, D. Romero-Morales

Methods of Multicriteria Decision Support Based on Reference Sets 282
A. Skulimowski

A Tabu Search Procedure to Solve Multiobjective Combinatorial Optimization Problem 291
X. Gandibleux, N. Mezdaoui, A. Fréville

Linking Production Theory and Multiobjective Fractional Programming as a Support Tool for Animal Diet Formulation 301
P. Lara

Multiobjective Energy-Environmental-Economy Model for Israel (ENMIS) 310
D. Soloveitchik

BayRes: A System for Stochastic Multiobjective Reservoir Operations 319
D. Ríos-Insúa, C. Bielza, J. Martín, K. Salewicz

Timber Harvesting Scheduling Problems: A Compromise Programming Approach 328
L. Díaz-Balteiro, C. Romero

Location of Semiobnoxious Facility: A Biobjective Approach 338
E. Carrizosa, E. Conde, D. Romero-Morales

A Hierarchical Location Model: Biobjective Programming vs FDH 347
M. Alminyana, F. Borrás, J.T. Pastor

Efficient Frontier Derived from Opinions and the Utility Optimization Problem 356
E. Ballestero

Finding Efficient Points in Multiobjective Quadratic Programming with Strictly Convex Objective Functions 365
A. Beato-Moreno, P. Luque-Calvo, R. Osuna-Gómez, A. Rufián-Lizana

Multiobjective Programming with a Concave Vector Value Function 374
A. Mateos, S. Ríos-Insúa

Modelling Best-Practice Frontiers When There Are Multiple Outputs 383
C. Tofallis

Introduction

This Volume contains 41 papers and comprises the refereed proceedings of the 2nd International Conference on Multi-Objective Programming and Goal Programming (MOPGP'96) that was held at the Hotel Meliá Costa del Sol, Torremolinos (15 km from Málaga), Spain, May 16-18, 1996.

The Conference was attended by 86 people from 22 countries and 63 papers were presented. The Conference was the Second in the series of MOPGP conferences devoted to optimization topics in multiple criteria decision making (MCDM), particularly multiple objective programming and goal programming. The First conference, MOPGP'94, organized by Mehrdad Tamiz, was held at the University of Portsmouth, Portsmouth, England, June 1-3, 1994. The proceedings volume for that conference was also published by Springer-Verlag. Its citation is: Mehrdad Tamiz (Ed.) (1996). "Multi-Objective Programming and Goal Programming: Theories and Applications," Lecture Notes in Economics and Mathematical Systems, Vol. 432, Springer-Verlag, Berlin, 359 pp. As of this writing, MOPGP'98, the Third conference in the series, is scheduled for the University of Laval, Quebec City, Canada, May 31-June 2, 1998.

MOPGP'96 was organized by the Departamento de Economía Aplicada (Matemáticas) of the University of Málaga, and chaired by Rafael Caballero. Other members of the Local Organizing Committee were Pablo Lara, University of Córdoba, and José M. Cabello, Mercedes González, Lourdes Rey and Francisco Ruiz, all also from the Departamento de Economía Aplicada (Matemáticas), University of Málaga.

The first paper of this Volume, entitled "On the Education of a Multi-Criteria Researcher: A Personal View" by Saul I. Gass, was the Conference's keynote address. In the MOPGP conference series, the keynote address is designated the Abraham Charnes Distinguished Lecture in memory of the many outstanding scientific contributions of Abraham Charnes (1917-1992) and their influence on the current practice of MOPGP. The remaining 40 papers in the Volume have been divided into three parts.

In Part 1 are eleven papers covering a range of interesting and innovative applications in the state-of-the-art use of goal programming. In Part 2 are fifteen papers directed at theoretical topics in multiple objective programming ranging from new ideas in combinatorial optimization, scalarizing functions, preference modelling, duality, and matrix games, to properties of efficient points. In Part 3 are fourteen papers on the implementation and practice of multiple objective programming covering heuristic searches, decision support systems, interactive algorithms, and applications.

MOPGP'96 was made possible by the Spanish Ministry of Education and Sciences, by resolution of the 'Secretaría de Estado de Universidades e Investigación' on May 9, 1996, grant No. CO94-0696. In addition to the University of Málaga, other institutions providing support to the Conference were the University of Córdoba, the Town Council of Torremolinos, the Town Council of Málaga, and the College of Economists of Málaga.

Those in attendance at MOPGP'96 were: Marcos Almiñana, Belaïd Aouni, Alejandro Balbas, Enrique Ballestero, Antonio Beato Moreno, Foued Ben Abdelaziz, Julio Berbel, Concepción Bielza, Rafael Blanquero Bravo, Willem Brauers, Rafael Caballero, José M. Cabello, Susana Calderón, Alberto Cambini, Analía Cano, Rafael Canero, Emilio Carrizosa, Emilio Cerdá, Eduardo Conde, Luis Diaz Balteiro, Cezary Dominiak, Matthias Ehrgott, Tomas Gal, Teodoro Galache, Xavier Gandibleux, Saul I. Gass, Trinidad Gómez, José Antonio Gómez-Limón, Mercedes González, Alfonso González Pareja, Jacinto González-Pachón, Henryk Gorecki, John Bowen Hughes, James P. Ignizio, Andrzej Jaszkiewicz, Pedro Jiménez Guerra, Dylan Jones, Valery Irikov, Ossama Kettani, Ewa Konarzewska-Gubala, Pekka Korhonen, and Pablo Lara.

Moussa Larbani, Margarita López de Pablo, Mariano Luque, Pedro Luis Luque Calvo, Lorenzo Mandow, Simon Mardle, Amparo M. Mármol, Laura Martein, Jean-Marc Martel, Francisca de Miguel, Eva Millán, S. K. Mir-Razavi, Luisa Monroy, María del Mar Muñoz, Adli Mustafa, Hirotaka Nakayama, Rafaela Osuna Gómez, Jesús Pastor, Anna Perekhod, Vasile Postolicã, Lourdes Rey, Mark A. Ridgley, Alfonso Mateos, María Victoria Rodríguez-Uría, Carlos Romero, Dolores Romero, Francisco Ruiz, Antonino Scarelli, Pascual Ruiz Canales, Mindia Salukvadze, Andrzej Skulimowski, David Soloveitchik, Ioan Stancu-Minasian, Ralph E. Steuer, Mehrdad Tamiz, Christiane Tammer, Tamaki Tanaka, Jorgen Tind, Christopher Tofallis, Angel Torrico, Tadeusz Trzaskalik, František Turnovec, Margaret Wiecek, and Ricardo Zamora.

Rafael Caballero, [1]
Francisco Ruiz, [1]
Ralph E. Steuer. [2]

1.- University of Málaga. Spain.
2.- University of Georgia. U.S.A.

Abraham Charnes
Distinguished Lecture

On the Education of a Multi-Criteria Researcher: A Personal View

The Abraham Charnesl Lecture
In Memory of Abraham Charnes

Saul I. Gass[1]

[1]College of Business and Management, University of Maryland
College Park, MD 20742 USA

Abstract

When the author first started working in the field of Operations Research and in the specific subfields of linear and mathematical programming, defined subject areas like MCDM, MOLP, GP, MAUT, and so on, did not exist. Unlike those early days, we now have a multitude of multi-criteria research monographs, texts, journals, and college courses. We also have a wide base of real-world applications and an ever expanding theoretical and computational framework for all aspects of MCDM. How did we get where we are today? The answer must, of course, deal with the work of Abraham Charnes. We owe much to his pioneering approach to solving unsolvable problems. In this paper, we review some aspects of the uncharted early days and Charnes' contributions, how the author became interested and involved in multi-criteria problems and research, and review some of the areas of multi-criteria research in which the author has been involved.

1 Introduction

It is an honor and a pleasure to have been asked to present this plenary paper in the memory of Abraham Charnes. I first met Abe over 40 years ago. He was a mathematician, a scientist and a friend whose research greatly influenced my own work and career, as well as the professional careers of all who work on multi-criteria problems .

problem. These models were proposed under the assumption that computers would eventually be available and new mathematical techniques like the simplex method would work. We must recognize the difficult methodological, data and computational challenges faced by the Project SCOOP staff. Their accomplishments seem common place now. Their hopes are captured in the following quotes from 1949 and 1953. They may seem somewhat quaint and naive today, but they are quite prescient:

"To compute programs rapidly with such a mathematical model, it is proposed that all necessary information and instructions be systematically classified and stored on magnetized tapes in the 'memory' of a large scale digital electronic computer. It will then be possible, we believe, through the use of mathematical techniques now being developed, to determine the program which will maximize the accomplishment of our objectives within those stated resource limitations," (Wood and Dantzig 1949).

"The work of the Planning Research Division with the three models (rectangular optimization model, square model of linear equations, triangular square model of linear equations) has given considerable impetus to the current interest in models of linear equations -- or 'linear models,' as the are coming to be known. Following a term contributed by the Division, this field is widely known among mathematicians as 'linear programming,' although 'activity analysis' is gaining favor. It is hard to say whether more attention is directed towards rectangular models or square models, but it is clear that many mathematicians view the rectangular model (the linear-programming model) as one with a great future. In a world where more and more attention is certain to be given to the efficient allocation of money and resources -- in various situations from the national level right down to the plant or process level -- the rectangular model is naturally exciting," (USAF 1953).

Although they were very insightful, the Project SCOOP staff had little information that things would work out mathematically and computationally. They took a chance when they decided to go with the linear programming model as the basis of Air Force planning. At this point in time, there was little consideration given to multi-criteria problems and methodologies for solving them.

2 The Simplex Method

Hand computations and some electronic aids helped to test the simplex method in the late 1940s and early 1950s. Stigler's diet problem, a (9 x 77) inequality system, took 120 person-days to solve on hand-operated desk calculators, (Dantzig 1982). A 40 origin and 60 destination transportation problem was solved by hand in nine person-days, and since only simple additions and subtractions are required, the process did not need to use desk calculators, (Dantzig 1949). For a (25 x 50) matrix, one simplex iteration would take about 4 hours on a desk calculator, 1 hour on an IBM 604 multiplier, and 25-30 minutes on an IBM-Card Programmed Calculator (Symonds 1955).

Was the simplex method the way to go? Orden describes three linear constraint problems: the linear inequality or linear equation linear program, the pure linear inequality problem (no optimization), and the zero-sum two-person game problem. Each problem can be converted to any of the others, and each problem type has a specific solution method associated with it: linear programming by the simplex method; inequality systems by relaxation; and zero-sum two-person games by fictitious play. In the early 1950's, it was unclear which method would be the best on a digital computer. A test run on the NBS SEAC computer produced the following typical results (Orden 1952a):

a. Simplex Method
- 10 x 10 zero-sum two-person game solved in 12 minutes
- 48 x 72 linear program solved in 20 hours
- A set of 28 simultaneous equations (converted to a linear-programming problem) solved in 2 1/4 hours.

b. Relaxation Method

- 32 x 16 set of linear inequalities solved in 14 minutes
- 6 x 6 zero-sum two-person game solved in 10 minutes

c. Fictitious Play Method

- 6 x 6 zero-sum two-person game solved in 10 minutes
- 10 x 10 zero-sum two-person game solved in 2 hours
- 3 x 4 linear programming problem solved in 10 minutes
- 48 x 72 linear-programming problem ran 8 hours and achieved accuracy to about 10%.

As we all know, the simplex method won out, but not without some help. A major advance in developing and maintaining the inverses required by the simplex method was Alex Orden's proposal to use the product form of the inverse (PFI). Orden, with reference to his Gauss-Jordan inversion work on the Whirlwind computer, stated:

"I noticed that it was neither necessary nor desirable to fully invert the matrix -- leaving it as the product form of the inverse cut the amount of computation in half. When I joined Project SCOOP, there was the simplex algorithm waiting for the same device," (Orden 1952a).

The PFI was used by Orchard-Hays on the Card-Programmed-Calculator (CPC) and in the design of his simplex code for the Rand Corporation's IBM 701 computer (Orchard-Hays 1954). Just about all present day methods of solving linear systems and linear-programming problems are based on the LU form of Gaussian elimination, augmented by sparse matrix techniques, with L being a product of lower-triangular matrices.

A simplex code for solving transportation problems was written for the SEAC in 1950, with a general simplex code developed in 1951, (Dantzig 1963). The latter code was used to successfully solve the first application on the SEAC, a U.S. Air Force programming problem dealing with the deployment and support of aircraft. This deployment model can be described as follows:

Given the D-Day availability a_0 of a specified type of combat aircraft, and the additional availabilities $a_1, a_2, \dots, a_{n-1}$ in the succeeding months, determine how to divide these availabilities between combat and training so as to maximize in some sense the sortie effort on one or more phases of the war, (Jacobs, 1951; Pollack 1952).

The system had 48 equations and 72 variables and was solved in 73 simplex iterations in 18 hours, with accuracy to five decimal places. The 18 hours includes the time to store and access data from the SEAC's new and novel magnetic tape system. The importance of this aircraft deployment problem is that it was one of the first problems to be solved under different objective functions, with possible objectives being maximize sorties; maximize early month sorties; maximize weighted sorties, with early months weighted more heavily; and simultaneously maximize sorties and minimize idle crews. The last objective leads to an early form of parametric programming.

The U.S. Air Force installed the second production unit of the UNIVAC in April 1952. The UNIVAC simplex code was written by the staff of the Air Force's Mathematical Computation Branch, under the direction of Emil Schell and could handle a (250 x 500) problem, (Schell 1953).

3 Charnes and the Early Days of Linear Programming

My first encounter with Abe Charnes was at a seminar he gave in 1953 or 1954 at the National Bureau of Standards, Washington, DC. Abe was doing linear programming research at Carnegie Institute under Air Force support for the study of intra-firm behavior. Although this work was not related to Project SCOOP, his important work with Cooper and Mellon on blending aviation gasolines was first reported at the Project SCOOP 1951 Symposium on Linear Inequalities and Programming (Orden and Goldstein 1952). In writing the first (1958) edition of my text, *Linear Programming*, I relied heavily on the mathematical discussion of linear programming given in Charnes, Cooper and Henderson, *An Introduction to Linear Programming*, John

Wiley, 1953. The material there, coupled with Dantzig's early papers on linear programming, provides the mathematical justification of linear programming and the simplex method. Charnes notes that W. W. Cooper introduced him to the subject.

My initial involvement in multi-criteria problems stemmed from the Air Force aircraft crew training problem. We looked at it as a production-inventory balancing problem: How many crews to train and graduate each period to meet demand for trained crews. Crews that were produced could meet the current demand or be put into a storage (waiting) pool. It is inefficient for a training facility to vary its production of crews from period to period, so its main objective is to have a "smooth" level of production. However, as the demand for trained crews fluctuates from time period to time period, smoothing the production usually produces a large cumulative inventory. Minimizing inventory is a competing objective to minimizing the fluctuations in production.

The situation is pictured in Figures 1 to 4 (Gass 1985). Figure 1 shows the fluctuating demand; Figure 2 shows that meeting the demand exactly yields a nonsmooth production pattern with a zero surplus (inventory); Figure 3 shows how a smooth production pattern will produce a rather large total monthly surplus; and Figure 4 shows a compromise solution in which the production has been somewhat smoothed, yielding a smaller total monthly surplus.

The mathematical structure for this problem is given by the following linear-programming model:

(P1) Minimize $a y_t + b s_t$
subject to
$x_t + s_{t-1} - s_t = d_t$
$x_t - x_{t-1} - y_t + z_t = 0$
$(x_t, s_t, y_t, z_t) \geq 0 \quad (a, b) > 0$

where
a = cost of a unit increase in production
b = cost of holding a unit in storage for one month
d_t = demand in month t
x_t = production in month t
s_t = storage at end of month t
y_t = increase in production from month t-1 to month t
z_y = decrease in production from month t-1 to month t
(Note that we are here concerned with just smoothing production increases.)

For = a/b, we give rise to the first instance of a parametric (multi-criteria) objective function:

$$\text{Min } y_t + s_t$$

The parametric version was posed by Walter Jacobs, who was Chief of a Project Scoop's Mathematical Formulation Branch. I was given the task to investigate how it could be solved it. This led to the parametric programming algorithm in which we balanced off the two competing linear objectives by the single parameter. We recognized that the scheme could be generalized to more objectives and described a multi-parameter approach in which two objectives meant one parameter; three objectives, two parameters and so on. We described this method for solving multi-objective linear programming in our early papers, but we failed to pursue it in terms of the parametric relationship between efficient solutions and the vector maximum problem, (Saaty and Gass 1954; Gass and Saaty 1955a, 1955b.)

I first heard of the vector maximum problem in a talk given by David Blackwell, a statistician who was at Howard University (in Washington, DC) and I was intrigued with the notion of multi-objective and multi-criteria optimization. It was clear that treated as a parametric problem (with linear constraints), we needed a method for finding and describing convex solution sets in parametric space that coincided to specific basic solutions that were optimal for ranges of the parameters. It really amounted to having an algorithm for efficiently solving inequality systems and being able to identify all the regions in parameter space that correspond to basic feasible solutions. To my knowledge, the vector maximum problem is first mentioned in the paper by Kuhn and Tucker "Nonlinear Programming" in which they developed the famous Kuhn-Tucker conditions (now Karush-Kuhn-Tucker conditions), (Kuhn and Tucker 1951). In their paper, Kuhn and Tucker note the relationship between T.C. Koopmans' efficient point in production activity analysis problems.

During this early period of mathematical programming--1950s and 1960s--there is little to be found with reference to multi-objective research and applications. The major event here is the goal programming (GP) innovation first presented in the paper "Optimal Estimation of Executive Compensation by Linear Programming," by Charnes, Cooper and Ferguson (1955). Other applications soon followed and GP became an important Operations Research modeling technique. Contributing to GP's acceptance were the books by Ijiri (1965), Lee (1972), and Ignizio (1976). It is interesting to note that Charnes and Cooper in their two volume *Management Models and Industrial Application of Linear Programming* (Charnes and Cooper 1955) spend only six pages on goal programming. In their view, GP is looked at as a way of determining a compromise (satisficing) solution to an infeasible linear programming problem (this is the way I still interpret GP).

The original GP problem posed by Charnes and Cooper (1955) is the following (Figure 5):

(P2) Max $z = x_1 + 0.5x_2$

st

(1) $3x_1 + 2x_2 \leq 12$

(2) $5x_1 \quad\quad \leq 10 \quad x_j \geq 0$

with the added constraints

(3) $x_1 + x_2 \geq 8$

(4) $-x_1 + x_2 \geq 4$

The added constraints produce an infeasible solution space.

In discussing the problem (P2), Charnes and Cooper ignored the original objective function (they did not assume some achievement level for z) and just considered the minimization of $|x_1 + x_2 - 8| + |-x_1 + x_2 - 4|$. For their objective function, Min $(y_3^- + y_4^+ + y_5^-)$, the optimal solution is $x_1 = 4/5$, x_2 24/5, $y_3^- = 12/5$, where y_3^- = the under-achievement variable for constraint (3), y_4^+ = the over-achievement variable for constraint (3), and y_5^- = the under-achievement variable for constraint (4). They discuss the first GP paper of Charnes, Cooper and Ferguson (1955) in a chapter separate from the goal programming section and do not call the reader's attention to the fact the problem formulation given in this paper is really one of goal programming. They looked at GP as an L_1 regression problem. It seems, at that time, as if they did not recognize the power of the GP formulation. I did not pay much attention to GP at first; I knew it mainly from the paper by Wagner (1975) that discussed the relationship between linear programming and L_1 and Chebychev regression. I included the L_1 discussion in the in 3rd edition of my linear-programming text, (1969), but did not include any GP material until the 4th edition (1975).

I really turned to GP in the mid 1970s when I became Director of Operations Research for Mathematica, a consulting organization in Washington. We did contract work mainly for the federal government. It seemed as if most problems encountered had goal-programming features--scheduling, allocation, planning, manpower planning, etc. Most of these problems could be structured as infeasible linear-programming problems and the GP formulation enabled planners to measure the difference between target goals and what could actually be accomplished.

My work on parametric programming and goal programming brought to the forefront the importance of multi-criteria decision analysis and the fact that we did not have many acceptable ways for solving such problems. My own formal training covered little that could be applied, including utility theory and game theory. In the 1950s and 1960s, utility theory was basically not known outside of the universities and had not been widely applied to real-world problems. The early OR texts hardly covered the topic: Churchman, Arnoff and Ackoff (1957) had a chapter on a cumbersome prioritization method, while Wagner (1969) did not mention multi-criteria problems.

4 Manpower Planning Models

My involvement in GP really began in earnest when I started to work on military manpower models. The first GP manpower applications were given in the book by Charnes, Cooper and Niehaus *Management Science Approaches to Manpower Planning*, Wiley (1969) in which they reported their seminal work done for the U. S. Navy. With caution, they noted

"... we propose to introduce a series of technical (mathematical) developments which will help to clarify some of the possibilities that are now present. This should be regarded, however, as only a beginning for such possibilities, It does not represent a final commitment (p 79)."

It is safe to say that all subsequent military manpower applications stemmed from their pioneering efforts.

The basic manpower policy planning problem can be stated as follows:

Given a work force at the beginning of the planning period, we want to determine the hiring, firing, promotion and training policies that should be used so that the ending work force best meets future work force requirements (forecasts).

Looked at from a goal programming perspective, the multi-year manpower problem can be stated as follows:

The total number of personnel in job type i from source j in time period t must meet a target goal for all manpower in time period t. Here, the goal is "soft" and the set of target constraints can be stated in terms of under- and over-achievement variables. Also considered are attrition, additions, budget constraints, and related conditions. The objective is to minimize a weighted sum of the achievement variables. Here the decision variables of the problem can be defined as:

X(g,s,y,t) = the number of individuals in grade g, with skill s, with years of service y in planning period t.

A combination of indices (g, s, y, t) is called a state and an individual can be in only one state at time period t, with the initial state conditions given by X(g,s,y,0). A typical constraint can be written (functionally) as

$$F[X(g,s,y,t)] + P(g,s,y,t) - N(g,s,y,t) = T(g,s,y.t)$$

where F[X(g,s,y,t)] is a function of decision variable, P(g,s,y,t) and N(g,s,y,t) are goal-programming under- and over-achievement deviation variables for the target constraint, and T(g,s,y.t) is the target for the function. Each P(g,s,y,t) and N(g,s,y,t) appears in the linear objective function multiplied by a weight WP(g,s,y,t) and WN(g,s,y,t), respectively.

One can usually formulate the conditions of a manpower model in terms of a standard goal programming model. But, quite often, the model takes on the form of a network with goal targets. The work of Glover and Klingman and their associates (see Glover, Klingman and Phillips, 1992) demonstrates how goal programming conditions can be incorporated into a network and still retain the important mathematical and computational features of linear networks. My paper given at first MOPGP conference discusses GP and networks in detail, Gass (1996).

My manpower research and applications are reported in a series of papers given in the references. In particular, the paper by Gass et al. (1988) demonstrates how GP and networks intersect from two perspectives: the network structure and need to determine weights for hundreds, if not thousands, of under- and over-achievement variables. This work highlighted the problem of determining objective function weights for goal-programming applications. For this situation, in which there are a multitude of weights, one needs a way of separating the deviation variables into classes and subclasses so that one can assign weights to the individual variables that is consistent and makes sense. This led to a proposal to use the Analytic Hierarchy Process (AHP) to set weights for the major goal attributes such as years and major manpower targets, and to develop priorities for these attributes that could then be translated into goal-programming weights, Gass (1986). The AHP, properly applied, is a recommended way of accomplishing the task. However, this important problem is still an open one, and its resolution probably depends on the specific application.

The weight problem will not go away. A related question is how does one trade-off between deviation variables when such variables represent different resources (constraints) and are measured in different units? Applying the Euclidean norm to each constraint is often suggested as the remedy. However, as I show in Gass (1987), one need not be concerned with such norming. The problem of weight determination and tradeoffs between deviation variables comes down to determining the objective function weights for the deviation variables.

5 The Interactive MOLP Problem with Partial Orders on the Variables

I have always had a strong interest in the MOLP (vector maximum problem with linear constraints). As we must deal with efficient (Pareto optimal) solutions, the computational problem is two-fold:

(1) how to enumerate the efficient solutions (or at least an acceptable sample), and

(2) how to select among the known set of efficient solutions.

Over the years, many techniques have been developed for enumerations and ideas such as interactive best-compromise procedures have been developed for selecting a specific solution (Steuer 1986). Often, some preferences were given with respect to the objectives, for example, minimize some measure of the deviation from goals or ideal solutions. According to Chankong and Haimes (1983), there are three types of preference information procedures:

- Global preference that uses utility functions, value functions, and other information to order alternatives completely;
- Preference through weights, priorities, goals and ideals;
- Preference through tradeoffs between objective functions in moving between efficient solutions.

I always thought something was missing in these approaches. Certainly, the DM was interested in the values of the objectives and deviations from goals. But I felt that the DM should also be interested in the variables that were positive in the optimal solution. That is, the selection of the efficient solution should be based on objective values, deviations from the goals, and the variables in the solution. For example, in a multi-objective project selection problem we would have objectives of cost, manpower, budget, with the variables, the projects to be selected, having a preference ordering; or in a diet (feed-mix) problem we could have objectives dealing with cost, calories, nutrient levels, and also have preferences in what foods (grains) should be included in the diet. That is, besides the usual preferences on objectives and goals, the DM also has preferences on the individual variables. This idea has been developed in a series of papers, Gass and Dror (1983); Dror, Gass and Yellin (1986); Dror and Gass (1987). The process does not require the DM to directly evaluate tradeoffs or to compare alternative solutions. The steps are:

1. The DM expresses a weak order preference structure on the objectives and a second on the variables.
2. The computational procedure (based on the standard Ecker and Kouada efficient solution procedure) determines a locally most preferred solution.
3. The DM is presented with the solution and, based on the DM's reactions, the process moves to an adjacent efficient solution.

The process continues until either an acceptable solution is obtained or the process cycles (for the latter, a compromise solution is determined based on the preference score of the efficient solutions in the cycle). An important contribution here is reducing the interactions with the DM. We end up with a candidate solution for the DM that is efficient, basic, and has maximal preference value among the basic efficient solutions. More recently, the idea has evolved into a multi-objective solution procedure with a graphical interface, Dror, Shoval and Yellon (1991); Anderson and Dror (1994).

6 Our Gratitude To Abraham Charnes

Much of my theoretical and applied research has been based on the seminal research of Abraham Charnes. He was pioneer in the field of Operations Research and Management Science; I am grateful for having known him as a friend. All of us who work in the areas of multi-objective programming, goal programming, and optimization, in general, are working in the light of his remarkable insight and contributions.

References

Anderson, R. K. and M. Dror. 1994. "An Interactive Graphic Presentation for Multi-objective Linear Programming," Department of Management Information Systems, University of Arizona, Tucson, Arizona.

Chankong, V. and Y. Y. Haimes, 1983. *Multi-objective Decision Making*, North-Holland, NY, NY

Charnes, A. and W. W. Cooper, 1961. *Management Models and Industrial Applications*, Volumes I and II, John Wiley and Sons, NY, NY.

Charnes, A., W. W. Cooper and R. J. Niehaus, 1969a. *Management Science Approaches to Manpower Planning*, John Wiley & Sons, NY, NY.

Charnes, A., W. W. Cooper and R. J. Niehaus, 1969b. "A Goal Programming Model for Manpower Planning," in *Management Science in Planning and Control*, J. Blood, Jr. editor, pages 79-93, Technical Association of the Pulp and Paper Industry, NY, NY.

Churchman, C. W., R. L. Ackoff and E. L. Ackoff, 1957. *Introduction to Operations Research*, John Wiley & Sons, NY, NY.

Collins, R. W., S. I. Gass, and E. E. Rosendahl, 1983. "The ASCAR Model for Evaluating Military Manpower Policy," *Interfaces*, 13, 3, 44-53.

Dantzig, G. B., 1949. "Programming of Inter-Dependent Activities II, Mathematical Model," Project SCOOP Report Number 6, Headquarters, USAF, Washington, DC.

Dantzig, G. B, 1957. "Concepts, Origins, and Use of Linear Programming," Report P-980, The Rand Corporation, Santa Monica, CA.

Dantzig, G. B., 1982. "Reminiscences about the origins of LP," *OR Letters*, vol 1 no 2, 1982, 43-48.

Dantzig, G. B., 1988. "Impact of Linear Programming on Computer Development," *OR/MS TODAY*, August, p. 12-17.

Dror, M. and S. I. Gass, 1987."Interactive Scheme for a MOLP Problem Given Two Partial Orders," *Applied Mathematics and Computation*, Vol. 24, No. 3, Part I.

Dror, M., S. I. Gass and A. Yellin, 1988. "Experiments with an Interactive Procedure for MOLP Given Weak Orders on Variables and Objectives," *European Journal of Operational Research*, Vol. 34, No.

Dror, M., P. Shoval and A. Yellin. 1991. "Multi-objective Linear Programming," *Decision Support Systems*, 7, 221-232.

Fossett, C., D. Harrison, H. Weintrob and S. I. Gass, 1991. An Assessment Procedure for Simulations Models: A Case Study, *Operations Research*, Vol. 39, No. 5, pp. 710-723.

Gass, S. I., 1984. "On the Development of Large-Scale Personnel Planning Models," *Proceedings of the 11th IFIP Conference on System Modeling and Optimization*, Springer-Verlag.

Gass, S. I., 1985. *Linear Programming*, 5th edition, McGraw-Hill Book Company NY, NY.

Gass, S. I., 1986. "A Process for Determining priorities and weights for Large-Scale Linear Goal Programs," *Journal of the Operational Research Society*, Vol. 37, No. 8.

Gass, S. I., 1987. "The Setting of Weights in Linear-Goal Programming Problems," *Computers & Operations Research*, Vol. 14, No. 3.

Gass, S. I., 1991. "Military Manpower Planning Models," *Computers and Operations Research*, Vol.18, No. 1, 65-73.

Gass, S. I., 1996. "Goal Programming in Networks," *Proceedings of First Conference in Multi-objective Programming and*, Goal Programming, M. Tamiz, editor, Springer-Verlag, Berlin, 212-234.

Gass, S. I., R. W. Collins, C. W. Meinhardt, D. M. Lemon,and M. D. Gillette, 1988. "Army Manpower Long Range Planning System," *Operations Research*, Vol. 36, No. 1.

Gass, S. I. and M. Dror, 1983. "An Interactive Approach to Multiple-Objective Linear Programming Involving Key Decision Variables," *Large Scale Systems*, vol. 5, pp. 95-103.

Gass, S. I. and T. L. Saaty, 1955a. "Parametric Objective Function - Part II: Generalization," *Operations Research,* vol. 3, no. 4, 395-401.

Gass, S. I. and T. L. Saaty, 1955b. "The Computational Algorithm for the Parametric Objective Function," *Naval Research Logistics Quarterly*, vol. 2, nos. 1 and 2.

Jacobs, W. W., 1951. "An Aircraft Deployment Model," Internal Report, Directorate of Management Analysis, Headquarters, USAF, Washington, DC.

Koopmans, T. C., ed., 1951. *Activity Analysis of Production and Allocation*, John Wiley & Sons, Inc., NY.

Kuhn, H. W. and A. W. Tucker, 1951. "Nonlinear Programming," in *Proceedings of the Second Berkeley Symposium on Mathematical Statistics and Probability*, University of California, Berkeley, California.

Orchard-Hays, W., 1954. "The Rand Code for the Simplex Method," RM1269, The Rand Corp., Santa Monica, California.

Orchard-Hays, W., 1990. "History of the Development of LP Solvers," *Interfaces*, 20, 4, 61-73.

Orden, A., 1952a. "Application of the Simplex Method to a Variety of Matrix Problems," pp. 28-55 in Orden and Goldstein 1952.

Orden, A., 1952b. "Solution of Systems of Linear Inequalities on a Digital Computer," (paper presented at the meeting of the Association for Computing Machinery, May 2, 1952, Pittsburgh, PA), Directorate of Management Analysis Headquarters, USAF, Washington, DC.

Orden, A., 1960. "Matrix Inversion and Related Topics by Direct Methods," Chapter 2, pp. 39 - 55 in *Mathematical Methods for Digital Computers*, A. Ralston and H. S. Wilf, eds., J. Wiley & Sons, NY.

Orden, A. 1993. "LP from the '40s to the '90s," *Interfaces*, 235, 5 2-12.

Orden A. and L. Goldstein (editors), 1952. *Proceedings of the Symposium on Linear Inequalities and Programming*, Project SCOOP Report No. 10, Headquarters, USAF, Washington, DC.

Pollack, S., 1952(?), "Optimization in Linear Programming," Internal Report, National Bureau of Standards, Washington, DC.

Saaty, T.L. and S. I. Gass, 1954. "The Parametric Objective Function - Part I," *Operations Research*, vol. 2, no. 3, 316-319.

Schell, E. D., 1953(?). "Application of the UNIVAC to Air Force Programming," Internal report, Headquarters, USAF, Washington, DC.

Steuer, R. E., 1986. *Multiple Criteria Optimization*, John Wiley & Sons, NY, NY.

Symonds, G. H., 1955. *Linear Programming: The Solution of Refinery Problems*, Esso Standard Oil Company, New York, New York.

USAF, 1953. "The Mathematical Computation Branch: Origins, Functions, and Facilities," Internal report, DCS/Comptroller, USAF, Washington, DC.

Wood, M. K. and G. B. Dantzig, 1949. "Programming of Inter-Dependent Activities I, General Discussion," Project SCOOP Report Number 5, Headquarters, USAF, Washington, DC. Also published in *Econometrica*, 17, 3 & 4, July-October, 1949, 193-199.

Wagner, H. M., 1959. "Linear Programming Techniques for Regression Analysis," *Journal of the American Statistical Association*, 54, 206.

Wagner, H. M., 1969. *Principles of Operations Research*, Prentice-Hall, Englewood Cliffs, NJ.

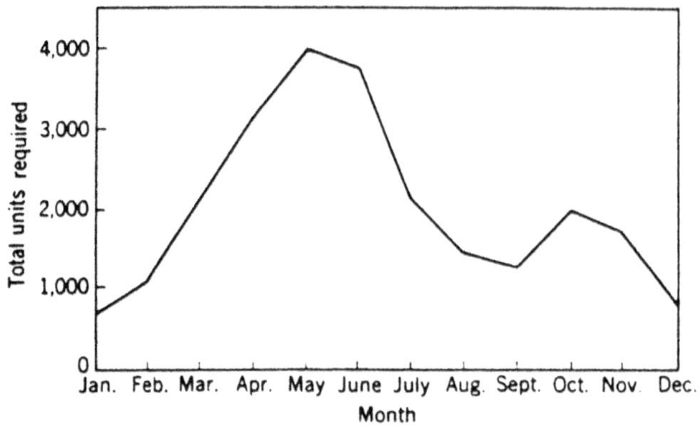

Figure 1. Smoothing Problem; Monthly Demand (Gass 1985)

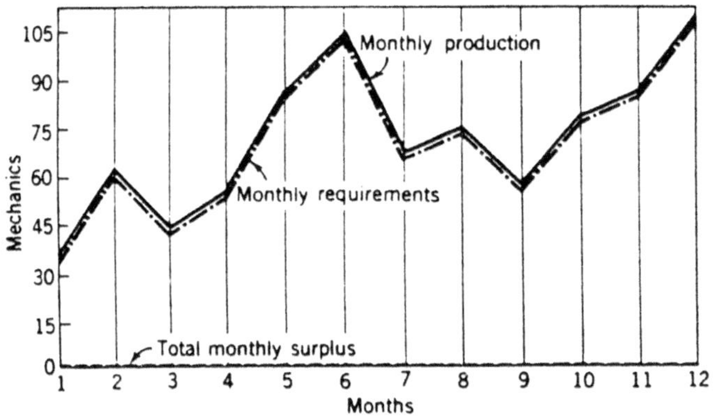

Figure 2. Smoothing Problem: Minimum Surplus Solution (Gass 1985)

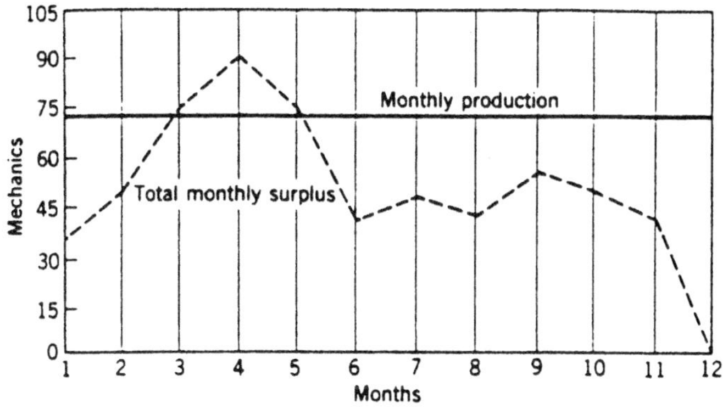

Figure 3. Smoothing Problem; Minimum Fluctuation Solution (Gass 1985)

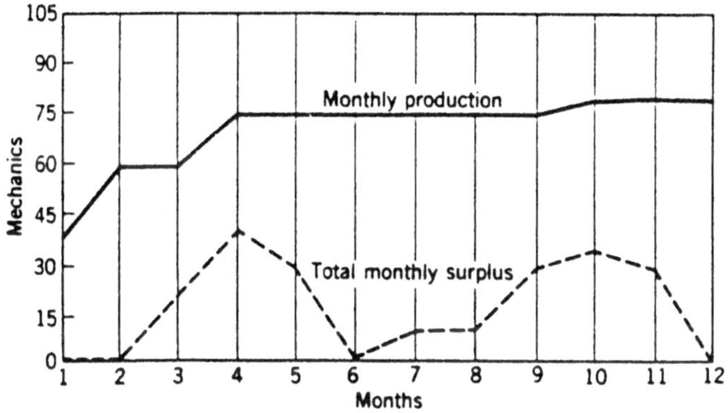

Figure 4. Smoothing Problem: An Intermediate Compromise Solution (Gass 1985)

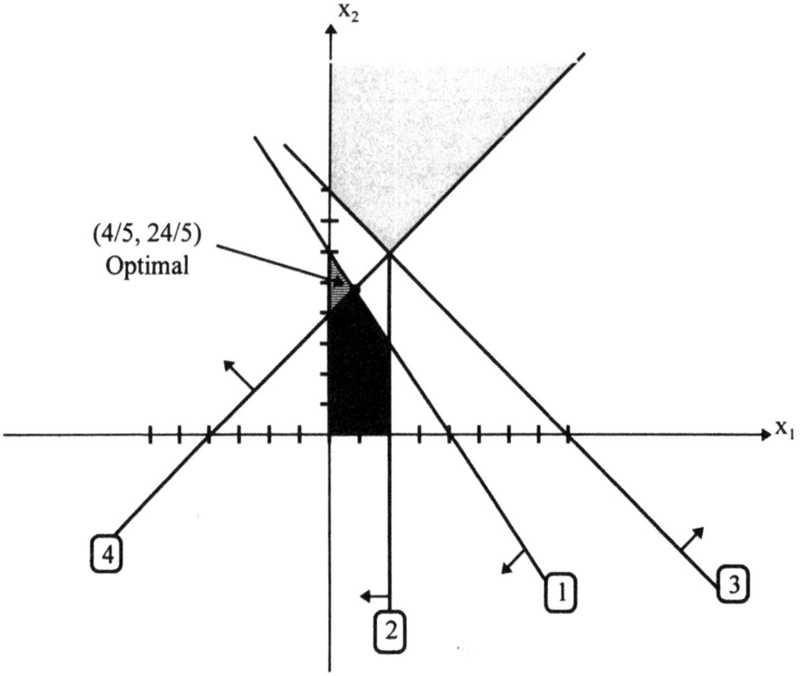

Figure 5. Goal Programming Example, Charnes and Cooper (1961)

Part 1

Goal Programming

An Example of Good Modelling Practice in Goal Programming : Means for Overcoming Incommensurability

M.Tamiz, D.F.Jones
University of Portsmouth [1] , UK.

Abstract

Much discussion has taken place over the need for good modelling practice when formulating and solving goal programmes (GP's). However, a considerable amount of poor or erroneous modelling still exists in the literature.

This paper examines one area where errors frequently arise, in the means of handling incommensurability in the GP. That is, coping with objectives that are measured in different units. Whilst many different means and techniques have been proposed for the resolution of incommensurability, the actual 'best' method is still an area of contention. However, the aim of this paper is to bring into reach of the practitioner practical means for dealing with incommensurabilty in GP, rather than proving one method superior to the others. Consequently, all the proposed methods are outlined in this paper. Their scope and robustness as well as extension to other GP variants is discussed. A hybrid algorithm which semi-automates the procedure of choosing a method is detailed.

Keywords: Goal Programming, Modelling

1 Introduction

Goal Programming is a powerful multi-criteria decision making technique. Instead of minimising a single objective as in linear programming, the sum of unwanted deviations from target values of a number of objectives are minimised. This minimisation may be as a vector of deviations, ordered in terms of the importance of their respective objectives. This is termed lexicographic goal programming(LGP). Alternatively, the deviations can be minimised as a weighted

[1] Address for correspondence: Dr M.Tamiz, SMS, University of Portsmouth, Mercantile House, Hampshire Terrace, Portsmouth. UK. E-mail TAMIZM@csovax.port.ac.uk

euclidean sum, known as weighted goal programming(WGP). The standard algebraic form for a WGP is given as:

$$MIN\ z = \sum_{i=1}^{q}(u_i n_i + v_i p_i) \tag{1}$$

Subject to,

$$f_i(\mathbf{x}) + n_i - p_i = b_i \quad i = 1 \ldots q$$
$$\mathbf{x} \in C_s$$
$$\mathbf{x}, \mathbf{n}, \mathbf{p} \geq 0$$

Where $f_i(\mathbf{x})$ is a linear function(objective) of $\mathbf{x}$, and b_i is the target value for that objective. n_i and p_i represent the negative and positive deviations from this target value. u_i and v_i are the respective weights attached these deviations in the achievement function z. These weights take a strictly positive value if the associated deviation is unwanted and a zero value otherwise. C_s is an optional set of hard constraints as found in linear programming.

Incommensurability in a WGP, or within a priority level of an LGP, occurs when deviational variables measured in different units are summed directly. This simple summation will cause an unintentional bias towards the objectives with a larger magnitude. This bias may lead to erroneous or misleading results.

The remainder of this paper is divided into four sections. Section two gives the suggested methods for the overcoming of incommensurability. Section three details a hybrid algorithm that chooses a method based on pre-analysis of the level of incommensurability in the GP. Section four details the suitability of the various methods in the case of other GP variants such as MinMax GP, Fuzzy GP and GP with non-standard utility functions. Finally, section five draws conclusions.

2 Methods for Dealing With Incommensurability

This section details suggested methods for the dealing with incommensurability in goal programmes. The strengths, weaknesses, and flexibility of each method are outlined

2.1 Use of a Lexicographic Structure

An early solution to the incommensurability problem was to use an LGP structure and only place deviations from commensurable objectives in the same priority level. This method will totally eliminate any incommensurability as summation of different units of measurement is banned. However, it also has the disadvantage of forcing every model into a lexicographic form, even when the underlying structure of the model may not be lexicographic. Furthermore, trade-offs between incommensurable objectives are not allowed even when required by the decision maker. In addition, the resulting lexicographic model must have at least as many priorities as there are different units of measurement in the model, thus sometimes violating the recommendations about LGP's having relatively few priority levels [7, 12].

These difficulties led to criticisms of LGP as inflexible and, in certain cases, unable to accurately reflect the decision maker's preferences [5]. It is clear that the use of a lexicographic structure is not the best tool to deal with incommensurability due to the rigid preference structure it imposes.

2.2 Implicit Scaling

Some authors argue that there is no need for normalisation providing the modeller bears in mind the different magnitudes of the objectives when setting their weights [13]. This process implies a kind of implicit scaling by the decision maker. If applied successfully, this process can deal with incommensurability. In some situations there exists a clear ratio between the different units (e.g. Pounds Stirling and American Dollars, where the ratio is the exchange rate). In other cases this ratio is a lot harder to estimate. This process requires the decision maker to clearly understand the nature of incommensurability in order to produce minimal distoration by the choice of appropiate scaling factors, as very little computerised automation is available due to the implicit nature of the process.

Gass [3] puts forward a similar argument, with the additional requirement that a small selection of optimal results from several different attempts should be presented to the decision maker.

2.3 Normalisation Techniques

The problem of incommensurability leads to the develop of normalisation techniques. A normalisation technique seeks to overcome incommensurability by the division of each objective by some constant pertaining to the magnitude of that objective. Thus under a normalisation technique the standard WGP model (1) is transformed to the following form

$$MIN\ z = \sum_{i=1}^{q}\left(\frac{u_i n_i + v_i p_i}{k_i}\right) \tag{2}$$

Subject to,

$$f_i(\mathbf{x}) + n_i - p_i = b_i \quad i = 1 \ldots q$$
$$\mathbf{x} \in C_s$$
$$\mathbf{x}, \mathbf{n}, \mathbf{p} \geq 0$$

Where k_i is the normalisation constant pertaining to the i'th objective. The exact nature of the normalisation constant is dependent on the type of normalisation used. There are several well-used methods for calculating the normalisation constant. None of these have been proved to dominate the others [8], and the type of normalisation able to be used depends on the nature of the GP in question. A review of the properties of the normalisation methods follows:

1. EUCLIDEAN MODEL - This is the first normalisation technique to be used in WGP applications. It was introduced by De Kluyver in 1979 [1]. In this method the normalisation constant is the euclidean norm of their technical coefficients in the objective. This has the advantage of reconciling geometric (L_2) distances with linear (L_1) deviations used in goal programming [12]. Wildhelm [18] and Hannan [4] both cite the euclidean normalisation method as a means of dealing with incommensurability. The Euclidean method is a robust technique, being able to be applied to every goal programme [8]. It does not, however, restore any significant meaning to the achievement function, it still being a sum of different units.

2. PERCENTAGE MODEL - Suggested and used by Romero [9, 12] during the past decade. This method expresses all deviations as percentages from their targets rather than as absolute values. Thus the normalisation constant is the target value of the objective divided by 100. The percent method has the advantage of restoring meaning to the achievement function, which now takes the form of a sum of percentage deviations from targets. The Percentage method is not as robust as the Euclidean method, failing to operate with zero target values and requiring modification for negative target values [8].

3. SUMMATION MODEL - Suggested by Jones [8]. Here the normalisation constant is the sum of the absolute values of the technical coefficients of the objective. As this method gives normalisation constants of greater magnitude than that of the Euclidean method, it helps deal with models which have particularly severe incommensurability. It has the same robustness as the Euclidean method and the same drawback of not restoring significant meaning to the achievment function.

4. ZERO-ONE MODEL - This method is popular amongst the MCDM community [17]. Each objective is scaled onto a zero-one range by establishing the minimal(best) and maximal(worst) deviations for each objective. The zero-one method is first used in a goal programming context, by Hwang and Masud [10]. The zero-one method has the advantage of restoring meaning to the achievement function and, where it can be used, is sucessful at overcoming incommensurability. It is not in general, however, a robust method due to problems with the calculations of the worst values

of each deviation [13]. It general it requires the existence of a closed feasible set of constraints. Even so, if there exist areas in the feasible region which are unattractive to the decision maker, these will serve to distort the normalisation constants by making some maximal deviations too large.

3 A Hybrid Algorithm

The results of the previous section show the Euclidean and Summation methods to be the only ones which can be robustly applied to all types of GP models. Therefore these are the logical choices for inclusion in a method that selects a normalisation procedure. An effective measure of incommensurability in a model is the logarithm of the maximum and minimum Euclidean norms in the objective set. If this index is below a certain threshold, Θ_1, then there is no requirement for a normalisation technique as the objectives are roughly commensurable. The Euclidean method produces good results for a large proportion of models and has some valuable theoretical properties [12], and should be used for models with incommensurability exceeding Θ_1. The Summation method produces normalisation constants of greater magnitude that of those given by the Euclidean method and hence there will exist a second incommensurability threshold Θ_2 above which it is more appropriate to apply Summation normalisation. Therefore a suggested automatic normalisation algorithm is given as:

1. Measure incommensurability index Θ of model.
2. If $\Theta \leq \Theta_1$ then Goto (5).
3. If $\Theta_1 \leq \Theta \leq \Theta_2$ then apply Euclidean normalisation. Goto (5).
4. If $\Theta \geq \Theta_2$ then apply Summation normalisation. Goto (5)
5. Solve model and output results.

The only parameters needed to be specified by the modeller are the threshold values Θ_1 and Θ_2. These values are dependent on the degree of incommensurability regarded as serious enough by the modeller to require the normalisation type beyond the threshold value. The values that the authors find most appropriate in their GP modelling are $\Theta_1 = 1, \Theta_2 = 3$. These values are used as defaults in the GPSYS [15] intelligent GP system.

3.1 Incommensurability Warning

For those modellers experienced in the setting of weights in GP, a less automatic normalisation process may be required. A further normalisation option in GPSYS is the production of an incommensurability warning if the incommensurability index exceeds the threshold at which Euclidean normalisation is applied in the hybrid algorithm detailed in section 6.1. The modeller can then alter their weights in accordance with this information to produce a modified version of the model which takes account of the incommensurability.

4 Incommensurability in Other GP Variants

Whilst the methods for dealing with incommensurability detailed in Section 2 were developed mainly for the weighted GP variant, the problem of incommensurability in fact exists throughout the GP paradigm. This section details particular issues that are of importance when applying incommensurability resolution techniques to other GP variants.

Applying normalisation within a lexicographic GP structure is a relatively straight-forward task. Here the only incommensurability of concern is that between objectives measured in different units that are placed in the same priority level. Objectives in different priority levels are not directly compared and thus incommensurability between them is not a problem. When examining a lexicographic GP each priority level should be examined in turn to see if incommensurability exists between the objectives in that priority level. If incommensurability is found, then a method outlined in Section 2. or the hybrid algorithm in Section 3, should be applied to the objectives within that priority level. Applying different normalisation techniques to different priority levels is, whilst possible, unadvisable because of the resulting confusion over the meaning of the achievement function.

Some further variants of GP are those that are based on the L_∞ distance metric. These being Chebyshev GP [2], MinMax GP, and Fuzzy GP [7]. These models seek to minimise the maximum deviation from amongst the set of unmet objectives, rather than to minimise the weighted sum (or lexicographic order of weighted sums) of deviations from unmet deviations as takes place in Weighted or Lexicographic GP. In this case normalisation is vital because any discrepency in units may cause one objective to have a larger deviation than the others. As only the maximal deviation is minimised, this objective will then be the only one considered. An effectual degeneration to a single objective (LP) model will have thus been caused by the incommensurability. Possibly because of the seriousness of this problem, the initial Chebyshev/Minmax GP formulation [2] is given with Percentage normalisation applied. This situation however limits the use of the technique to models with non-zero RHS's. It is possible to use any of the techniques mentioned in Section 2 in a Chebyshev/Minmax GP model. However, some kind of normalisation is strongly recommended for the reason listed above.

The fuzzy GP model [7] is normally stated with zero-one normalisation applied. This allows the deviation from each of the objectives to be measured on a scale of zero to one. Thus interpretation of the situation can be given from the viewpoint of fuzzy set theory [19]. The deviation then acts as a fuzzy membership function of the objective. Further details of fuzzy GP formulations are given by Ignizio [7].

A final GP variant to mention is Preference-Function GP [14, 16]. In this case the relationship between the deviation from the target and the penalty imposed is not the standard linear one, it may be piecewise linear, discontinuous, or even non-linear. This GP variant is used to model objectives with non-standard underlying utility functions. The result of the modelling of this case is a series of objectives with the same left-hand side but differing right-hand sides. All these objectives derive from a single criterion and thus have the same unit of measurement. In this case, the use of normalisation method based on the right-hand-side (ie. Percentage or Zero-one) distorts the preference structure being modelled. In these cases a method based on the left-hand-side, such as Euclidean or Summation normalisation, or the hybrid method given in Section 3, should be preferred.

5 Conclusion

This paper has discussed the practical difficulties arising from the base issue of incommensurability in goal programmes. Upon the detailing and examination of the different methods used to overcome the incommensurability, it becomes clear that no method can be said to 'dominate' the other methods. Some methods cannot be applied to all models, but at least one method can be applied to any particular model that may have been built.

It is clear that the methods detailed do alleviate a large amount of the problems. Therefore, it is advisable that those aiming to apply GP in an accurate manner to their real-world GP problems use at least one of the techniques mentioned in Section 2. The hybrid algorithm mentioned in Section 3 goes someway to choosing a method based on the properties of the model in question. The hybrid algorithm, as well as all of the normalisation techniques mentioned in sectioned 2, are best used in an automated manner as part of a goal programming software package. To this end, they have been incorporated into GPSYS, an intelligent GP solver, being developed at the University of Portsmouth, UK [15].

When dealing with most other GP variants, the problem of incommensurability persists. In this case, the means of dealing with the incommensurability needs to be chosen in accordance with the variant used as well as the nature of the particular model.

References

1 DE KLUYVER, C.A.(1979) An exploration of various goal programming formulations - with application to advertising media scheduling, *Journal of the Operational Research Society*, **30**, 167-171.

2 FLAVELL, R.B.(1976) A New Goal Programming Formulation. *Omega*, **4**, 731-732.

3 GASS, S.I.(1987) The settings of weights in linear goal programming, *Computers and Operations Research*, **14**, 227-229.

4 HANNAN, E.L.(1985) An assessment of some of the criticisms of goal programming, *Computers and Operations Research*, **12**, 525-541.

5 HARRALD, J., LEOTTA, J., WALLACE, W.A. and WENDELL, R.E.(1978) A note on the limitations of goal programming as observed in resource allocation for marine environmental protection. *Naval Research Logistics Quarterly*, **25**, 733-739.

6 IGNIZIO, J.P.(1994) *Linear Programming*, Lexington, Mass.: Heath (Lexington Books).

7 IGNIZIO, J.P.(1982) *Linear Programming in Single and Multiple Objective Systems*, Prentice-Hall Inc., Englewood Cliffs, New Jersey.

8 JONES, D.F.(1995) *The Design and Development of an Intelligent Goal Programming System*, PhD. Thesis. University of Portsmouth, UK.

9 LARA, P. and ROMERO, C.(1992) An interactive multigoal programming model for determining livestock rations: an application to dairy cows in Andalusia, Spain. *Journal of the Operational Research Society*, **43**, 945-953.

10 MASUD, A.S. and HWANG, C.L.(1981) Interactive Sequential Goal Programming, *Journal of the Operational Research Society*, **32**, 391-400.

11 ROMERO, C.(1985) Naive weighting in non-preemptive goal programming-letter to the Editor. *Journal of the Operational Research Society*,**36**, 647-648.

12 ROMERO, C.(1991) *Handbook of Critical Issues in Goal Programming*, Pergamon Press.

13 SUTCLIFFE, C., BOARD, J. and CHESHIRE, P.(1985) Naive weighting in non-preemptive goal programming: a reply, *Journal of the Operational Research Society*, **36**, 648-649.

14 TAMIZ M. and JONES D.F.(1995) 'Improving the flexibility of goal programming via preference modelling techniques', *Omega*, **23**, 41-48.

15 TAMIZ, M. and JONES, D.F.(1995) GPSYS, Specifications for an intelligent goal programming system. user guide, version 1.1. University of Portsmouth, UK.

16 TAMIZ. M. and JONES, D.F.(1995) 'Algorithmic extensions to the method of Martel and Aouni', *Journal of the Operational Research Society*, **46**, 254-257.

17 VINCKE, P.(1992) *Multicriteria Decision Aid*, J. Wiley & Sons.

18 WILDHELM, W.B.(1981) Extensions of goal programming models. *Omega*, **9**, 212-214.

19 ZIMMERMAN, H.J.(1983) Using Fuzzy Sets in Operational Research, *European Journal of Operational Research*, **13**, 201-206.

Pattern Classification by Linear Goal Programming and its Applications

Hirotaka Nakayama and Naoko Kagaku

Department of Applied Mathematics, Konan University
8-9-1 Okamoto, Higashinada, Kobe 658, JAPAN

Abstract. For pattern classification, the multi-surface method proposed by Mangasarian is attracting much attention, because it can provide an exact discrimination function even for highly nonlinear problems without any assumption on the data distribution. However, the method produces often too complicated discrimination surfaces, which cause poor generalization ability.
In this paper, several trials in order to overcome this difficulty of the multi-surface method will be suggested: One of them is the utilization of the goal programming in which the auxiliary linear programming problem is formulated as a goal programming in order to get as simple discrimination curve as possible. Another one is to apply the fuzzy programming by which we can get fuzzy discrimination curves with gray zones. In addition, using the suggested methods, the additional learning can be easily made. These features of the methods make the discrimination more realistic. The effectiveness of the methods are shown on the basis of some applications.

Keywords. Pattern classification, linear goal programming, fuzzy goal programming

1 Introduction

Machine learning for pattern classification is one of most important subjects in artificial intelligence. Several approaches to this topic have been developed so far: in particular, statistic discrimination analysis, artificial neural network (ANN), mathematical programming approach are popular in many real fields.

In ANN, the back propagation method is widely applied for learning. However, unless several parameters are decided moderately in advance, it tends to be trapped in local minima. Moreover, it is a problem how the structure of ANN should be decided.

Although statistic discrimination analysis has been developed remarkably, it requires some assumption on data distribution. This assumption restricts the application.

On the other hand, mathematical programming approach has been attracting many researchers' attention due to its applicability to wide range of real problems and the simplicity of its algorithm. In particular, Multisurface Method (MSM) is very attractive because it is very simple and can provide an exact discrimination function even for highly nonlinear problems without any assumptions on the data distribution.

The method finds a piecewise linear discrimination function by solving linear programming iteratively. One of most prominent features of MSM is that 1) there is no parameters to be decided in advance, 2) relatively short computing time is required, and 3) the resulting solution is guaranteed to be the global optimal. Mangasarian *et al.* (1992) reported its good performance in application to medical diagnosis problems.

However, we observed that the method tends to produce too complicated discrimination sufaces, which cause a poor generalization ability because it is affected by noise. In order to overcome this difficulty, we proposed a multi-objective optimization approach (Nakayama *et al.* 1994). On the other hand, Benett-Mangasarian proposed a method called Robust Linear Programming Discrimination (RLPD) whose idea is based on the goal programming. Although these new methods provide simpler discrimination functions than by the original MSM, they still do too complicated ones in many cases. This is because that they try to find an exact discrimination surface which classifies all data correctly. It has been observed that such a perfect learning gives a poor generalization ability in many cases. This phenomenon is known as the over learning.

In order to avoid the over-learning, in this paper, we suggest methods which do not require the perfect classification for given data, but allow unclassified data. This idea can be performed by applying fuzzy mathematical programming techniques which produce "gray zones". The effectiveness of the modified methods can be proved by several applications to artificial classification problems. Next, these methods are extended to trinary classification problems. Finally, we apply them to a stock portfolio problem.

2 Multisurface Method (MSM)

Suppose that given data in a set X of n-dimensional Euclidean space belong to one of two categories $\mathcal{A}$ and $\mathcal{B}$. Let A be a matrix whose row vectors denote points of the category $\mathcal{A}$. Similarly, let B be a matrix whose row vectors denote points of the category $\mathcal{B}$. For simplicity of notation, we denote the set of points of $\mathcal{A}$ by A. The set of points of $\mathcal{B}$ is denoted by B similarly. MSM suggested by Mangasarian (1968) finds a piecewise linear discrimination surface separating two sets A and B by solving linear programming problems iteratively. The main idea is to find two hyperplanes parallel with each other which classify as many given data as possible:

$$g(\boldsymbol{u}) = \boldsymbol{x}^T \boldsymbol{u} = \alpha$$

$$g(\boldsymbol{u}) = \boldsymbol{x}^T\boldsymbol{u} = \beta$$

This is performed by the following algorithm:

Step 1 . Solve the following linear programming problem at k-th iteration (set $k = 1$ at the beginning):

$$\text{(I)} \qquad \begin{array}{l} \text{Maximize} \quad \phi(A, B) = \alpha - \beta \\ \text{subject to} \\ \quad A\boldsymbol{u} \geq \alpha\mathbf{1} \\ \quad B\boldsymbol{u} \leq \beta\mathbf{1} \\ \quad -\mathbf{1} \leq \boldsymbol{u} \leq \mathbf{1} \end{array}$$

where, $\mathbf{1}^T = (1, \ldots, 1)$.

In order to avoid a trivial solution $\boldsymbol{u} = 0$, $\alpha = 0$, $\beta = 0$, we add a constraint which is a linear approximation to $\boldsymbol{u}^T\boldsymbol{u} \geq \frac{1}{2}$. Namely,

$$\boldsymbol{u}^T\boldsymbol{u} \cong \boldsymbol{p}^T\boldsymbol{p} + 2\boldsymbol{p}^T(\boldsymbol{u} - \boldsymbol{p}) \geq \frac{1}{2}.$$

which finally yields

$$\boldsymbol{p}^T\boldsymbol{u} \geq \frac{1}{2}(\frac{1}{2} + \boldsymbol{p}^T\boldsymbol{p}).$$

Now our problem to be solved is given by

$$\text{(I')} \qquad \begin{array}{l} \text{Maximize} \quad \phi_i(A, B) = \alpha - \beta \\ \text{subject to} \\ \quad A\boldsymbol{u} \geq \alpha\mathbf{1} \\ \quad B\boldsymbol{u} \leq \beta\mathbf{1} \\ \quad -\mathbf{1} \leq \boldsymbol{u} \leq \mathbf{1} \\ \quad \boldsymbol{p}_i^T\boldsymbol{u} \geq \frac{1}{2}\left(\frac{1}{2} + \boldsymbol{p}_i^T\boldsymbol{p}_i\right) \end{array} \qquad (1)$$

where $\boldsymbol{p}_i$ is given by one of

$$\boldsymbol{p}_1^T = (\frac{1}{\sqrt{2}}, 0, \ldots, 0)$$

$$\boldsymbol{p}_2^T = (-\frac{1}{\sqrt{2}}, 0, \ldots, 0)$$

$$\vdots$$

$$p_{2n}^T = (0, \ldots, 0, -\frac{1}{\sqrt{2}})$$

After soving LP problem (I') for each i such that $1 \leq i \leq 2n$, we take a hyperplane which classify correctly as many given data as possible. Let the solution be $\boldsymbol{u}^*, \alpha^*, \beta^*$, and let the corresponding value of objective function be $\phi^*(A, B)$.

If $\phi^*(A, B) > 0$, then we have a complete separating hyperplane $g(\boldsymbol{u}^*) = (\alpha^* + \beta^*)/2$. Set $\tilde{A}^k = \{\boldsymbol{x} \in X | \ g(\boldsymbol{u}^*) \geq (\alpha^* + \beta^*)/2\}$ and $\tilde{B}^k = \{\boldsymbol{x} \in X | \ g(\boldsymbol{u}^*) < (\alpha^* + \beta^*)/2\}$. $\tilde{A}^k$ and $\tilde{B}^k$ represent subregions of the category $\mathcal{A}$ and $\mathcal{B}$ in X, respectively, which is decided at this stage. Go to Step 3.

Otherwise, go to Step 2.

Step 2 . First, remove the points such that $\boldsymbol{x}^T\boldsymbol{u}^* > \beta^*$ from the set A. Let A^k denote the set of removed points. Take the separating hyperplne as $g(\boldsymbol{u}^*) = (\beta^* + \tilde{\beta})/2$ where $\tilde{\beta} = \text{Min}\ \{\boldsymbol{x}^T\boldsymbol{u}^* | \ \boldsymbol{x} \in A^k\}$. Let $\tilde{A}^k = \{\boldsymbol{x} \in X | \ g(\boldsymbol{u}^*) > (\beta^* + \tilde{\beta})/2\}$. The set $\tilde{A}^k$ denotes a subregion of the category $\mathcal{A}$ in X which is decided at this stage. Rewrite $X \backslash \tilde{A}^k$ by X and $A \backslash A^k$ by A.

Next, remove the points such that $\boldsymbol{x}^T\boldsymbol{u}^* < \alpha^*$ from the set B. Let B^k denote the set of removed points. Take the separating hyperplne as $g(\boldsymbol{u}^*) = (\alpha^* + \tilde{\alpha})/2$ where $\tilde{\alpha} = \text{Min}\ \{\boldsymbol{x}^T\boldsymbol{u}^* | \ \boldsymbol{x} \in B^k\}$. Let $\tilde{B}^k = \{\boldsymbol{x} \in X | \ g(\boldsymbol{u}^*) < (\alpha^* + \tilde{\alpha})/2\}$. The set $\tilde{B}^k$ denotes a subregion of the category $\mathcal{B}$ in X which is decided at this stage. Rewrite $X \backslash \tilde{B}^k$ by X and $B \backslash B^k$ by B.

Set $k = k + 1$ and go to Step 1.

Step 3. Construct a piecewise linear separating hypersurface for A and B by adopting the relevant parts of the hyperplanes obtained above.

Remark At the final p-th stage, we have the region of $\mathcal{A}$ in X as $A^1 \cup A^2 \cup \ldots \cup A^p$ and that of $\mathcal{B}$ in X as $B^1 \cup B^2 \cup \ldots \cup B^p$. Given a new point, its classification is easily made. Namely, since the new point is either one of these subregions in X, we can classify it by checking which subregion it belongs to in the order of $1, 2, \ldots, p$.

3 Reformulation as Multi-objective Programming and Goal Programming

3.1 Multi-objective Programming Approach

It has been observed that MSM yields sometimes too complex discrimination hypersurfaces. One of reasons for this phenomenon may be that the discrimination surface is constructed by two parallel hyperplanes solving a LP problem at each iteration.

In order to overcome the difficulty of MSM, we can consider a hyperplane which has the set B on its one side and classifies correctly as many points of A as possible. This is performed by solving the following LP problem:

$$\text{(II)} \qquad \begin{array}{ll} \text{Maximize} & (A\boldsymbol{u} - \beta\mathbf{1}) \\ \text{subject to} & \\ \end{array}$$

$$\begin{array}{rcccl} & & B\boldsymbol{u} - \beta\mathbf{1} & \leqq & \mathbf{0} \\ -\mathbf{1} & \leqq & \boldsymbol{u} & \leqq & \mathbf{1} \end{array}$$

The problem (II) is a multi-objective optimization problem. We can take a scalarization of linear sum of the objective functions. For this scalarization, we have

$$\text{(II')} \qquad \begin{array}{ll} \text{Maximize} & \sum_{i=1}^{m}(y_i^+ - y_i^-) \\ \text{subject to} & \end{array}$$

$$\begin{array}{rcccl} & & A\boldsymbol{u} - \beta\mathbf{1} & = & \boldsymbol{y}^+ - \boldsymbol{y}^- \\ & & B\boldsymbol{u} - \beta\mathbf{1} & \leqq & \mathbf{0} \\ -\mathbf{1} & \leqq & \boldsymbol{u} & \leqq & \mathbf{1} \\ & & \boldsymbol{y}^+,\ \boldsymbol{y}^- & \geqq & \mathbf{0} \end{array}$$

More generally, considering the symetric role of A and B, we solve another LP problem in addition to (II'):

$$\text{(III)} \qquad \begin{array}{ll} \text{Minimize} & (B\boldsymbol{u} - \alpha\mathbf{1}) \\ \text{subject to} & \end{array}$$

$$\begin{array}{rcccl} & & A\boldsymbol{u} - \alpha\mathbf{1} & \geqq & \mathbf{0} \\ -\mathbf{1} & \leqq & \boldsymbol{u} & \leqq & \mathbf{1} \end{array}$$

from which

$$\text{(III')} \qquad \text{Minimize} \quad \sum_{i=1}^{m}(z_i^+ - z_i^-)$$

$$\begin{array}{rcccl} \text{subject to} & & & & \\ & & A\boldsymbol{u} - \alpha\mathbf{1} & \geqq & \mathbf{0} \\ & & B\boldsymbol{u} - \alpha\mathbf{1} & = & \boldsymbol{z}^+ - \boldsymbol{z}^- \\ -\mathbf{1} & \leqq & \boldsymbol{u} & \leqq & \mathbf{1} \\ & & \boldsymbol{z}^+,\ \boldsymbol{z}^- & \geqq & \mathbf{0} \end{array}$$

In oder to avoid a tivial solution $\boldsymbol{u} = 0,\ \boldsymbol{y} = 0,\ \boldsymbol{z} = 0,\ \alpha = 0,\ \beta = 0$, we add the constraint (1) to each LP problem above.

At the beginning, set k=1. Let A^k be the set of points of A such that $\boldsymbol{x}^T\boldsymbol{u}^* > \beta^*$ for the solution to (II'), and let n_A denote the number of elements of A^k. Similarly, let B^k be the set of points of B such that $\boldsymbol{x}^T\boldsymbol{u}^* < \alpha^*$ for the solution to (III'), and let n_B denote the number of elements of B^k.

If $n_A \geqq n_B$, then we take a separating hyperplane as $g(\boldsymbol{u}^*) = (\beta^* + \tilde{\beta})/2$ where β^* is the solution to (II') and $\tilde{\beta}$=Min$\{\boldsymbol{x}^T\boldsymbol{u}^* | \ \boldsymbol{x} \in A^k\}$. Let $\tilde{A}^k = \{\boldsymbol{x} \in$

$X|\ g(\boldsymbol{u}^*) > (\beta^* + \tilde{\beta})/2\}$. The set $\tilde{A}^k$ denotes a subregion of the category $\mathcal{A}$ in X which is decided at this stage. Rewrite $X\backslash\tilde{A}^k$ by X and $A\backslash\tilde{A}^k$ by A.

If $n_A < n_B$, then we take a separating hyperplane as $g(\boldsymbol{u}^*) = (\alpha^* + \tilde{\alpha})/2$ where α^* is the solution to (III') and $\tilde{\alpha}$=Max$\{\boldsymbol{x}^T\boldsymbol{u}^*|\ \boldsymbol{x} \in B^k\}$. Let $\tilde{B}^k = \{\boldsymbol{x} \in X|\ g(\boldsymbol{u}^*) < (\alpha^* + \tilde{\alpha})/2\}$. The set $\tilde{B}^k$ denotes a subregion of the category $\mathcal{B}$ in X which is decided at this stage. Rewrite $X\backslash\tilde{B}^k$ by X and $B\backslash\tilde{B}^k$ by B.

Setting $k = k+1$, solve the problem (II') and (III') until we have $A\backslash\tilde{A}^k = \emptyset$ and $B\backslash\tilde{B}^k = \emptyset$.

Remark At the final p-th stage, we have the region of $\mathcal{A}$ in X as $A^1 \cup A^2 \cup \ldots \cup A^p$ and that of $\mathcal{B}$ in X as $B^1 \cup B^2 \cup \ldots \cup B^p$. Given a new point, its classification is easily made. Namely, since the new point is either one of these subregion in X, we can classify it by checking which subregion it belongs to in the order of $1, 2, \ldots, p$.

3.2 Goal Programming Approach

Unlike the method by multi-objective optimization above, we can also consider another method obtaining a hyperplane which has one of the sets on its one side and minimizes the sum of distances between misclassified data of another set and the hyperplane at each iteration. This is formulated as a goal programming, and was originally given by Benett-Mangasarian (1992), who named it the robust linear programming discrimination (RLPD). The algorithm is summarized as follows:

In order that the separating hyperplane minimizes the sum of distances between misclassified data in the class $\mathcal{A}$ and the hyperplane, we can solve the linear programming problem

$$
\text{(IV)} \qquad \begin{array}{ll} \text{Minimize} & \sum_{j=1}^{k} y_j \\ \text{subject to} & \end{array}
$$

$$
\begin{array}{rcl}
-A\boldsymbol{u} + \beta\mathbf{1} & \leqq & \boldsymbol{y} \\
B\boldsymbol{u} - \beta\mathbf{1} & \leqq & \mathbf{0} \\
-\mathbf{1} \ \leqq \ \boldsymbol{u} & \leqq & \mathbf{1} \\
\boldsymbol{y} & \geqq & \mathbf{0}
\end{array}
$$

Similarly, for the class $\mathcal{B}$, we have

$$
\text{(V)} \qquad \begin{array}{ll} \text{Minimize} & \sum_{i=1}^{m} z_i \\ \text{subject to} & \end{array}
$$

$$
\begin{array}{rcl}
A\boldsymbol{u} - \alpha\mathbf{1} & \geqq & \mathbf{0} \\
B\boldsymbol{u} - \alpha\mathbf{1} & \leqq & \boldsymbol{z} \\
-\mathbf{1} \ \leqq \ \boldsymbol{u} & \leqq & \mathbf{1} \\
\boldsymbol{z} & \geqq & \mathbf{0}
\end{array}
$$

In order to avoid a trivial solution $\boldsymbol{u} = 0,\ \boldsymbol{y} = 0,\ \boldsymbol{z} = 0,\ \alpha = 0,\ \beta = 0$, we add the constraint (1) to each LP problem above. We can make the same procedure as the multi-objective optimization stated above taking (IV) in place of (II') and (V) in place of (III').

3.3 Classification of New Data

A point $\boldsymbol{x} \in R^n$ is classified into $\mathcal{A}$ or $\mathcal{B}$ as follows: The judgment is made by hyperplanes obtained as solutions to corresponding LP problems. Suppose that the given new point is not classified definitely for the first k-1 hyperplanes, and that it is classified definitely by the k-th hyperplane (i.e., it is in the subset removed by the k-th hyperplane).

- If the k-th hyperplane is obtained by solving (II') or (IV), then the point is judged to be in $\mathcal{A}$.
- If the k-th hyperplane is obtained by solving (III') or (V), then the point is judged to be in $\mathcal{B}$.

4 Application of Fuzzy Mathematical Programming

The multi-objective programming approach and the goal programming approach make a perfect classification for given teacher's data as well as MSM. As a result, we have too complex discrimination surfaces which cause a poor generalization ability due to the high sensitivity to the noise of data.

In order to overcome this difficulty, we can introduce fuzzy mathematical programming into the methods so that some critical points may be unclassified. In the following, a case for the goal programming approach will be stated.

In the problem (IV), the constraints $B\boldsymbol{u} - \beta\mathbf{1} \leqq \mathbf{o}$ may be fuzzified into

$$B\boldsymbol{u} - \beta\mathbf{1} \lesssim \mathbf{o}$$

where $\lesssim$ implies "almost $\leqq$". The degree of satisfaction of the inequality $B\boldsymbol{u} - \beta\mathbf{1} \leqq \mathbf{o}$ can be represented by the membership function $M_B(\boldsymbol{x})$ denoting the degree of membership of $\boldsymbol{x}$ in B:

$$M_B(\boldsymbol{x}) = \begin{cases} 1, & -\frac{\boldsymbol{x}^T\boldsymbol{u}-\beta}{e} \geq 1 \\ -\frac{\boldsymbol{x}^T\boldsymbol{u}-\beta}{e}, & 0 < -\frac{\boldsymbol{x}^T\boldsymbol{u}-\beta}{e} < 1 \\ 0, & -\frac{\boldsymbol{x}^T\boldsymbol{u}-\beta}{e} \leq 0 \end{cases}$$

Similarly, the membership function $M_A(\boldsymbol{x})$ of the point $\boldsymbol{x}$ in A is given by

$$M_A(\boldsymbol{x}) = \begin{cases} 0, & \frac{\boldsymbol{x}^T\boldsymbol{u}-\alpha}{e} - 1 \geq 0 \\ \frac{\boldsymbol{x}^T\boldsymbol{u}-\alpha}{e} - 1, & -1 < \frac{\boldsymbol{x}^T\boldsymbol{u}-\alpha}{e} - 1 < 0 \\ -1, & \frac{\boldsymbol{x}^T\boldsymbol{u}-\alpha}{e} - 1 \leq -1 \end{cases}$$

Using the usual technique in fuzzy mathematical programming, we have the following LP problems taking into account the fuzzines stated above:

(VI) Minimize $\sum_{j=1}^{k} y_j - w\lambda$
subject to

$$\begin{aligned} -A\boldsymbol{u} + \beta\mathbf{1} &\leq \boldsymbol{y} \\ B\boldsymbol{u} - \beta\mathbf{1} + e\lambda\mathbf{1} &\leq e\mathbf{1} \\ -\mathbf{1} \leq \boldsymbol{u} &\leq \mathbf{1} \\ \boldsymbol{y} &\geq \mathbf{0} \end{aligned}$$

(VII) Minimize $\sum_{i=1}^{m} z_i - w\lambda$
subject to

$$\begin{aligned} A\boldsymbol{u} - \alpha\mathbf{1} - e\lambda\mathbf{1} &\geq -e\mathbf{1} \\ B\boldsymbol{u} - \alpha\mathbf{1} &\leq \boldsymbol{z} \\ -\mathbf{1} \leq \boldsymbol{u} &\leq \mathbf{1} \\ \boldsymbol{z} &\geq \mathbf{0} \end{aligned}$$

Remark The parameter e represents the width of the "gray zone" of the separating hyperplane, and w does the weight for considering the membership function. The fuzziness of the boundary between the set A and B varies according to the value of these parameters. They should be decided by users on the basis of their experiences.

5 Additional Learning and Application to Portfolio Selection

The obtained discrimination surface follows just the teachers' data given in advance. In general, however, when the discrimination surface can not classify an additional new data correctly, it needs to be modified. In other words, the machine learning also needs the additional learning like human beings so that it can follow adaptivley the change of situation. In the stated methods, the additional learning can be easily made by restarting the procedure from the stage at which the new data is misjudged. The effect of additional learning can be shown along an example of application to portfolio selection as follows:

Our problem is to judge whether a stock is to be sold or to be bought. Considering in addition whether to be hold, our problem becomes trinary classification. We can make trinary classification by repeating binary classification for one and the others among three categories. In our stock portfolio problem, seven economic indices are taken into account. We have the data in the 119 periods in the past for which it is already known to be sold or to be bought. We obtained a discrimination function by the stated methods taking the first 50 data as the teachers' ones, and examined the ability of classification for the rest 69 data. The following figures compare the result without additional learning and the one with additional learning. Flags with thick line represent misclassified data, and those with thin line do data in a gray zone in the fuzzy approach.

It can be seen from Fig. 1, for example, that the additional learning of the misclassified 52-th data enables correct judgment for the data until 64-th period while the method without additional learning misclassified many of them. On the other hand, many of data after 99-th period were misclassified by the method with additional learning, but not by the one without additional learning. One of reasons for this phenomenon may be that the rule around here is not affected by the near past, but by the far past. Another reason may be that the discrimination surface became more and more complex with additional learning. This suggests the necessity of not only additional learning but also moderate forgetting. The word "moderate forgetting" means "forgetting and recalling if necessary". This will be a very attractive subject of machine learning in the future.

Generally speaking, our results above show the effectiveness of additional learning in the original MSM, Fuzzy MSM and Fuzzy RLPD as is seen in Figs. 1-3.

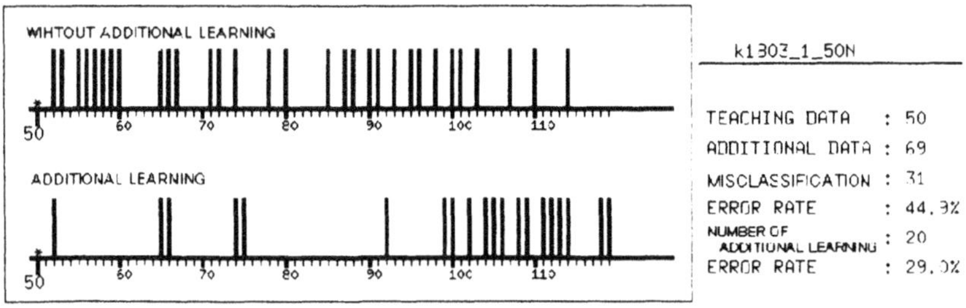

Fig. 1. MSM (trinary classification)

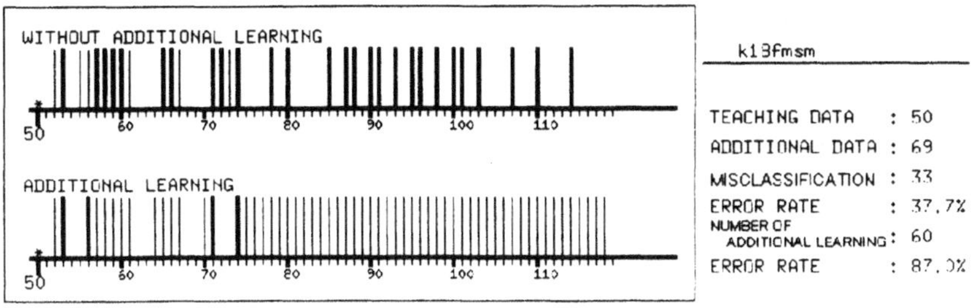

Fig. 2. Fuzzy MSM (trinary classification)

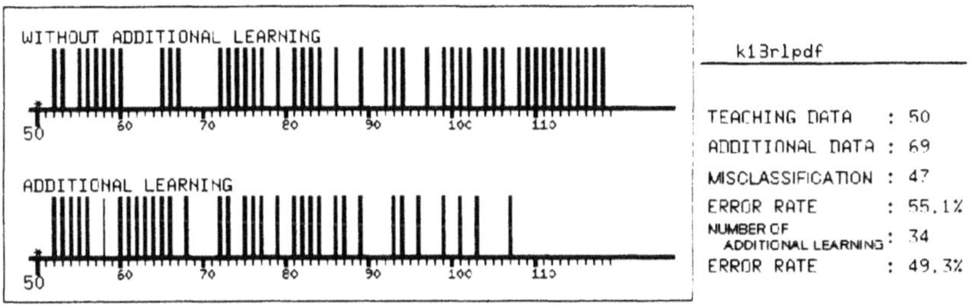

Fig. 3. Fuzzy RLPD (trinary classification)

References

[1] K. P. Benett and O. L. Mangasarian, Robust Linear Programming Discrimination of Two Linearly Inseparable Sets, *Optimization Methods and Software*, Vol. 1, pp. 23-34, 1992.

[2] O. L. Mangasarian, Multisurface Method of Pattern Separation, *IEEE Transactions on Information Theory* Vol.IT-14, No. 6, pp. 801-807, 1968.

[3] H. Nakayama and K. Nasu, Pattern Classification by Linear Programming, presented at XI-th International Conf. on Multiple Criteria Decision Making, Coimbra/Prtugal, 1994 1978.

Goal Programming and Heuristic Search

L. Mandow and E. Millán.
Dept. de Lenguajes y Ciencias de la Computación.
Universidad de Málaga. Campus de Teatinos. MÁLAGA 29071. SPAIN.
FAX: +34-5-2131397. E-mail: {lawrence, eva}@lcc.uma.es.

Abstract. The A* algorithm belongs to the class of *best first* graph-search algorithms and is frequently used to find minimum-cost paths. Recently, A* has been generalised to solve multiple-objective search problems, returning the set of non-dominated solution paths. This paper presents METAL-A*, an admissible heuristic search algorithm for search problems with lexicographic goals that returns the set of non-dominated paths that best achieves a set of lexicographic goals.

Keywords. Lexicographic goals, heuristic search, goal programming

1 Introduction

Most real world problems involve the achievement of multiple objectives or goals. In Artificial Intelligence, problem solving is frequently defined as search in a graph or state space. However, search procedures, most notably A* [3], are formulated in terms of just a single objective function.

Recently, A* has been extended to support multiobjective search in trees [2] and graphs [4]. Although multiobjective approaches to decision making play an important role in multicriteria decision theory, most practical applications often come from goal-based methods.

This paper describes a heuristic search procedure for lexicographic goals. We refer to this algorithm as METAL-A* which stands for METAs Lexicográficas A* (Spanish translation for lexicographic goals A*). This research is motivated by our desire to introduce more practical and realistic decision criteria in Artificial Intelligence problem-solving methods.

The next section introduces notation and some details about search with several criteria. Then the algorithm is presented. Finally, an example is used to illustrate the algorithm with the help of a general technique that provides admissible heuristic functions for any finite graph.

2 Search with lexicographic goals

The following notation is used: lower case letters stand for constants, real-valued functions and nodes; capital letters for sets and capital boldface for vectors.

The *search problem with lexicographic goal satisfaction* can be defined in the following terms: Let G be a graph, s an initial node and Γ a set of goal nodes. Each arc in the graph (n_i, n_j) is labelled with a vector of positive costs

$$\mathbf{K}(n_i, n_j) = (cost_1(n_i, n_j), cost_2(n_i, n_j),..., cost_q(n_i, n_j)),$$

where $cost_u(n_i, n_j)$ (u = 1,..., q) refers to the cost of the associated transition in space u. A path C within the graph G is a sequence of nodes $(n_1, n_2,..., n_m)$ such that for two consecutive nodes n_i and n_{i+1} there is an arc between them. The problem consists in finding a path from the initial node to some goal node which satisfies a set of goals, one for each cost measure in the cost vector.

Definition

Let a *type 1 search problem with (lexicographic) goal satisfaction* be a problem which consists of finding a non-dominated solution path C that satisfies a set of lexicographic goals with just one goal for each priority level, i.e.:

LEVEL 1: $cost_1(C) \leq t_1$
LEVEL 2: $cost_2(C) \leq t_2$
:
LEVEL q: $cost_q(C) \leq t_q$ $\quad\quad \forall i \ t_i \geq 0$ ♣

[1] shows how any search problem with lexicographic goals may be expressed in this form. METAL-A* has been designed to find just one or the set of non-dominated solution paths to lexicographic goal satisfaction problems. A set of path evaluation functions is used to guide the search, one for each additive cost function.

Definition

Let G be a graph where an initial node s and a set of goal nodes Γ have been selected. The function $\mathbf{F}^*(n) = (f_1(n), f_2(n), ..f_q(n))$ gives the cost of a efficient path that goes through node n from s to a goal node, where

$$f_i^*(n) = g_i^*(n) + h_i^*(n), \quad\quad i = 1,2,...,q.$$

$g_i^*(n)$ is the optimal cost in space i for a path from s to n; and $h_i^*(n)$ is the optimal cost in space i from n to any goal node. $f_i^*(n)$ is generally not known but can be approximated by a *path evaluation cost function* $f_i(n)$ of the type:

$$f_i(n) = g_i(n) + h_i(n) \quad\quad i = 1,2,...,q.$$

where $g_i(n)$ is cost in space i of the best path known from s to n, and $h_i(n)$ is an *admissible heuristic function*, i.e. $\forall n \ \forall i \ \ h_i(n) \leq h_i^*(n)$. ♣

A significant difference between A* and METAL-A* is the use of several cost evaluation functions $(f_1(n),..., f_q(n))$ for each path.

Definition

Let $\mathbf{G}(n)$ be the vector of cost $(g_1(n), g_2(n), ..g_q(n))$ of an *efficient path* from s to n.

Let $\mathbf{H}(n)$ be a vector $(h_1(n), h_2(n), .. h_q(n))$ of an *optimistic estimate* of the cost of a path from n to some goal node.

Let $\mathbf{F}(n)$ be a vector $(f_1(n), f_2(n), .., f_q(n))$ of the *estimated values* of the cost of a path that goes through n from the initial node s to some goal node. ♣

1. CREATE

   an empty list of nodes OPEN and open the initial node s,
   an empty list of nodes CLOSED and
   an acyclic search graph SGRAPH with the initial node s.

   GOALN $\leftarrow \varnothing$;expanded goal nodes
   COSTS $\leftarrow \varnothing$;efficient cost vectors of known paths to goal nodes
   CONSTRAINTS $\leftarrow \varnothing$;achievement of previous priority levels
   pr-level $\leftarrow$ 1 ;current priority level

2. NODE SELECTION.

   Include in CANDIDATES all pairs (n, **F**) from OPEN such that **F** is non-dominated by any cost vector in COSTS and satisfies the set of CONSTRAINTS.
   Select a pair (n_1 **F**) from CANDIDATES with the best value for: $f_{pr\text{-}level}(n)$.
   If there are several such pairs, select one with non-dominated **F**, e.g. using the values $f_{pr\text{-}level+1}(n)$, $f_{pr\text{-}level+2}(n)$,..., $f_1(n)$,...,$f_{pr\text{-}level-1}(n)$ to break the tie.
   If several pairs still have equal vector **F**, choose arbitrarily preferring goal nodes.

3. BODY.

   3.1. If CANDIDATES is empty, then go to 4 (TERMINATE).
   3.2. If n_1 is a goal node
   - COSTS $\leftarrow$ eff(COSTS $\cup$ G-SET(n_1)).
   - If an efficient path for *pr-level* has not been found yet,
     then add "$f_{pr\text{-}level}(n) \leq \max(t_{pr\text{-}level}, f^{*r}_{pr\text{-}level}(n))$" to CONSTRAINTS
   - While there are more priority levels left and $f_{pr\text{-}level+1}(n) \leq t_{pr\text{-}level+1}$
     * *pr-level* $\leftarrow$ *pr-level* +1
     * Add "$f_{pr\text{-}level}(n) \leq t_{pr\text{-}level}$" to CONSTRAINTS
   - If there are more priority levels to satisfy
     then *pr-level* $\leftarrow$ *pr-level* +1
     else If only one solution path is needed,
          then go to 4 (TERMINATE).
          else close(n_1) and add n_1 to GOALN.

   3.3. If n_1 is not a goal node,
   - Close(n_1)
   - For each node n_2 successor of n_1 (without cycles).
     If n_2 is in SGRAPH, and new efficient paths to n_2 have been found then
       * update SGRAPH, i.e. add the new efficient paths and delete dominated ones
     If n_2 is closed,
       then open(n_2),
       else update F-SET(n_2) and OPEN.
     If n_2 is a new node (is not in SGRAPH)
       * Open(n_2)
       * Add n_2 to SGRAPH with the new paths found.

   3.4. Go back to 2 (NODE SELECTION)

4. TERMINATE, using SGRAPH, COSTS, GOALN and CONSTRAINTS to return the set of efficient solution paths to the goals problem (or n_1, F and SGRAPH if only one efficient solution path is needed).

Fig. 1. METAL-A*

3 Metal-A* algorithm

The algorithm METAL-A* is shown in figure 1. It is an iterative procedure that explores and makes explicit a portion of a graph G. The basic operation is node expansion, i.e. the generation of all the successors of a node and updating of the data structures. The following data structures are used:

SGRAPH: an acyclic search graph that records efficient paths found to nodes.

OPEN: a list of paths in SGRAPH that can be further expanded.

CANDIDATES: a list of paths in SGRAPH that may lead to efficient solutions to the problem.

CLOSED: a list of nodes already expanded.

GOALN: a list of goal nodes reached by efficient solution paths to the problem.

COSTS: a list of cost vectors of all paths found to goal nodes (efficient of not).

CONSTRAINTS: a set of constraints to ensure goal achievement.

Definition

Let G-SET(n) be the set of all non-dominated cost vectors **G**(n) corresponding to paths from the initial node s to node n included in SGRAPH.

Let H-SET(n) be the set of all non-dominated estimated cost vectors **H**(n) from n to goal nodes.

Let F-SET(n) be the set obtained adding each vector in G-SET(n) to each vector of H-SET(n), and removing the dominated ones, that is;

$$\text{F-SET}(n) = \text{eff}(\ \{\mathbf{G}(n)+\mathbf{H}(n)/\ \mathbf{G}(n) \in \text{G-SET}(n) \text{ and } \mathbf{H}(n) \in \text{H-SET}(n)\})$$

where eff(A) gives the set of efficient vectors of the set A without repetitions. ♣

Some properties of METAL-A* are presented in [1] including a proof about the admissibility of the algorithm provided all heuristic functions are admissible, i.e. the algorithm terminates with the set of efficient solution paths to the lexicographic goals problem (or optionally, a single efficient solution path).

This section describes some important features of METAL-A*. First, the use and structure of OPEN and CLOSED lists is discussed, then the procedure for goal satisfaction is described and finally, the node selection strategy is presented.

3.1 OPEN and CLOSED lists in multicriteria search

In METAL-A* the list OPEN is formed by pairs of the form

$$(n, \mathbf{F})\ , \text{ where } \mathbf{F} \in \text{F-SET}(n),$$

i.e., any generated but not yet expanded node will appear in OPEN as many times as vectors **F** it has in F-SET(n). Thus each pair (n, **F**) represents a different path that can be further expanded in the graph. The operations OPEN(node) and CLOSE(node) are defined in the following way,

- To open a node n_1, add to OPEN a pair (n_1, **F**) for each vector **F** in F-SET(n_1).

- To close a node n_l, delete all pairs (n_l, ·) from OPEN and add node n_l to CLOSED. Note that in any iteration any node n_l may have one (or several, but not necessarily all) vector in F-SET(n_l) corresponding to a possibly efficient solution path in G. Although the algorithm seeks to extend a single path at each iteration, the expansion of a node produces the expansion of all its associated paths.

Note also that when new efficient paths to a node n are found, G-SET(n) changes, and therefore F-SET(n) has to be updated. These changes are reflected in OPEN too, meaning the addition of new paths and subtraction of dominated ones.

3.2 Goal satisfaction

The solution of a type 1 problem involves the achievement of a set of goals according to their priority. METAL-A* uses the functions $f_i(n)$ to estimate the cost of paths in each space. Before trying to find paths that fulfil $f_{i+1}(n) \leq t_{i+i}$ we have to find paths that satisfy $f_i(n) \leq t_i$.

When using A* with a function f(n) and initial node s, the existence of a solution path with cost $f^*(s) \leq t$ cannot be assured until an optimal path is found, since A* always uses an *optimistic* estimation of the cost of the paths. To satisfy lexicographic goals, METAL-A* tries to optimise each function $f_i(n)$ in turn. Once the optimal value for a function $f^*_i(s)$ is found, there are several possibilities:

- If $f^*_i(s) \leq t_i$, then the corresponding goal can be satisfied, and therefore any node n candidate for expansion should have $f_i(n) \leq t_i$.
- If $f^*_i(s) > t_i$, then the corresponding goal cannot be satisfied, but at least the best value found can be preserved. Any node n expanded while searching for paths for level i+1 will have to satisfy $f_i(n) \leq f^*_i(n)$.

The satisfaction of the goal corresponding to level i+1 is constrained to search only those paths that preserve the achievement of goals from previous levels.

The set CONSTRAINTS, used when the algorithm is searching priority level i+1, can be defined as the set of constraints of the form:

$$f_k(n) \leq \max(t_k, f^*_k(s)) \qquad k=1, 2, \ldots, i$$

3.3 Node selection strategy in METAL-A*

Perhaps the most important feature in any best-first search algorithm is the way nodes are selected for expansion. This mechanism is used to determine which path to explore next taking the constraints into account. METAL-A* always expands the node n from the pair (n **F**) selected from CANDIDATES. The pairs (n, **F**) from OPEN that are CANDIDATES to lead to a solution must satisfy two conditions,

- **F** satisfies the set of CONSTRAINTS, i.e., it preserves previous goal fulfilment.
- **F** is non-dominated by any vector of COSTS, i.e., it represents a possibly efficient path to a goal node.

If the current priority level is i, then the pair (n, **F**) selected from CANDIDATES is such that the i-th component of vector **F**, $f_i(n)$, has the *best* value among all the vectors **F** in CANDIDATES. It is possible that several nodes have the same (smallest) value for that component. In this case the values $f_{i+1}(n)$, $f_{i+2}(n),\ldots, f_q(n), f_1(n),\ldots, f_{i-1}(n)$ are used to break the tie. Thus, a path with non-dominated vector **F** from CANDIDATES is selected.

Note that in general lexicographic problems with several goals in each priority level, $f_i(n)$ measures the deviation from a satisfactory solution in a weighted goal model. However, the achievement of the goal $f_i(n) \leq 0$ does not guarantee an efficient solution to the weighted goals problem, and non-dominance has to be ensured over an extended vector **F**, including all original goals from that weighted goal model. [1] discusses this simple extension in more detail.

At termination the efficient solution paths can be found in the search graph, selecting from COSTS those vectors that satisfy the goal set of CONSTRAINTS and tracing back in SGRAPH, from the nodes stored in GOALN, paths with those costs.

4 Using Euclidean distance as a heuristic function

Any graph can be reduced to an Euclidean space in such a way that preserves the *maximum information* possible, i.e., the graph is *minimally collapsed.* Then this Euclidean distance between nodes in the collapsed graph can be used as an admissible heuristic for the original graph. The problem of collapsing the graph can be modelled as a quadratic programming problem. E.g., the coordinate x_i for each node i in a unidimensional Euclidean space is given by the model:

$$\text{Max.} \sum_{i,j \in E} (x_i - x_j)^2$$

$$\text{subject to } |x_i - x_j| \leq c_{ij}, \qquad \text{for all } (i, j) \in E$$

where E is the set of arcs in the graph and c_{ij} the cost associated with each arc (i,j). Note that any feasible solution to the model can be used to obtain a heuristic. The better the solution, the more informed the heuristic will be.

5 Example

This section shows how the algorithm METAL-A* and the graph reduction technique presented above can be applied to a hypothetical search problem. Let us consider that the graph in figure 2 represents a structured environment for a mobile robot on a forbidden planet, i.e. nodes stand for significant locations and arcs denote navigation is possible between nodes. Each arc is labelled with a cost vector (c_1, c_2) where c_1 is the cost in fuel units of navigating through the arc, and c_2 the time in minutes needed for navigation..

Robby, the robot, is currently working at location S and three refuelling stations are located at γ_1, γ_2 and γ_3. Robby is given a break and decides to pick up some fuel. Currently he has 4 fuel units and according to his scheduled work load should not take more that 6 minutes to get to the fuel station. Which path should Robby take?

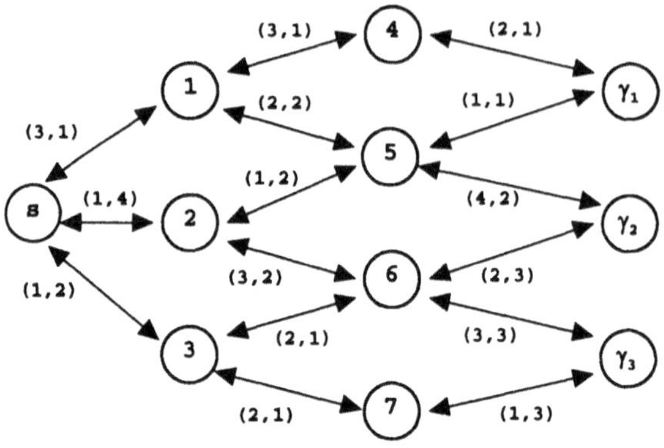

Fig. 2. Example.

Let the goals be:

PR. LEVEL 1: $cost_1(C) \leq 4$ PR. LEVEL 2: $cost_2(C) \leq 6$

For convenience, we may think of the problem as composed of two different graphs, with the same nodes and arcs, but with different scalar costs associated with each arc. The graphs corresponding to each space have been collapsed using the model in section 4. The results obtained are shown in figure 3. The Euclidean distance according to these coordinates will be used as a heuristic for both spaces.

It can be shown that the set of efficient paths to goal nodes for this problem is,

(s 3 7 γ_3) : cost (4, 6) (s 3 6 γ_2) : cost (5, 5) (s 2 5 γ_1) : cost (3, 7)
(s 1 5 γ_1) : cost (6, 4) (s 1 4 γ_1) : cost (8, 3)

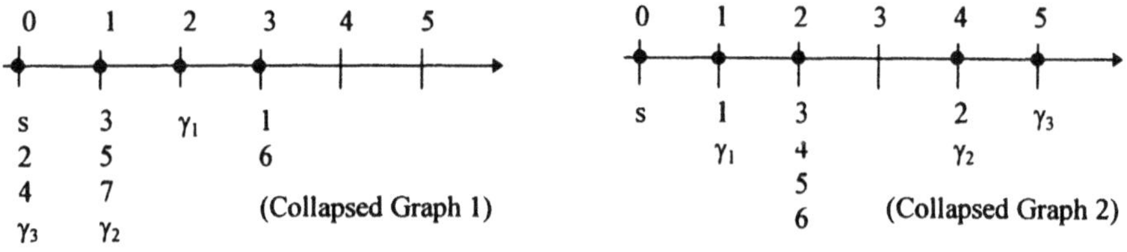

Fig. 3. Coordinates of nodes for the collapsed graphs.

Let $f_1(n)$, $f_2(n)$ be the evaluation functions used by METAL-A* for $cost_1(C)$ and $cost_2(C)$. The sets OPEN, CLOSED, COSTS, CONSTRAINTS and the priority level and node selected for expansion are shown for each iteration in table 1. All vectors **F** of nodes in OPEN shown underlined are not in CANDIDATES.

It.	OPEN node :F-SET(node)	CLOSED	Pr. level	Constr.	Costs	Sel. Node
0	S :(2 1)(1 4)(0 5)	∅	1	∅	∅	S
1	3 :(1 4)(2 3) 2 :(1 5)(2 4) 1 :(4 1)	S	1	∅		3
2	7 :(3 5)(4 4) 6 :(4 4) 2 :(1 5)(2 4) 1 :(4 1)	3 S	1	∅	∅	2
3	5 :(2 8)(3 7) 7 :(3 5)(4 4) 6 :(4 4) 1 :(4 1)	2 3 S	1	∅	∅	5
4	γ_2:(6 8) γ_1:(3 7) 7 :(3 5)(4 4) 6 :(4 4) 1 :(4 1)	5 2 3 S	1	∅	∅	7
5	γ_3:(4 6) γ_2:(6 8) γ_1:(3 7) 6 :(4 4) 1 :(4 1)	7 5 2 3 S	1	∅	∅	γ_1
6	γ_1:(3 7) γ_3:(4 6) γ_2:(6 8) 6 :(4 4) 1 :(4 1)	7 5 2 3 S	2	$f_1(n) \le 4$	(3 7)	1
7	5 :(2 8)(3 7)(5 5)(6 4) 4 :(6 5)(7 4)(8 3) γ_3:(4 6) γ_2:(6 8) γ_1:(3 7) 6 :(4 4)	1 7 2 3 S	2	$f_1(n) \le 4$	(3 7)	6
8	γ_2:(5 5) γ_1:(3 7) 5 :(2 8)(3 7)(5 5)(6 4) 4 :(6 5)(7 4)(8 3) γ_3:(4 6)	6 1 7 2 3 S	2	$f_1(n) \le 4$	(3 7)	γ_3
9	γ_2:(5 5) γ_1:(3 7) 5 :(2 8)(3 7)(5 5)(6 4) 4 :(6 5)(7 4)(8 3)	γ_3 6 1 7 2 3 S	2	$f_1(n) \le 4$ $f_2(n) \le 6$	(3 7) (4 6)	-

*Table 1. Trace of METAL-A**

Initially, the heuristic function $h_1(n)$ takes METAL-A* to the path s-2-5-γ_1 (only 6 nodes are expanded instead of 9 if $h_1(n) = 0$ were used). Satisfaction of the first goal is enforced through the set CONSTRAINTS, and $h_2(n)$ guides the search to satisfy the second goal. Although the best paths for $f_2(n)$ lead to γ_1 and then γ_2, the algorithm quickly turns to γ_3, since paths to γ_1 and γ_2 do not satisfy the first goal. Finally, the efficient solution path S-3-7-γ_3 is found. After the corresponding constraint is added to CONSTRAINTS the algorithm finds there is no open node that can possibly lead to a solution (CANDIDATES is empty) and terminates. The solution paths can be traced from COSTS, GOALN and SGRAPH. In this problem there is only one efficient solution path: S-3-7-γ_3 to the lexicographic goals problem. Note that although a non-dominated path was found to γ_1 it was later discarded as a solution to the problem, since it did not satisfy the time goal.

6 Conclusions

Previous work in the field of heuristic search has laid the foundations of search for non-dominated paths in graphs with multiple objectives. Although non-dominance of solution paths is used as a decision criteria, many Pareto-optimal solutions may exist. Most practical applications include goals to further establish the desired properties of a solution. We have devised and implemented a new algorithm, METAL-A*, that extends the search paradigm to find solution paths for a set of goals, and proved its admissibility, i.e. the algorithm terminates with the set of non-dominated solution paths to lexicographic goals problems. The algorithm is also admissible if we want to terminate with the first non-dominated solution path found. An algorithm like METAL-A* has great potential for applications in Artificial Intelligence problems, since search is widely used as a problem solving paradigm. The application of METAL-A* to problems like design and robot path planning is the subject of our current research.

REFERENCES

1. Mandow, L. *Búsqueda Heurística con Satisfacción de Metas.* Tech. Rep. Dpto. Lenguajes y Ciencias Computación. University of Málaga, 1997.
2. Navinchandra, D. *Exploration and Innovation in Design: Towards a computational model.* Springer-Verlag, New York, 1991.
3. Pearl, J. *Heuristics.* Addison-Wesley, Reading, Mass. 1984.
4. Stewart, B. and White, C.C. *Multiobjective A**, Journal of the ACM, Vol 38, No 4, October 1991, pp 775-814.

Goal Programming Model for Evaluating Hospital Service Performance

ARENAS PARRA, M. MAR
LAFUENTE ROBLEDO, ESTHER
RODRIGUEZ URIA, M.VICTORIA
Departamento de Matemáticas . Universidad de Oviedo*. Spain.

Abstract. The purpose of this paper is to present a G.P. model to design the real performance of a surgical service at a local general hospital. Through the G.P. approach we intend to offer the decision centre a suitable methodology that permits, on the one hand , to analyse the inner coherency of the goals and, on the other hand to know whether it is possible or not to improve the running of the service, taking into account all the real constraints, e.g. spatial occupation, staff availability and financial support.

Keywords: Goal Programming, decision making, hospital management.

1 Introduction

This paper attempts to explicity recognise both, financial and nonfinancial objectives, presenting a Lineal Goal Programming model for clinic and surgical operations. The model permits the decision maker to evaluate their operation plans in terms of providing quality service, minimizing costs and maintaining resource allocation.

Hospitals in general function in a highly complex environment in which there are conflicting opinions about how best to meet the primary missions of that institution. There is a set of values that stresses the quality of care and a set of pressures to meet the community medical service needs. At the same time, there is pressure to maintain a financially viable institution. That is why Administration of virtually every Hospital is not a unique management problem.

The Spanish National Health Service is involved in a process of radical changes, searching for a way to improve efficiency and efficacy and to control public health costs. Public hospitals will have more autonomy than at present from central/regional government, especially in the areas of operational management and human resource policy.

---

* e-mail:VRODRI@hp845.ECONO.UNIOVI.ES

Evaluation criteria will be introduced to measure the effectiveness and efficiency of providers. There will be fewer layers of management in hospitals and hospital boards will become more business oriented in their outlook.

Even though hospital administration has become a very complex management process, an efficient delivery of health services through a more scientific management could be developed using GP techniques.

This study attempts to apply a management science technique to improve the efficiency of Hospital Administration. More specifically, it applies G.P. to design an efficient set of products offered by our local Hospital.

2 Formulation of a Goal Programming Model

Multiple Criteria Decision Making -M.C.D.M.- attempts to give the decision maker a mathematical support to decision making processes that involve several conflicting objectives subjected to a set of constraints. In real world problems, the concept of goal may be more useful than the concept of optimization of an objective function because, as we said in the introduction, many corporations follow a satisfaction criteria, which can be described by an acceptable level of achievement for any attributes considered, that is, by setting goals.

Lee (1971- 1973) has applied GP formulation to problems in the health care field and has shown that the flexibility of choosing priorities of the model is a special advantage because it permits the decision maker to make different choices to find the one that represents the best option in each circumstance.

The G.P. Model described here was developed for the COT[1] service of a local stated-owned Hospital, to plan one year ahead, because the economical previsions are also made on a yearly basis. To simplify the model design, we will omit a detailed description of the Hospital.

The model is formulated according to the following steps:

2.1 Program decision variables.

We are intending to adequate the activity developed by the COT service through a GP model and in order to do that, we have considered necessary to link the products offered by this in the following groups:

a.- **Surgery Activity (X).**
b.- **External Activity (Y).**
c.- **Hospitalization (Z).**

[1] - Traumatology and Orthopedic Surgery Cirugía - Ortopédica y Traumatología.

2.2 Data

In order to use the algorithm we have collected certain data related to the problem's variables and we have presented them in several tables along the paper. They have been filled in using the information returned by the decision makers

2. 3 Goals and their priorities..

The objectives to be considered by the GP model in order to design the optimum performance of the hospital service were arrived at through discussions with hospital decision makers. These are:

- Keeping staffing requirements and spatial availability.
- Minimizing operating costs.
- Decreasing waiting lists.
- Providing quality service.

Priority One

- *Assure the necessary personnel to give satisfactory services to the patient:* The prior objective is to ensure the personnel needed to offered the patient a satisfactory service; the decision centre estimates that the actual staff will be enough to keep the demand in.

Table 2.1 shows the staff requirements:

Table 2.1

Var.	Products	Physicians	Nurses
x_{11}	Scheduled operation with admission	200′	200′
x_{21}	Emergency Operation with no admission	200′	200′
x_{12}	Ambulatory scheduled Operation	90′	90′
x_{22}	Ambulatory Operation Non-scheduled	90′	90′
y_{11}	First Appointments	20′	
y_{12}	Following Appointments	10′	
y_{13}	Ambulatory Emergency	32′	
z	Hospitalisation	15′	

Continuing with this priority, next we will show table 2.2 containing the global time available for Doctors, Surgery Nurses, Hospitalisation Nurses, and the Supervisor Nurse.

Table 2.2

	Physicians	Hospitalisation Nurses	Supervisor	Surgical Nurses
Time Resource[2]	23.333	24.480	493,5	6.120
Discount		560		140
Total Amount	23.333	23.920	493,5	5.980

Following with this priority, the spatial requirements are also a hard restriction; the hospital is not intending to make any kind of extension and the resources are fixed, at least, in the space time considered in the model. Spatial requirements are shown in table 2.3.

Table 2.3

	Beds	Consultory Rooms	Operating Rooms
Available	66	10.920 h/annum	2.912 h/annum

These are the restrictions corresponding to this priority:

Human Resources

Physicians:

$$200x_{11} + 200x_{21} + 90x_{12} + 90x_{22} + 20y_{11} + 10y_{12} + 32y_{13} + 15z \leq 1.400.000$$

Surgical Nurses: $200x_{11} + 200x_{21} + 90x_{12} + 90x_{22} \leq 358.800$

Consultory Nurses: $(20/3)y_{11} + (10/3)y_{12} \leq 218.400$

Hospitalisation Nurses: $70,28z \leq 1.435.200$

Spatial Resource:

Consultory Room: $20y_{11} + 10y_{12} \leq 655.200$ [3]

Operating Room: $100x_{11} + 100x_{21} + 45x_{12} + 45x_{22} \leq 174.720$

Priority two:

-Minimise cost:

We have, as second priority, taken into account the need of decreasing or, at least maintaining COT service costs. Therefore, we will carry out the service costs analysis in order to minimise them. Our base-year has been 1994 and the decision centre consider that costs will have to stay or decrease but they are not willing to increase.

---

[2] In hours

[3] Data in restrictions are expresed in minutes.

Once having made the cost evaluation we have developed a model to allocate it into each of the variables. In order to do that we have applied a costs assignment technique used for most of the Spanish Hospitals. Table 2.4 shows costs corresponding to each variable:

Table 2.4

Var .	Activity Level actual	Activity Level Desirable	Cost per unit
x_{11}	1.230	1320	19.437,57
x_{21}	66	300	19.437,57
x_{12}	27	260	9.718,78
x_{22}	8	20	9.718,78
y_{11}	14.187	10.000	9.719,47
y_{12}	22.663	20.000	5.830,5
y_{13}	16.539	10.000	11.663,25
z	20.655	20.850	38.875,15

Here is the corresponding equation[4]:

$$19{,}43757\,x_{11} + 19{,}43757\,x_{21} + 9{,}71878\,x_{12} + 9{,}71878\,x_{22} + 9{,}71947\,y_{11} + 5{,}8305\,y_{12} + 11{,}66325\,y_{13} + 38{,}87515\,z \leq 1.291.423$$

Priority 3

This third group includes some aspects related to quality service. One of the main problems of the National Health Service are waiting lists in some medical and surgical services, specially in surgical ones. Due to strictness and impossibility of doing any change in a short term we have been obliged to restrict the model with the following bounds - upper and down- which show the need of spanding activities in order to reduce waiting lists. The desired activity levels were shown in table 4:

Equations can be expressed as:

$1.230 \leq x_{11} \leq 1.320$

$66 \leq x_{21} \leq 360$

$27 \leq x_{12} \leq 130$

$8 \leq x_{22} \leq 20$

$10.000 \leq y_{11} \leq 14.187$

$20.000 \leq y_{12} \leq 22.663$

$10.000 \leq y_{13} \leq 16.539$

$20.655 \leq z \leq 20.850$

Priority four

In this priority we set constraints referred to the quality of C.O.T. service, through the most frequent activity indexes:

4 In thousand pesetas

* **Occupation Index**: (Patient-days / Annual Average of beds working x 365days) x 100:

In 1994, the number of patient-days was 20.655, with 66 beds: **O.I.** = (20.655/(66x365)) x 100 = 85,74%.

* **Turnover**: Number of Admissions / Number of beds:

In 1994, the number of admitted patients was 1.779, then : **R.I.** = 1.779 / 66 = 26,95.

* **Average length of stay** : Patient-days / Number of Admissions (scheduled and emergency):

In 1994: **A.L.S.** = 20.655 / 1.779 = 11,65.

Next table shows data and pretensions:

Table 2.5

	Actual	Minimum	Maximum
Occupation Index	85,74%	75	85
Turnover	26,95	27	30,4
Average Length of stay	11,65	9	11,5

As we have seen previously, the Occupation Index appears in the first goal, anyway Average Length of Stay as well as Turnover are matched and they could be placed in the same equation. An increase in the length of stay will imply a decrease in turnover. So this is the constraint referred to this last priority:

$$27 \leq \frac{\frac{x_{11} + x_{21}}{0,73}}{66} \leq 30,4$$

$$1.301 \leq x_{11}+x_{21} \leq 1.465$$

The right hand side has been obtained from historical data got from the Hospital Decision Centre, so Total Admissions could be separated into Surgical and Clinical Admissions, the last ones being about 27% of the total amount.

Once all equations have been developed, we may model the G.P. program. In order to solve it we will use the lexicographic method. Therefore, the modelization will be as follows:

$$Lexmin\ a = (p_1 + n_1 + p_2 + p_3 + p_4 + p_5 + p_6 \cdot p_7 \cdot n_8 \div p_9 + n_{10} + p_{11} + n_{12} + p_{13} + n_{14} + p_{15} + n_{16} + p_{17} + n_{18} + p_{19} + n_{20} + p_{21} + n_{22} + p_{23} \cdot n_{24} + p_{25})$$

Subject to:

$$200x_{11} + 200x_{21} + 90x_{12} + 90x_{22} + 20y_{11} + 10y_{12} + 32y_{13} + 15z + n_1 - p_1 = 1.400.000$$

$$.200x_{11} + 200x_{21} + 90x_{12} + 90x_{22} + n_2 - p_2 = 358.800$$

$$(20/3)y_{11} + (10/3)y_{12} + n_3 - p_3 = 218.400$$

$$70,28\, z + n_4 - p_4 = 1.464.810$$

$$20y_{11} + 10y_{12} + n_5 - p_5 = 655.200$$
$$100x_{11} + 100x_{21} + 45x_{12} + 45x_{22} + n_6 - p_6 = 174.720$$
$$19{,}43757x_{11} + 19{,}43757x_{21} + 9{,}71878x_{12} + 9{,}71878x_{22} + 9{,}71947y_{11} + 5{,}8305y_{12} +$$
$$11{,}66325y_{13} + 38{,}87515z + n_7 - p_7 = 1.291.423$$

$x_{11} + n_8 - p_8 = 1.230$ $\quad$ $x_{11} + n_9 - p_9 = 1.320$

$x_{21} + n_{10} - p_{10} = 66$ $\quad$ $x_{21} + n_{11} - p_{11} = 360$

$x_{12} + n_{12} - p_{12} = 27$ $\quad$ $x_{12} + n_{13} - p_{13} = 130$

$x_{22} + n_{14} - p_{14} = 8$ $\quad$ $x_{22} + n_{15} - p_{15} = 20$

$y_{11} + n_{16} - p_{16} = 10.000$ $\quad$ $y_{11} + n_{17} - p_{17} = 14.187$

$y_{12} + n_{18} - p_{18} = 20.000$ $\quad$ $y_{12} + n_{19} - p_{19} = 22.663$

$y_{13} + n_{20} - p_{20} = 10.000$ $\quad$ $y_{13} + n_{21} - p_{21} = 16.539$

$z + n_{22} - p_{22} = 20.655$ $\quad$ $z + n_{23} - p_{23} = 20.850$

$x_{11}+x_{21} + n_{24} - p_{24} = 1.301$ $\quad$ $x_{11}+x_{21} + n_{25} - p_{25} = 1.465$

3 Model runs and results

We have pretended to offer DM several possible optimal solutions, so we have formulated the above G.P model which has been solved in four different ways:

First of all we have solved the problems **P1**and **P2** with the previously mentioned preemtive priorities in two different alternative hypothesis:

P1 spending all the Physicians time and **P2** not spending all the time.

Secondly, we have taken into account the possibility of trading off Priorities second and third, giving more weight to decrease the waiting list instead of minimising costs.

P3 is the option of spending all the Physicians time, like in **P1**. In **P2**, as in **P4**, we have left free Physicians time.

In any case, all the goals have been reached, and the decision variables values are very similar, as it is shown in the next figure:

It may be remarked, according with figure 3.1, that decision variable values in **P1** and **P3** are almost the same, as well as **P2** and **P4**. Main differences between all problems are in deviation variables, although they scarce significance for Decision Centre.

After presenting all the options to the Decision Maker, it has been chosen the second alternative (**P2**), -or its corresponding Pareto Optimums- because it allows some surplus in physicians time which could be used for teaching and clinical investigation.

Table 3.1

Var.	Desirable	Activity level	P1	P2	P3	P4	First Pareto Optimum **P1/P3**	Second Pareto Optimum **P1/P3**	First Pareto Optimum **P2/P4**	Second Pareto Optimum **P2/P4**
x_{11}	1.320	1.230	1.235	1.235	1.235	1.235	1.230	1.235	1.230	1.235
x_{21}	300	66	66	66	66	66	71	66	71	66
x_{12}	260	27	260	27	260	27	27	27	27	27
x_{22}	20	8	20	8	20	8	8	8	8	8
y_{11}	10.000	14.187	10.000	10.000	10.000	10.000	10.000	10.000	10.000	10.000
y_{12}	20.000	22.663	20.000	20.000	20.000	20.000	20.000	20.000	20.000	20.000
y_{13}	10.000	16.539	12.561	10.000	12.561	10.000	13.338	13.338	10.000	10.000
z	20.850	20.655	20.842	20.842	20.842	20.842	20.655	20.655	20.655	20.655
n_1			0	104.013	0	104.013	0	0	106.825	106.825
p_1			0	0	0	0	0	0	0	0
n_2			73.400	94.450	73.400	95.450	95.450	95.450	95.450	95.450
n_3			85.068	85.068	85.068	85.068	85.067	85.067	85.067	85.067
n_4			0	0	0	0	13.177	13.177	13.177	13.177
n_5			255.200	255.200	255.200	255.200	255.200	255.200	255.200	255.200
n_6			32.020	32.020	32.020	43.045	43.045	43.045	43.045	43.045
n_7			92.848	92.848	92.848	125.102	93.455	93.455	132.391	132.391

Once being chosen the optimum solution we will discuses all the priorities:

Priority One: It has been achieved. There are some surpluses in Physicians, nursery, and local Consultory 's time. It is significant the evidence of surgery rooms time which has been totally spent, Sensitivity Analysis has shown that increasing in more than minutes surgery room time, global surgical activity could be increased without acting on any other constraint.

Priority two: The optimum activity level could be reached without spending all many available in the service. It could mean some inefficient allocation of resources. Actual level of activity is lower than optimum one and costs are bigger.

Priority three: Some of the desirable levels of activity have been reached in spite of the other ones. Scheduled operation with admission -x_{11}-, has increased in 0,40%; however, Emergency operation with no admission keeps in its level. External activity represented by variables y_{11}. , y_{12} and y_{13} reached their desirable values. The wished days-of-stay (Hospitalisation - z) has been almost achieved. Ambulatory surgery keeps in their actual levels of activity, which could be increased growing surgical rooms time availability.

Priority four: Turnover has been reached in its minimum desirable value. Despite this, average length of stay has not been achieved neither in its minimum nor in its maximum value.

Solution Average length of stay: 20.842/1.782[5] = 11,69 greater than actual level.

Solution Turnover : 1.782/66 = 27.

[5] This has been obtained : (1.235+66)/0,73.

4 Conclusions

Once being observed all the solutions and after exposing them to the Decision Centre, we could conclude that with actual technical restrictions it is impossible to reduce significantly waiting lists. These constraints could be eliminated by producing health services in a more efficient way, having in mind than there exists an important economic, as well as personnel, surplus which should be a Decision Centre task.

It is necessary to reorganise the COT service by expanding surgical activities decreasing the external and clinical ones. There should be some reduction on the number of External Health Services through a joint-elaborated program between Ambulatory Health Services and Hospital Services.

5 References

ANDERSEN CONSULTING (1993): "The Future of European Health Care". *A study by Andersen Consulting in Co-operation with Burson-Marsteller.*

ARENAS, M.; BILBAO, A.; RODRIGUEZ, M.V. (1992): "Aplicación de la programación por objetivos a la toma de decisiones de un centro de salud". *VI Reunión Anual ASEPELT* - ESPAÑA. Granada.

BITRAN, G.R.; VALOR-SABATIER, J. (1987): "Some Mathematical Programming Based Measures of Efficiency in Health Care Institutions". *Advances in Mathematical Programming and Financial Planning.* Vol 1: 61-84.

CHARNES, A., COOPER, W. (1961). *Management Models and Industrial Applications of Linear Programming".* vols. 1 and 2 / New York Wiley.

LEE, S. (1973): "An Aggregative Resource Allocation Model for Hospital Administration". *Socio-Economic Planning Science.* Vol. 7, 381-395.

ORTÙN RUBIO, V; (1990) *La economia en sanidad y medicina: Instrumentos y limitaciones.* Ed. Euge, Barcelona.

WATCH, R.F.; WHITFORD, D.T. (1976): "A Goal Programming Model for Capital Investment Analysis in Non-profit Hospitals". *Financial Management* 5: 37-56.

An Application of Interactive Multiple Goal Programming on the Warsaw Stock Exchange

Cezary Dominiak
Department of Econometrics, Academy of Economics
ul. Bogucicka 14, 40-226 Katowice, Poland

Keywords. portfolio selection, multicriteria decision aid, interactive multiple goal programming

1 Introduction

The first research concerning stock selection and evaluation methods goes back to the 1929 great depression. Initially there were two main approaches: fundamental and technical analysis. In the 50's a new trend in stock market analysis started up, namely portfolio analysis. Its origins and development were connected with Markowitz's [4], [5] and Sharpe's [12] papers. According to them portfolio selection is conceived as a problem of finding an efficient set of portfolios. The probabilistic estimates of future risks and returns of securities are applied. Then an investor chooses one portfolio from the efficient set.

Theoretical research on portfolio analysis, as well the empirical works on the efficiency of capital market and the serious revaluations in decision science brought about the consideration of portfolio analysis as a multicriteria decision making problem. In the multilicriteria approach prospective models are considered rather than normative ones. This enables us to take into consideration an individual investor's goals and preferences. Many scholars divide the portfolio selection problem into two phases: security evaluation and selection and the determination of the proportions of stocks in the portfolio. In the security selection phase the discrete multicriteria decision support methods are applied eg.: Electre [6], [14], [15], AHP [11]. To support the determination of the portfolio structure Lee and Lerro [3], Hallerbach and Spronk [2] apply goal programming, Zouponidis [14], [15] presents the decision support system using ADELAIS method, Nijhoff [8] and Hallerbach and Spronk [2] consider the use of Interactive Multicriteria Goal Programming.

In this paper we propose the multicriteria model of portfolio selection which is based on Sharpe's single index model. Moreover we consider additional objectives that relate to the market and financial attributes regarding the Decision Maker's (DM) preferences. Finally, we consider an occurence of additional constraints. This enables us to take into account an individual's restriction set upon the proportions of the chosen stock groups in the portfolio. We apply Interactive Multiple Goal Programming (IMGP) in the decision support process. In the first part of this paper the proposed model and optimizing problems under consideration are described. Then an example of the portfolio selection support session is presented. In the third paragraph we consider two strategies of portfolio selection and we present the results of a numerical example based on the real world data from the Warsaw Stock Exchange. Finally some concluding remarks are made.

2. Multicriteria model for portfolio selection

We consider the multicriteria model using the expected return and risk according to Sharpe's single index model. The next two formulae describe portfolio expected return and variance, respectively.

$$R_p = \sum_{i=1}^{n} x_i * \alpha_i + R_M * \sum_{i=1}^{n} \beta_i * x_i$$

$$V_p = V_M \left(\sum_{i=1}^{n} \beta_i x_i \right)^2 + \sum_{i=1}^{n} x_i^2 v_i$$

where n denotes the number of stocks, x_i -the weight of i-th stock in the portfolio. R_M denotes the expected market rate, V_M - the variance of the market rate. The parameters α and β are estimated using ordinary least squares from the
equation:

$$R_{M,t} = \alpha + \beta * R_{i,t} + \zeta_t,$$

where R_{Mt} is the market return in period t, R_{it} is the return of the i-th stock in period t, v_i denotes the residual variance of i-th stock. Morover we consider the set of additional objectives that are the linear functions of weights x_j. These functions are as follows:

$$F_j = \sum_{i=1}^{N} x_i * c_{i,j} \rightarrow opt \quad j = 1,\dots,k$$

where j is the number of goal, $c_{i,j}$ is the value of j-th attribute of i-th stock. As objectives we consider market indicators (valuation measures) such as P/E - price earning ratio, P/BV price of stocks to book value, MC/S - market capitalization to sales. Many scholars (e.g. [1], [10]) report that return of stocks with low P/E ratio exceed the return of stocks with high P/E ratio on average in the long time even after adjusting for betas. Morover Reingunum

[10] shows that P/BV is one of the most important criteria for security selection. We propose to take into account objectives based on financial indicators such as ROE - return on stockholder's equity that reflects the profitability of assets included in the portfolio, and DR - debt ratio that can be treated as an additional measure of risk. We assume that the DM chooses the objectives that he/she is interested in. We do not consider short sales and borrowing money, so the basic constraints can be written as follows:

$$\sum_{i=1}^{n} x_i = 1 \quad x_i \geq 0$$

We propose to take into account the additional constraints on grouping stocks regarding the DM preferences. The stock groups may reflects for instance different industries or single stocks.

$$Ax \leq b$$

where **A** is 0-1, (k x n) matrix, k detones the number of groups considered by the DM and **b** is a vector of maximum weights of each group in the portfolio. The element $a_{i,j} = 1$ dentones that j-the stocks belongs to i-th group.

In the portfolio selection we apply the IMGP method. IMGP consists of the following main steps:

Step 1: solve single criterion problems,
Step 2: compute the potency matrix,
Step 3: the DM evaluates the results on the basis of the potency matrix,
Step 4: modify the single criterion problems and go to the step 1.

Some important advantages are connected with the IMGP. First of all the DM does not have to give his preference information on a priori basis but has to consider all kinds of choices and trade-off questions which may be relevant (see.: Nijhoff 1980 p.104). Another important advantage of IMGP is its relatively simply and easy to understand idea. Finally the DM during an interactive procedure has to answer the simple questions:

1. Is the given solution acceptable or not?
2. Which goal value needs to be improved?
3. How much should this goal value be improved at least?
4. Do you accept the consequences of the proposed improvement of the value of the indicated goal variable? (see.: [8] p.250)

Thus in the presented multicriteria model of portfolio selection we consider the following single criterion problems:

Problem 1.
Maximizing the return:

$$R_p = \sum_{i=1}^{n} x_i * \alpha_i + R_M * \sum_{i=1}^{n} \beta_i * x_i \rightarrow MAX$$

st.

$$Ax \leq b$$

$$\sum_{i=1}^{n} x_i = 1, \ x_i \geq 0$$

Problem 2
Minimizing the risk:

$$V_p = V_M(\sum_{i=1}^{n} \beta_i x_i)^2 + \sum_{i=1}^{n} x_i^2 v_i \rightarrow MIN$$

st.

$$Ax \leq b \quad \sum_{i=1}^{n} x_i = 1, \ x_i \geq 0$$

Problem j-th (j=3,...,k)
Optimizing the additional goals.

$$F_j = \sum_{i=1}^{n} x_i * c_{i,j} \rightarrow opt$$

where opt=max or opt=min regarding the choosen goal.
st.

$$Ax \leq b \quad \sum_{i=1}^{n} x_i = 1, \ x_i \geq 0$$

From the second iteration according to DM's choices one of the following constraints is added in each iteration.

$$R_p = \sum_{i=1}^{n} x_i * \alpha_i + R_M * \sum_{i=1}^{n} \beta_i * x_i \geq r_p$$

where r_p denotes an acceptable value of the expected portolio return.

$$V_p = V_M(\sum_{i=1}^{n} \beta_i x_i)^2 + \sum_{i=1}^{n} x_i^2 v_i \leq \mathrm{var}_p$$

where var_p is an accepted portfolio risk (return variance).

If F_j -> MAX then

$$F_j = \sum_{i=1}^{n} x_i * c_{i,j} \geq f_j$$

If F_j -> MIN then

$$F_j = \sum_{i=1}^{n} x_i * c_{i,j} \leq f_j$$

where f_j denotes an acceptable value of the j-th criterion. If the DM wants to improve an acceptable goal value that was improved in one of the previous iterations, then the right-hand side value is changed. Since the considered problems are not linear we solve them using the optimizing package based on numerical optimizing methods Solver for MS Excel 5.0. We use the conjugate gradient method for computing the search direction, quadratic extrapolation from a tangent vector and central differencing for estimates of partial derivates.

3. An examle of the decision support session

In this paragraph we present an example of the decision support session using the model and method described above. The portfolio selection support system is written as the Excel macro. It consists of two subprocedures: the security selection procedure and the procedure of setting the portfolio structure. The security selection procedure uses the multicriteria discrete analysis method based on the idea of reference solution. Evaluated stocks are compared to each other regarding their position in relation to the reference system. The reference system consists of two sets of stocks: the desirable ones (extremely good) and the undesirable ones (extremely bad). The reference sets are constructed using the past data and the proposed weights of the criteria and veto levels are calculated on the basis of the distribution analysis of the considered criteria within the reference sets. For the purpose of this example let us assume that the DM has selected 10 stocks that may be included in the portfolio. The DM has decided to consider the following objectives: Return, Risk, P/BV, P/E and ROE. The parameters of the single index model were computed on the basis of weekly returns. Finally, let us assume that the DM does not cosider additional constraints on the weights of stocks. In the first iteration we obtain the optimal solutions that are presented in table 1. The first potency matrix contains values shown in table 2. The DM chooses the goal whose value should be improved and inputs a new acceptable value. Let us assume that the DM wants to reduce the risk to maximum 3.50%. After the optimizing process the next potency matrix is presented to the DM. Let us assume that the DM accepts the new optimistic values of objectives and decides to reduce the acceptable P/BV of the portfolio to 1.20. The obtained results and the DM's decision for the further iterations are shown in table 3.

Table 1. The optimal solutions of single criterion problems (Iteration 1)

Problem Criterion	Max: Return	Min: Risk	Min: P/E	Min: P/BV	Max: ROE
Return	**2.35%**	1.60%	1.80%	1.40%	1.83%
Risk	3.91%	**2.91%**	6.38%	4.47%	6.32%
P/E	7.23	7.81	**6.08**	11.20	6.08
P/BV	1.07	1.29	1.74	**0.97**	1.74
ROE	22.00%	24.00%	52.40%	9.10%	**52.45%**

Table 2 The optimistic and acceptable values of criteria

	Return	Risk	P/E	P/BV	ROE
optimistic	2.35%	2.91%	6.08	0.97	52.45%
acceptable	1.40%	6.38%	11.20	1.74	9.10%

Table 3 An example of the DM's session

Iteration	Solution	Return	Risk	P/E	P/BV	ROE
1	opt.	2.35%	2.91%	6.08	0.97	52.45%
	acc.	1.40%	6.38%	11.20	1.74	9.10%
	decision		**≤3.50%**			
2	opt.	2.20%	2.91%	6.46	1.01	34.00%
	acc.	1.40%	3.50%	9.87	1.62	13.14%
	decision				**≤1.20**	
3	opt.	2.17%	3.00%	6.95	1.01	26.00%
	acc.	1.60%	3.50%	9.88	1.20	13.40%
	decision					**≥25%**
4	opt.	2.15%	3.16%	6.96	1.17	26.00%
	acc.	2.00%	3.50%	7.46	1.20	25.00%
	decision		**not**	**accepted**		
			≤3.10%			
4"	opt.	1.82%	3.00%	7.76	1.15	22.80%
	acc.	1.60%	3.10%	8.63	1.20	19.40%
	decision	**≥1.82%**				
5	opt.	1.82%	3.10%	7.76	1.20	22.12%
	acc.	1.82%	3.10%	7.76	1.20	22.12%

When the optimistic and acceptable values of the criteria of considered goals are equal to each other (with desired tolerance limit) then the procedures stops. The obtained portfolio structure is presented in table 4.

Table 4. Final solution - portfolio structure

Stock	Weight
BRE	18.61%
Stalexport	47.22%
Dębica	23.04%
MostalZab	0.41%
Novita	4.62%
Sokołów	0.77%
Wólczanka	5.33%

If the DM accepts the results then the decision support process is finished. If not, then the DM is able to carry out the procedure again, revising his/her decisions.

4. A comparison between simple multicriteria strategies and the market index

We carried out the comparison between the market index and two simplified strategies based on the multicriteria model. In both of them a goal quasi-hierarchy is assumed. The first strategy assumes that the DM after the first iteration reduces the risk to the optimistic value of 15% of tolerance stage and subsequently improves the P/E to the optimistic value obtained in the second iteration. The second strategy is the same in the first step but in the second iteration of the IMGP the P/BV is improved to the optimistic value. We carried out the simulation based on the monthly data from December 1994 till March 1996. In each month 30 stocks were randomly selected for the portfolio. The obtained average returns of the strategies are presented in table 5.

Table 5 The rates of return

	Market Index	Strategy1	Strategy2
Average return	3.12%	6.58%	6.22%
STDEV of Return	15.12%	19.57%	19.13%

The average deviation from the index return are 3.46% and 3.10% for the first and the second strategies respectively. The comparison between the capital invested in the market index and invested according to the considered strategies during 16 months is shown in the Figure 4.

Figure 4. A comparison of between assumed strategies and market index

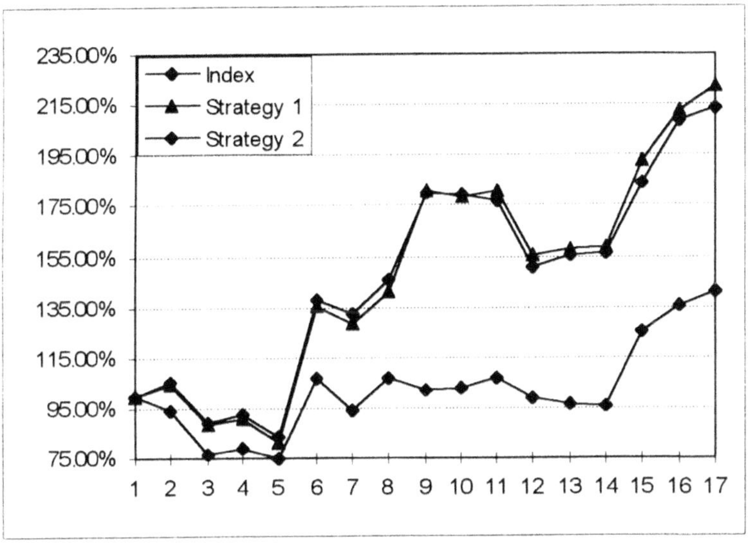

5. Conlusions

In this paper the multicriteria model of portfolio selection is presented. The model uses the risk return measures according to Sharpe's single index model and the set of some additional objectives and constraints. We applied the considered model in the decision support system which uses Interactive Multiciteria Goal Programming. An example of the decision support session is shown in the second paragraph. In the last part of the paper the results of the two simplified multicriteria strategies are shown.

The values of alpha parameters, as well as the residual variance and their significant differentiation obtained for the public companies quoted at the Warsaw Stock Exchange qualify to use Sharpe's non-simplified model. Thus we take into account alphas, betas and residual variances rather than only betas. However, regarding the complex character of optimizing problems we apply the numerical methods while solving them.

The results of the simulation for real data shown in the fourth paragraph of this paper, though obtained for a very short period of time confirm the use of the multi objective approach in the portfolio selecion problem.

References

[1] Basu S.: The investment performance of Common Stocks in Relation to their Price-Earning Ratios: A Test of the Efficient Marke Hypothesis, Journal of Finance vol. 32, 1977 pp 663-682

[2] Hallerbach W., Spronk J.; Portfolio management for the non-average investor, Proceedings of the XIIth International Conference on MCDM, Hagen, June 1995

[3] Lee, S.M., Lerro, A.J.: Optimising the Portfolio Selection for Mutual Funds, Journal of Finance, Vol.28, No.5, pp 1087-1101, 1973

[4] Markowitz H.M: Portfolio Selection, "Journal of Finance", 1952 vol. 7, s. 77-91

[5] Markowitz, H.M.: Portfolio Selection, Willey, New York 1959

[6] Martell J.M. Khoury N.T, Bergeron M.: An Application of a Multicriteria Approach to Portfolio Comparisons, Journal of Operations Research No 7,1988 pp. 617-628

[7] Nicholson S. F.: Price Earnings Ratios, Financial Analysist Journal, September /October 1980, pp 43-45.

[8] Nijhoff M.,: Interactive Multiple Goal Programming for Capital Budgeting and Financial Planning, Boston 1980

[9] Nijkamp P., Spronk J.: Interactive Multiplie Gcal Programming: an evaluation and some results. W Multiple Decision Makin Theory and Application, Ed. G. Fandel, T. Gal, Springer, Berlin 1980, s. 278-293

[10] Reingunum M.: The Anatomy of a Stock Market Winner, Financial Analysts Journal, March-April 1988

[11] Saaty T.L: Decision Making for Leaders: The Analitical Hierarchy Process for Decision in Complex World, RWS Publications, 1988

[12] Sharpe W.F: Capital Asset Prices: a theory of market equilibrium under conditions of risk, "Journal of Finance", 1964, vol. 19, s. 425-442

[13] Sim L.S.M., Bector C.R., Turtle H.J.: A fuzzy programming formulation for portfolio analysis, ASAC Conference, Lake Louis 1993

[14] Zouponidis C.: On the Use of the Minora Multicriteria Decision Aiding System to Portfolio Selection and Management, Journal of Information Science and Technology, Vol 2, January 1993

[15] Zouponidis C.: On the Use of Multicriteria Decision Aid Methods to Portfolio Selection, Technical University of Crete, Working Paper 94-06, 1994

Methodology for an Analysis of Decision Criteria: An Application to Irrigated Farms in Southern Spain

A. Rodríguez-Ocaña[1]; J. Berbel[2]; P. Ruiz Avilés[3]

[1] Consejería de Agricultura y Pesca. Delegación Provincial. Avda. de la Borbolla, 1. 41004 Sevilla. Spain.
[2] ETS Ingenieros Agrónomos y Montes, Universidad de Córdoba, Apdo. 3048. 14080 Córdoba, Spain.
[3] Centro de Investigación y Fomento Agrario. CIFA Córdoba. Alameda del Obispo, s/n. 14071 Córdoba, Spain.

Abstract. An analysis of a decision-making process in agricultural enterprises is approached in this paper by cluster analysis which allows a classification of firms according to production behaviour. The socioeconomic type decision-making is modellized by weighted goal programming applied to the defined clusters. This methodology is applied to two neighbouring irrigation units in Southern Spain and we found that there was an important degree of hetereogeneity in production plans explained by differences in the objective weights of farmers.

Key words: Multicriteria analysis, Irrigated agriculture, Goal programming, Cluster analysis.

1 Introduction

There is a general consensus between economists and managers that in real decision-making economic and institutional agents take into account many conflicting criteria. This hypothesis is the basis for the development of a multicriteria decision theory.

According to Barnett et al. (1982), multiple objective research can be characterized as "descriptive", "operational" or "combined". The descriptive approach concerns whether or not decision makers possess multiple objectives, developing relative rankings or objectives. The operational approach utilizes hypothesized objective weights, examining their impact upon a decision model and finally the combined approach embodies an attempt first, to discover objectives and their weights and then to use them in a decision model. A multicriteria decision theory has extensively developed the normative or "operational" approach meanwhile the descriptive and combined approaches are not so well developed.

The present research may be included in the "combined approach" type.

Combined approach may attempt to identify weights and use them in a model simultaneously as in Candler and Rochie (1971) for capital budgeting decisions or iteratively as in Barnett et al. (1982) which is the type of application we intend to explain.

Our interest is in the descriptive power of the Multiple Criteria Decision Theory and in our research we have developed a methodology to test and to analyze the decision criteria of the economic agents. We have selected as a field work ananalysis of farmer' decisions in an irrigated agricultural system. Farmers' decisions are an interesting empirical field because a relatively homogeneous type of decision makers is found, with a similar cultural, economic and physical background, devoted to the same industry and with a "price-taker" nature, therefore in a relatively "perfect competition" world.

In this paper we will describe the first stage of our research which is the development of a descriptive analysis of a decision making process in an economic (agrarian) system. The second phase of the complete project is the development of a normative model of the agrarian system analyzed.

2 Goals and values of farmers in previous works

There are works on empirical research in decision criteria of farmers some years ahead of the MCDM theory. A pioneering work is the research conducted by Gasson (1967) among farmers in Cambridge (U.K.). Gasson argues that we should distinguish between values defined as observable subjective beliefs related to the way of life of an economic agent and management criteria that are observable values which guide the management decision.

Values are cultural products possessed by members of a social system. Values do not exist in isolation but are organised in systems of "value orientations". They are a more permanent property of the individual, less liable to change with time and circumstances. Values are abstract criteria and they can only be approached indirectly through observed behaviour or verbal responses. Gasson defined 'for convenience' value orientation under the following four headings:

- *Instrumental*: making a maximum income; making a satisfactory income; safeguarding income for the future; expanding the business; providing congenial working conditions (hours, security,...).
- *Social*: gaining recognition; belonging to farming community; continuing farming tradition; working with other members of the family; maintaining good relations with workers.
- *Expressive*: feeling pride of ownership; gaining self-respect for doing a worthwhile job; exercising special abilities; meeting a challenge.
- *Intrinsic*: enjoyment of work task; preference for healthy, outdoor job; value of hard work; independence-freedom; control of variety of situations.

In the research about farmers in Cambridgeshire, Gasson found that farmers' value orientations are related to business size. We find the setting of value orientation types a very interesting approach to simplify the second stage of the analysis, i.e. the development of a decision model for each of the important types found in the descriptive analysis.

The paper of Patrick and Kliebestein (1983) makes a more complete review of the subjetc, methodologies and cases. Patrick (1978) developed a simulation model that uses the "goals orientation" based upon a behavioral model previously developed by Patrick & Eisgruber (1968). The concept of "goals orientation" is similar to Gasson's "values orientation", and this type of analysis based upon a types definition will be the approach of our analysis. The reason for the determination of types is that it is quite difficult to establish a single goal hierarchy for a heterogeneous group of farmers or decision makers, even in the case of a common cultural background and similar resource availability.

3 The case study

The first stage of our analysis was the establishment of the types of decision makers by using cluster techniques. The term *"cluster analysis"* embraces a loosely structured body of algorithms which are used in the exploration of data that arise from the measurement of a number of characteristics for a collection of individuals. Cluster analysis is concerned with the discovery of the groups. The word "cluster" or "group" should be interpreted as a collection of "similar" individual objects.

The sample is based upon a random choice of 115 farmers in the Medium Guadalquivir Valley (Southern Spain) from two neighbouring irrigation sectors (68 farmers in area 1, *Campiña Baja*, and 47 farmers in area 2, *Las Colonias*). The number of variables were 101 and the variables were grouped into ten categories. In this paper, we only show the results of 7 categories. Each of these categories was formed with the answer (observations) taken from each individual farmer. We grouped the farmers by cluster technique in the following clusters or groups as shown in Table 1.

The second stage was the application of a Group Analysis to the different categories. An analysis of this information is quite useful to understand the relationship between the individual socioeconomic characteristics and the values and criteria orientation.

An interesting innovation in our research was the questioning about "desired crop plan" i.e. the final output of the decision making process with and without water constraints. These questions were motivated because a severe drought was been experienced in the area thus limiting irrigation possibilities and affecting the real crop plan decisions as against "desired" or "unrestricted" farmer's types and orientations.

Table 1. Results of cluster analysis

Variable	G	%	Definition
Irrigation Sector	Z1	59%	Campiña Baja
	Z2	41%	Las Colonias
Socioeconomic	A1	17%	Farm income under 50%, older low qualif.
	A2	63%	Full-time, medium age, low qualification
	A3	10%	Full-time, younger, high qualification
	A4	10%	Farm income around 50%,medium age,low qualification
Structure	B1	43%	Medium farms, irrigated below 20%
	B2	35%	Small farm, 100% irrigated, no wells
	B3	12%	Small farm, mostly irrigated, own wells
	B4	10%	Large farms, mostly irrig., non propietary water supply
Decision	D1	26%	Cotton around 50%, rest sunflower and various
	D2	17%	Cotton around 50%, rest mainly corn
	D3	57%	Cotton almost 100%
Criteria	F1	17%	1) Max profit; 2) Max production (sales)
	F2	14%	1) Max profit; 2) Max resource use
	F3	12%	1) Max profit; 2) Max family labour
	F4	37%	1) Max profit; 2) Min cost
	F5	21%	1) Max production (sales); 2) Max profit
Values I "Why are you farmer?	H1	43%	Instrumental: 1) income; 2) independence
	H2	36%	Expresive-intrinsic: 1) to own land; 2) indep.
	H3	11%	Expresive: 1) creativity; 2) independence
	H4	10%	Intrinsic: 1) farm work; 2) creativity
Values II	I1	30%	Intrinsic-instrumental: 1) max. yield; 2) max. profit
"A good yield	I2	19%	Instrumental-family: 1) max. profit; 2) max.
farmer is?"	I3	17%	Instrumental-intrinsic: 1) max. yield; 2) no debts
	I4	11%	Family: 1) family labor; 2) work himself
	I5	12%	Mixed: 1) max. profit; 2) well informed
	I6	10%	Intrinsic-family: 1) conserve the land; 2) work himself

Table 2 shows the Chi Square Pearson Test of Independence. It shows that variables *y* and *j* are not independent when $\alpha < 0,05$, with 95% confidence level. Based upon the paired comparison of groups, we built the system of relationships which determinated the decision, and it is shown in Figure 1.

The third stage, once we had done the clustering and clumping, was to test the dependence between pairs of nominal variables.

Table 2. Chi Square Pearson Test of Independence

	A	B	D	F	H	I	Z
A: Socioecon.	--	**0,000**	**0,003**	0,082	0,190	0,012	**0,000**
B: Structure	**0,000**	--	0,768	0,118	**0,026**	0,05	**0,002**
D: Decision	**0,003**	0,768	--	0,697	**0,015**	0,369	**0,001**
F: Criteria	0,082	0,118	0,697	--	0,093	0,508	0,150
H: Values I	0,190	**0,026**	**0,015**	0,093	--	0,081	**0,013**
I: Values II	0,012	0,055	0,369	0,508	0,081	--	**0,002**
Z: Sector	**0,000**	**0,002**	**0,001**	0,150	**0,013**	**0,002**	--

Figure 1.- Complete system of relationship between the DECISION variable and the others variables.

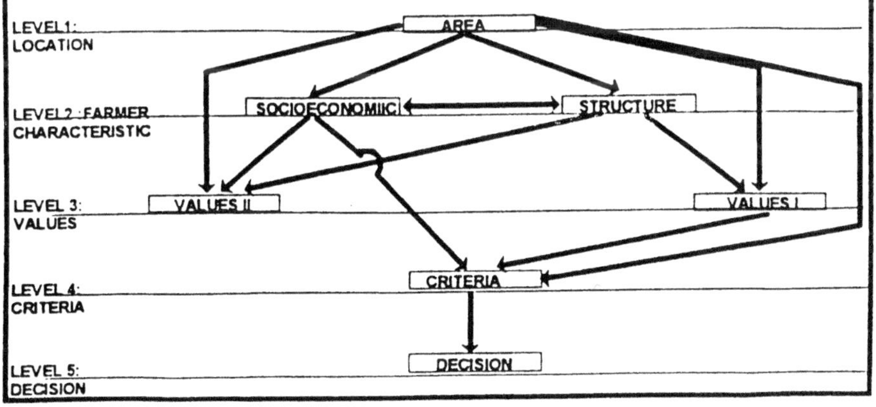

After building the complete model, we studied the relationship between the variables inside it to identify interdependences. We studied groups of variables with the help of log-linear models. In this sense, the general model in figure 1 is well explained by a closed subsystem that is selected because it is the one that minimizes the probability of error type I with the maximum likelihood ratio (Caridad, 1996). The result is the model shown in figure 2. We can see the double effect and triple effect and how Decision is affected by the Area, Socioeconomic Type of Farmer and Values Orientation I.

An important result in our research was that Criteria does not affect Decision. This result could be explained by the fact that farmers understood more easily the verbal questioning about cultural beliefs than questioning on management criteria (profit/risk/labour, etc.). More intensive research should be conducted in the analysis of management criteria with the help of direct methods. A similar conclusion was reached by Deffontaines and Petit (1985) who affirmed that "farmers' criteria are better observed by indirect methods than by direct verbal questioning".

Figure 2.- Relationships system between the DECISION variable and the others variables by the log-lineal model

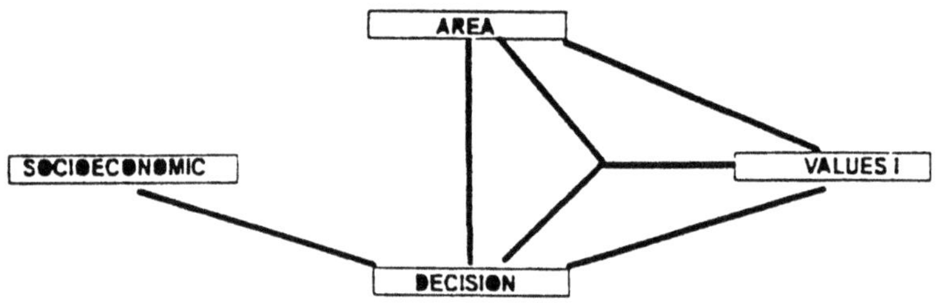

MODEL SOCIOECONOMIC-DECISION, AREA-DECISION-
LIKELIHOOD RATIO= 58 14 DEGREE OF FREEDOM=39 PROBABILITY= 2 49 %

4 Multicriteria analysis

As we mentioned before, it was not possible to find a relationship between decisions and criteria elicited, because we suppose that there were some linguistic and psychological problems that we have not yet solved. Therefore, we shall seek the objective weights with the help of a weighted goal programming (WGP) methodology as explained in Sumpsi et al. (1996).

The final output of the Sumpsi et al. algorithm gives the objective weights shown in Table 3 for farmer groups D1, D2 and D3. When applying the WGP to the three cluster distributions we found that for Sector 1 group D1 seems to take into account three criteria in decision making: maximizing income (56%), minimizing working capital (35%) and minimizing hired labor (8%). Group D2 is quite different as the main criteria is minimizing labor (60%), minimizing risk (31%) and thirdly maximizing profit (6%). Finally, group D3 seems to seek a risk minimization (64%) and, secondly, minimizing hired labour (35%).

Three orientations may explain the production plans: D1 instrumental values orientation (maximizing income, expanding business); D2 and D3 seem to be expressive or intrinsic orientations, with different degrees of risk aversion. When we proceed to apply the methodology to Sector 2, results are similar, and differences may be explained by different technical characteristics (soil quality, water supply, climate) and socioeconomic characteristics.

Table 3. Objective weights by type of decision cluster and County

Sector 1: Campiña Baja	D_1	D_2	D_3
w_1 (MAX income)	56,5%	6,0%	0,0%
w_2 (MIN hired labour)	8,1%	60,7%	35,4%
w_3 (MAXIMIN)	0,0%	31,1%	64,6%
w_4 (MIN working capital)	35,4%	2.2%	0,0%
Sector 2: Las Colonias	D_1	D_2	D_3
w_1 (MAX income)	51,7%	5,0%	0,0%
w_2 (MIN hired labour)	6,1%	59,5%	36,6%
w_3 (MAXIMIN)	0,0%	33,0%	63,4%
w_4 (MIN working capital)	42,1%	2,4%	0,0%

Relative weights given to decision criteria in clusters D1 to D3 were remarkably similar in both counties, showing that the WGP method when applied to a region where a previous categorization of farm operators has been made, may be an appropriate methodology for analyzing real decision making.

Table 4 shows for Sector 1 the result of applied weights to objectives and the projected real area of crops. It can be seen that crop distribution is fairly heterogeneous according to farm-type. On the other hand, the results are remarkably close to real crop plans. Sector 2 has a similar behaviour. More information on this research can be found in Rodríguez Ocaña (1996).

Table 4. Projected and Real Crop Area

	Simulated Crop Plan			Actual Crop Plan		
	D1	D2	D3	D1	D2	D3
Income	273,079	196,697	317,722	254,007	234,107	314,337
Hired Labour	42,349	23,860	31,204	54,786	40,861	37,012
Maximin	271,363	197,432	323,366	238,530	216,508	316,539
Working Capital	0	0	49,550	0	15,992	646,490
Soft Wheat	0,0%	33,2%	0,0%	6,9%	0,0%	0,0%
Durum Wheat	5,4%	1,4%	0,0%	5,4%	1,4%	0,0%
Corn	6,2%	0,0%	0,0%	2,7%	41,2%	0,0%
Potatoes	7,1%	0,0%	0,0%	4,7%	0,5%	0,0%
Sugar Beet	6,0%	1,8%	0,0%	6,0%	1,8%	0,0%
Cotton	59,3%	48,6%	100,0%	51,3%	44,3%	98,4%
Onion	0,0%	0,0%	0,0%	3,2%	1,2%	1,6%
Watermelon	0,3%	0,0%	0,0%	4,1%	1,6%	0,0%
Sunflower	11,6%	7,6%	0,0%	11,2%	0,5%	0,0%
Set-A-Side	4,1%	7,4%	0,0%	4,6%	7,6%	0,0%

5 Concluding remarks

The analysis of the decision-making process in agricultural enterprises is approached in the paper with the development of a methodology based upon a combination of techniques that allow a deep understanding of the influence of socioeconomic and technical heterogeneity on production decisions. A major advantage of the proposed method is that it is composed of two stages in the first of which we broaden our knowledge of the system under study by a cluster analysis of the farm enterprises.

Once the farms are classified according to production behaviour, we may also reach a deeper understanding of the agricultural system analyzed. The result of this stage is to classify farms according to crop pattern, which is the basis for the second stage. WGP is selected as the methodology that approaches the weights given by different farmers to the objectives of the decision process.

As we applied this methodology to two different irrigation units or counties, we found that there was an important degree of heterogeneity in production plans, and that the variation of crop plans could be explained by differences in objective weights caused by differences in the farmers' value orientation.

6 References

Aickin, M. 1981. *Linear Statistical Analysis of Discrete Data*. New York. Ed. John Wiley and Sons.

Barnett, D., B. Blake and B. McCarl, 1982. Goal Programming via Multidimensional Scaling Applied to Senegalese Subsistance Farms, *American Journal of Agricultural Economics*, nº 9, pp. 720-727.

Barnett, D. 1979. *Farmer's Goals and Constraints: Their effects on the Cultivation of Crops in Sine Saloum, Senegal*. Unpublished M.S. Thesis, Purdue University.

Deffontaines, J.P. and M. Petit, 1985. Comment étudier les explotations agricoles d'une région. Presentation d'un ensemble méthodologique. *Etudes et recherches*, 4, INRA. Dijon.

Gasson, R. 1973. Goals and Values Farmers. *Journal of Agricultural Economics*, 24, pp. 521-537.

Patrick, G.F. 1978. Some Impacts of Inflation on Farm Firm Growth. *Southern Journal of Agricultural Economics*, 10, pp. 9-14.

Patrick, G.F. and L.M. Eisgruber, 1968. The Impact of Managerial Ability and Capital Structure on Farm Firm Growth. *American Journal of Agricultural Economics*, 50, pp. 891-907.

Patrick, G.F. and J.B. Kliebenstein, 1983. *Multiple Goals in Farm Decision-Making: A Social Perspective*. Department of Agricultural Economics. Agricultural Experimental Station. Mimeo. Pardue University.

Rodríguez Ocaña, A. 1996. *Propuesta metodológica para el análisis de la toma de decisiones de los agricultores: Aplicación al caso del regadío extensivo cordobés*. Tesis Doctoral. ETSIAM. Universidad de Córdoba.

Romesburg, H.C. 1984. *Cluster Analysis for Researchers*. Belmont: Lifetime Learning Publication.

Romero, C. 1994. *Teoría de la Decisión Multicriterio: Conceptos, técnicas y aplicaciones*. Madrid: Ed. Alianza Universidad Textos.

Sumpsi, J.M., F. Amador and C. Romero. 1993. A Research on the Andalusian Farmers' Objectives: Methodological Aspects and Policy Implication. In: *Aspects of the Common Agricultural Policy*. VIIth EAAE Congress, Stressa, Italy, vol D.

Sumpsi, J.M., F. Amador and C. Romero. 1996. On Farmers' Objectives. A Multicriteria Approach, *European Journal of Operational Research*, Vol. 96, nº 1.

Socioeconomic Impact Evaluation of the Drought in Irrigated Lands in Southern Spain: A Multicriteria Decision Making Approach

Gómez-Limón, J.A.[1]; Lara, P.[1]; Rodríguez, A.[2]; Sánchez, F.J.[2].

[1] Dpt. Agricultural Economics. ETSIAM, University of Córdoba. Apdo. Correos 3048. 14080 Córdoba. SPAIN.
[2] Dpt. Agricultural Economics, Centro de Investigación y Desarrollo Agrario de Córdoba. 14080 Córdoba. SPAIN.

Abstract. The severe drought present in Andalusia (Southern Spain) since 1993 determined the complete absence of a water supply for agricultural lands for the following two agricultural years. This situation forced farmers to decide to sow crops with lower hydric demand requirements in irrigated lands. This long period of drought was superimposed by the beginning of the Common Agricultural Policy (CAP) Reform and this implies that the degree of the socioeconomic impact of the latter cannot be judged just by comparing the income and the labour generated between the pre-drought period (1986 to 1992) and the actual dry period (1994 and 1995). This paper aims to establish a methodology, based upon weighted goal programming, which lets us simulate the farmers' sowing decision if they had disposed of the normal water supply for irrigation, bearing in mind the new CAP institutional environment installed in the European Union. Once this simulation is obtained, we will be able to compare the theoretical sowing decision (without water constraints) for the drought biennium with the ones which occurred in reality. In this way, we will be able to determine, just by difference, the drought effects on income and labour generated. This methodology will be implemented in a practical way in La Campiña Baja district (Andalusia).

Keywords: Weight Goal Programming, Drought, Socioeconomic Impact, Irrigated Lands, Southern Spain.

1 Introduction

The severe drought present in Andalusia (Southern Spain) since 1992 determined an important reduction of the water supply in Andalusian irrigated lands for 1992, and the total absence of water for the following two agricultural years. This situation forced the farmers to decide to cultivate crops with lower hydric demands, which coincide with those that generate less income and demand less labour.

These decreases in income and generated employment have been evaluated by the different agrarian organizations and by the Administration in very different ways. Most of them have been carried out comparing the agricultural income and employment in the period before the drought (1986 to 1992) with that of the drought years (1994 to 1995). However, we should mention that during this period of drought the institutional environment changed, a circumstance that none of these evaluations have borne in mind. Actually, during this period of drought the Common Agricultural Policy (CAP) Reform has been developed, meaning that its protection system for farmers by the artificial maintenance of agricultural prices has been changed into a system of aid with direct payments for surfaces. This coincidence of the drought with the CAP Reform signifies that the mechanism of evaluation previously expounded of mere comparison between both periods can not be considered as being the most adequate one.

Because of this situation, the principal objective of this paper is a methodological development which evaluates the socioeconomic impact of the drought using the Multicriteria Decision Making theory. In this way we attempt to simulate farmers' sowing decisions if they had used the normal water supply for irrigated lands (6000 m^3/ha), taking into account the new agricultural policy established by the European Union. Once obtained, this simulation will enable us to compare the decisions of a theoretical crop coverage without water restrictions for the drought biennium (93-94) with those occurring in reality, establishing the effects of the drought on the loss of income and employment by simple difference.

Lastly it should be indicated that this methodology will be implemented in an agricultural district called La Campiña Baja (Córdoba), in which we shall quantify this impact on income and employment.

2 Analytical procedure: the Weighted Goal Programming

Once our objective has been clearly defined, the first step will be to determine, within the paradigm of Multicriteria Decision Making, the methodology that allows us to simulate what would have occurred during the years 93-94 if the drought had not existed. For this purpose we will suppose, as a kick-off hypothesis, that in this period the only structural factor that did not change was the farmers' objectives. In this way, our methodology will be based on the determination of these objectives and the calculation of their weights for agricultural producers. This study, carried out with the data available of the period before the drought, will serve to explain the behaviour of these farm managers. Once these weights have been obtained, it will be posible to simulate what would have been the behaviour of these managers during the years 93-94 if they had had the habitual water supply.

2.1 Multicriteria Decision Making and Multiobjective Programming

The farmers, like any economic agents, at the time of making their decisions, attempt to search for an equilibrium or compromise between a group of objectives, usually in conflict. In this context, Multicriteria Decision Making and its mathematical implementation, multiobjective programming, constitutes a focus of a great potentiality when the decision context is defined for a series of objectives to be optimized that should satisfy a determined group of restrictions.

So the first step in our analysis consists of defining a tentative set of objectives $f_1(X)$... $f^i(X)$... $f_n(X)$ which seek to represent the real objectives followed by the farmers.

Once these objectives had been defined, the second step is the calculation of the pay-off matrix. This will be of great use, because it will allow us to quantify the level of existing conflict between the different objectives that we are considering. This matrix will have the following formulation:

	$f_1(X)$	$f_2(X)$...	...$f_i(X)$...	...$f_q(X)$
$f_1(X)$	f^*_1	f_{12}	f_{1i}	f_{1q}
$f_2(X)$...	f_{21}	f^*_2	...	...
...$f_i(X)$...	f_{j1}	...	f_{ji}	...
...$f_q(X)$	f_{q1}	...	...	f^*_q

[1]

2.2 Weighting objectives

Once the pay-off matrix is obtained, we will be able to form the following system of equations:

$$\begin{vmatrix} f_{11} = f_1^* & f_{12} \;\cdots\cdots & f_{1i} & \cdots\cdots & f_{1q} \\ f_{i1} & f_{i2} \;\cdots\cdots & f_{ii} = f_i^* & \cdots\cdots & f_{iq} \\ f_{qi} & f_{q2} \;\cdots\cdots & f_{qi} & \cdots\cdots & f_{qq} = f_q^* \end{vmatrix} \begin{vmatrix} w_1 \\ w_i \\ w_q \end{vmatrix} = \begin{vmatrix} f_1 \\ f_i \\ f_q \end{vmatrix} \quad [2]$$

If this system has a non-negative solution, this will represent the set of weights possessed by each one of the real objectives and thus, the actual behaviour followed by the farmer is reproduced. However, in most cases, no exact solution exists. In other words, there is no set of weights w_1... w_i...w_q actually capable of reproducing the real behaviour revealed by farmers. In those cases, the best possible solution will be sought. For that purpose, a *weighted goal programming* can be formulated (Romero, 1991). This solution will be obtained by means of the resolution of the following linear programme (Model [3]):

$$\text{Min} [(n_1 + p_1)/f_1 + ... + (n_i + p_i)/f_i + ... + (n_q + p_q)/f_q]$$

Subject to:

$$\begin{array}{llllllll} w_1 f_{11} + & .... & + w_i f_{1i} + & .... & + w_q f_{1q} & + n_1 & - p_1 & = f_1 \\ w_1 f_{i1} + & .... & + w_i f_{ii} + & .... & + w_q f_{iq} & + n_i & - p_i & = f_i \\ w_1 f_{qi} + & .... & + w_i f_{qi} + & .... & + w_q f_{qq} & + n_q & - p_q & = f_q \\ w_1 + & .... & + w_i + & .... & + w_q + & & & = 1 \end{array} \quad [3]$$

Where p_i are positive deviational variables (i.e. the measurement of the overachievement of the i-th objective with respect to a given target), and n_i are negative deviations concerning values reached in reality for the i-th objective.

By solving this goal programming we will obtain a first approach of the solution of our problem, with an initial set of weights $w_1 ... w_i ... w_q$.

Briefly, this methodology can be summarized in an algorithmic manner as follows:

1º) Tentatively establish a set of objectives supposed to be most important ones for farmers.
2º) Determine the pay-off matrix for the previous objectives.
3º) With the previous matrix resolve de model calculating weighted goal programming (Model [3]).

This calculation of the objectives will be carried out independently for the three years prior to the drought (1989, 1990 1991). With them we aim, on one hand, to check if indeed these objectives are a structural characteristic, and furthermore, in the case of our supposition being correct, to calculate the mean of these objectives for their utilization in the simulation.

3 Applying the methodology to the farmers from La Campiña Baja

3.1 Decision variables

For our model the decision variables are the surfaces covered by different crops in the irrigated lands we studied. We decided to choose the most representative ones in this district, these being ten crops all of them representing over 90% of the total agricultural surface in this district. The following are the crops which build the X vector of decisions:

Variable	Crops
X_1	Surface covered by soft Wheat
X_2	Surface covered by hard Wheat
X_3	Surface covered by Corn
X_4	Surface covered by Potato
X_5	Surface covered by Beet
X_6	Surface covered by Cotton
X_7	Surface covered by Sunflower
X_8	Surface covered by Watermelon
X_9	Surface covered by Garlic
X_{10}	Surface covered by Onion
X_{11}	Surface without any crop (set aside)

3.2 Objectives

Different authors who have studied farmers' behaviour in irrigated lands in Andalusia (Sumpsi *et al.*, 1993 and Gómez-Limón & Berbel, 1995) have established different objectives that these producers take into consideration for establishing their sowing decisions. The same objectives will be considered as being the most relevant in order to explain the farmers' behaviour in La Campiña Baja, and they are therefore those which will be used for the calculation of the pay-off matrix. These are:

- Maximize the prospective Gross Margin (GM)
- Minimize Labour (L)
- Minimize the Working Capital (K)
- Minimize the Risk (MOTAD)
- Maximize the ratio GM/K.

Once the objectives to consider have been revealed, next we shall analyze each of them, studying how they are calculated and how they can be inserted in our multicriteria model.

a. Maximize the prospective Gross Margin (GM). The gross margin is an indicator of absolute profitability and is measured in pesetas. The calculation of the gross margin (GMi) of each of the crops considered for every year will be calculated through the following expression:

$$GMi = [price * yield] + direct\ payments\ (C.A.P.\ aids) - variable\ costs \quad [4]$$

b. Minimize Labour (L). Labour refers to the amount of employment required by the different crops yearly and is measured in hours.

c. Minimize the Working Capital (K). All productive cycles need an amount of capital that finances the temporal immobilizations in production. This capital is denominated working capital and is measured in pesetas.

d. Minimize Risk (MOTAD). Risk is a usual objective considered in many studies of farmer's behaviour. In this case it has been included by the MOTAD method (Hazell, 1971).

e. Maximize GM/ K. This gross margin/working capital ratio is measured in relative profitability. The maximization of this ratio is not calculated for the lineal code. Therefore, for this purpose we have used a parametric method.

3.3 Constraints

a. Utilization of the surface. The sum of the surfaces covered by each crop must be 100 hectares. In this way the coverage of each of the Xi activities will be given to us in percentages. Therefore, this constraint is formulated hence:

$$\Sigma X_i = 100 \quad [5]$$

b. Market limits. Horticultural crops (potato (X4), watermelon (X8), garlic (X9) and onion (X10)) are very perishable. This characteristic causes the maximum amount of hectares cultivated with this crop to depend on the demand for these products in the markets. We suppose that the maximal capacity of the market for these crops (Maxi) is equal to the maximal historic surface from 1985. Thus, this kind of constraints are like this:

$$Xi \leq Maxi \quad [6]$$

The same happens with sugar beet (X5) and hard wheat (X2), subject to CAP quotas.

4 Model results

4.1 Pay-off matrix

Once the model is defined, we shall successively solve the different objectives: Maximize GM and GM/K, and Minimize L, K and MOTAD, for each year (89-91), obtaining the pay-off matrixes for each year.

4.2 Weighted goal programming

Once the pay-off matrixes have been obtained, we shall formulate the model [3], according to the third step of the methodology. By solving this model, the set of weights is obtained.

Table 1. Weighted objectives

GM	L	MOTAD	K	GM/L
30,25%	16,74%	53,01%	0,00%	0,00%

In this way we can conclude by affirming that the farmers' utility function have the following formulation:

$$U = 0{,}3025\ GM - 0{,}1674\ L - 0{,}5301\ MOTAD \qquad [7]$$

This expression, which the producers will attempt to maximize, will be employed for the simulation carried out next.

5 Simulation results

The simulation of the crop coverage in the case of there not being any drought has been calculated using fictitious price values and fictitious direct payments, in accordance with the new CAP.

With these changes we will run our model again with a maximization of the utility function shown above. The results can be observed in Figure 1.

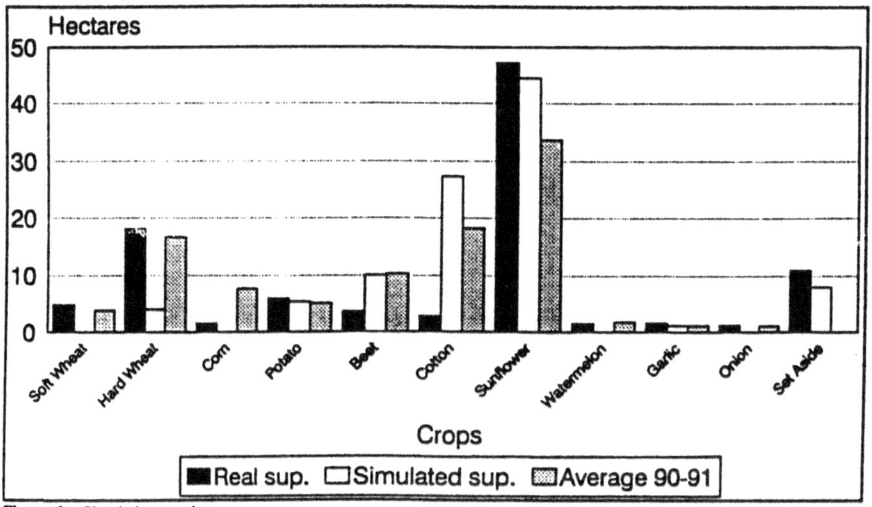

Figure 1.- Simulation results.

6 Final impact socioeconomic evaluation of the drought in La Campiña Baja

From the outputs of the previous simulation, the differences in the gross margin and labour caused in a year by a change in the planning of crops triggered by the drought can be calculated. We believe that the differences in these variables are the most adequate indicators for a quantification of the economic impact (decrease of income) and social impact (loss of working hours) caused by the absence of water in La Campiña Baja.

However, the previous outputs would only be for a farm of 100 hectares, which was the one we used in our model. In order to measure the global impact on the district, this amount should be multiplied by 51,369, the total amount of agricultural hectares which formed La Campiña Baja. By doing this calculation the final outputs can be seen in Table 2.

Table 2. Global impact of the drought in La Campiña Baja

Surface (Ha.)	Gross margin (million pts.)	Labour (thousand hours)
51.369	-5.215	-1.909

7 Concluding remarks

The methodology expounded of the determination of objectives of farmers can be useful in order to predict farmers' behaviour, and thus, an estimate of crop coverage. That is why this methodology can be applied efficiently in order to calculate the losses occasioned by drought on incomes (economic impact) and employment (social impact).

References

Alaejos, A. M., 1990. *Aplicación de modelos de programación en contexto de riesgo para la selección de planes de cultivos en la zona regable del Bembézar. Trabajo Profesional de Fin de Carrera*. Universidad de Córdoba.

Amador F.; Barco A.; Romero C., 1985. Maximización del empleo frente a estabilidad de la mano de obra en un programa de reforma agraria en Andalucía: una aplicación de la programación compromiso. In: *Trabajos elaborados por autores españoles para el XIX Congreso Internacional de Economistas Agrarios*. Banco de Crédito Agrícola, Madrid, pp. 19-27.

Cabrera de la Colina J., 1985. Las transformaciones recientes de la agricultura de regadío en la Valle Media del Guadalquivir. In: *Trabajos elaborados por autores españoles para el XIX Congreso Internacional de Economistas Agrarios*. Banco de Crédito Agrícola, Madrid, pp. 43-50.

Consejería de Agricultura y Pesca, Junta de Andalucía & E.T.E.A., 1993. *Estudio de Márgenes Brutos Estándar de las explotaciones agrarias de Andalucía*. Servicios de Publicaciones de la J. A., Sevilla.

Gómez-Limón, J.A.; Berbel, J., 1995. Aplicación de una metodología multicriterio para la estimación de los objetivos de los agricultores del regadío cordobés. *Investigación Agraria: Economía*, vol. 10 (1), 1995, pp. 142-157.

Guerrero, A., 1987 y 1992. *Cultivos herbáceos extensivos*. Mundi-Prensa, Madrid.

Hazell, P.B., 1971. A linear alternative to quadratic and semivariance programming in farm under uncertainty. *American Journal of Agricultural Economic*. (53), pp. 53-62.

Junta de Andalucía, Consejería de Agricultura y Pesca, 1991. *Anuario de estadísticas agrarias y pesqueras Andalucía*. Servicios de Publicaciones de la J. A., Sevilla.

Loring, J., 1989. *Estudio de Costes Agrarios en la provincia de Sevilla*. Diputación de Sevilla, Sevilla.

M.A.P.A, 1986, 1987, 1988, 1989, 1990, 1991, 1992. *Anuario de Estadística Agraria*. Secretaría General Técnica, Madrid.

Millán, D., 1994. *Aplicación del modelo de índice simple al regadío de la provincia de Córdoba*. Trabajo Profesional de Fin de Carrera. Universidad de Córdoba.

Romero, C., 1993. *Teoría de la decisión multicriterio: conceptos, técnicas y aplicaciones*. Alianza Universal Textos, Madrid.

Sumpsi, J. M.; Amador, F.; Romero, C., 1993. A research on the andalusian farmers' objectives: methodological aspects and policy implications. In: *Aspects of the Common Agricultural Policy*. VIIth EAAE Congress, Stresa Italy, vol. D.

Zionts, S.; Wallenius, J., 1976. An interactive programming method for solving the multiple criteria problem. *Management Science*, vol. 22, nº 6, February 1976, pp. 652-663.

Goal Programming in Distribution System Design

Ewa Konarzewska-Gubała and Andrzej Zajączkowski

Faculty of Management and Informatics, Wroclaw University of Economics, 53-345 Wroclaw, Poland

Abstract. In distribution system design we are dealing with multiple goals concerning three main sets of requirements for competitive distribution network: (a.) physical and organizational transportation of the demanded quantities of goods and/or services (b.) clients` expectations to the product/service delivery time (c.) budget restrictions for organizing and maintaining the distribution network. The purpose of this paper is to present the Goal Programming (GP) approach to the designing of the distribution network for non consumer goods in one of Polish enterprises.

Keywords. Goal Programming, discrete location, distribution system

1 Introduction

While location is not the only factor influencing the success or failure of a firm, it is critical in many cases. The ability of a firm to produce and market its products effectively or/and to deliver high-quality services is dependent in part on the location of the firm's facilities in relation to other facilities and to its customers.

In distribution system design the commonly occurring problem is the optimal location of intermediate distribution centers between plants and customers.

The simplest version of the real-life problem to be modeled is this. The problem consists in re-designing the distribution and service network for relatively inexpensive, small in size and weight goods, mostly of industrial application. A good example are pneumatic tools. There is a known demand for tools as well as for tool repair services at each of a number of customer zones. The important requirement is that the deliveries (goods or services) should be done strictly on order and within a limited time. The production plant is single and its area of operating is relatively large. The existing, traditional way of satisfying customers` demand is based on the combined system of independent dealers, buying and selling products on their own account. All inventory and transportation costs are included in the price, so none or little concern is given to the rationalization of these processes. The dealers are also unable to improve their organization since their resources are too scarce. The management of tools factory have realized that organization and optimization of their own intermediate distribution centers system can improve their competitiveness against the western companies offers.

The problem just presented as many other real world facilities location problems can be posed as discrete location problem. We are dealing with discrete location problem in the case of selection of sites from some finite set of candidate locations (e.g., in case of a network location problem we must locate facilities on a network, *Daskin, 1995*).
In mathematical modeling of discrete location problems there are four classical models: covering, center, median and fixed charged discrete location models. The recent survey and general comparative analysis of these models and their applications were given by Daskin *(1995)*. The integer linear programming model is commonly used as a modeling and solution searching tool (see *Geoffrion and Graves 1971, Erlenkotter 1977, Garfinkel and al. 1977, Daskin 1995*). The particular class of intermediate distribution location ILP models is formed by models using the hub-and-spoke system (see *Campbell 1994, Mirchandani and Francis 1990*).

Facility location problems are inherently strategic in nature (*Witkowski 1995*). As such, multiple objectives (usually conflicting) must be considered in location analysis. In distribution systems we are dealing usually with goals concerning three main sets of requirements for competitive distribution network:

a. physical and organizational transportation of the demanding quantities of goods or/and services
b. clients` expectations concerning the product/service delivery time
c. budget restrictions for organization and maintaining the distribution network.

Therefore multiobjective programming may be effectively applied in network design. The purpose of this paper is to present the Goal Programming approach to designing distribution system of goods and services. This area has received so far limited attention in discrete location literature. In Daskin monograph we can find a section devoted to multiple objectives location decisions using the fundamental approach of nondominated solutions. The GP facility location problem was formulated by *Taylor and al. (1985)* for allocation and deployment of state highway patrol units. However the characteristics of this problem (the network location) do not correspond to the problem outlined above.
The paper presents the model formulation stage of the distribution system design project. The research effort was focused on the recognition and specification of the crucial model assumptions and on adopting the appropriate mathematical formulation. In section 2. we describe briefly the initial full-scale problem of re-designing the distribution system of a major Wroclaw firm. The integer goal programming location model is developed in section 3. The properties of this model are discussed in the section 4. The final section of the paper is a brief summary.

2 Problem description

ARCHIMEDES S.A. is a manufacturing company. Its only factory is placed in Wroclaw. Production is diversified: pneumatic tools (over fifty catalogue positions plus spare parts constitute about sixty percent of total sales) and hydraulic equipment for agricultural machines (sold mainly to Ursus Tractor factory in Warsaw). The company has an overwhelming market share: more than 50% in the tools market and about 80% in hydraulic components. Unfortunately, this is nothing more than a remnant of socialist monopoly and the shares are constantly decreasing. The customers are production plants, workshops, and public institutions. About seventy percent are big state owned firms but the number of private customers is increasing. The customers are located all over Poland; however some regionalizing is clearly visible, in accordance with uneven dislocation of industry. Most important areas are Silesia, South Eastern Poland, Warsaw, Coastline esp. shipyards in Gdańsk and Szczecin. Export is not significant.

Competition threat from foreign companies--Ingersoll-Rand, Atlas Copco or Bosch to name a few--had caused increasing interest in improvement of operations. The problem the company is facing now is the quality. Much like the rest of Polish industry ARCHIMEDES is lagging behind its competitors. Efforts to improve the quality are being made but significant improvement requires some time. Therefore quick and decent repair, maintenance and spare parts delivery service may--to some extent, of course--balance the lacks in quality. The target outbound logistics strategy is a network of independent dealers. While building the network, company also sells and delivers directly using combined transportation routes: road shipments via independent carrier companies, using its own or its customer's vehicles, and packages via post or DHL. This complex system makes all cost calculations difficult but according to estimations transportation is about 25% of production cost for the average product. The network is supposed to decrease logistics costs by concentration of shipments and ceding the inventory costs on the dealers. It is also supposed to be easier to manage; however, the rules of cooperation must be required from the dealers very strictly, since they are often reluctant to provide company with current marketing survey from operations area.

The idea of the project related in this paper is to create an alternative model of distribution, with dealers replaced by company's distribution centers (DC) serving the inventory/transportation needs as well as repair and maintenance of already sold tools. Considering the size and weight of products, the inventory space and personnel needed to stock the monthly demand of several assigned customer zones is small and does not require any special equipment. (In this case customer zone is an area where distances between customers are insignificant--a city, for example). Organization of a service workshop is much more expensive but also does not require a large space. Basic assumption is that both types of centers can be placed in the same site to minimize the fixed costs of possession. Therefore a distribution center can exist in three forms:

- center for products transshipment,
- center for repair services,
- combined center for both transshipment and repair operations.

The transportation is divided into two kinds of shipments. The "primary" ones are conducted from the origin factory to the distribution center, using a large transport vehicle (truck). Its purpose is to supply a transshipment center, so the shipments are scheduled, bulk and relatively rare. It is assumed that these services are performed by an independent transportation company. The other kind, the "secondary" shipments from a DC to the customer zone are often and stochastic. Their purpose is to carry the ordered items (mostly in small numbers) straight to the customer. The most proper vehicle seems to be a small delivery minivan. There is a certain number of such vehicles at the company's disposal. Those cars can serve both transportation or repair needs without significant modification.

Therefore the fixed location cost of establishing a distribution center is composed of three primary components:

- fixed cost of possession (maintenance, lease) for the entire site,
- cost of transshipment equipment, staff, organization,
- cost of repair workshop equipment, staff and organization.

The possession costs differ across the location sites.

The operating costs are distance related for all pairs: (origin factory - DC) and (DC - customer zone). There are two operation unit costs: for shipment from any DC and/or for repair service from any DC. (Repair travel mostly requires the presence of a repair specialist taken to the customer and back, thus it is more expensive than a regular shipment. The repair car may be also equipped with some amount of tools and spare parts.)
In order to achieve a better marketing position in satisfying the customer demand all distribution centers should be located close enough to the customer zones to assure the quality standard time of service (about three hours at most for goods deliveries; in case of repair travels this time is extended to six hours).

The budget preliminated for re-designing the existing distribution network is limited.

3 Integer GP location model formulation

3.1 Decision variables

The mathematical formulation of the problem uses the following notation:

indices

i - index for possible distribution center (DC) sites,
j - index for customer demand zones,
$k1$, $k2$ - indices for goal constraint deviational variables,

input values

fcp_i - fixed cost of possession for a DC i per month,
$ptcs_i$ - primary transportation cost for shipments per month from factory to DC.
It is assumed that primary shipments are conducted once per month. The cost is calculated as distance from factory to DC multiplied by cost per kilometer for a track.
$stcd_{ij}$ - secondary transportation cost for shipment from a DC i to customer zone j,
$stcr_{ij}$ - secondary transportation cost for repair service from a DC i to customer zone j,
This cost is also calculated as distance from DC to customer zone multiplied by cost per kilometer for a delivery or repair vehicle.
dd_j - demand for deliveries in zone j,
dr_j - demand for repairs in zone j,
Demand is calculated as an expected average number of vehicle travels to zone j per month.
tt_{ij} - travel time from DC i to customer zone j, in minutes,
$fcdu$ - fixed cost of delivery operations in a DC per month,
$fcru$ - fixed cost of repair operations in a DC per month,
$fcvu$ - fixed cost of a transportation vehicle maintenance,
msc - maximum capacity of a single vehicle in shipments,
mrc - maximum capacity of a single vehicle in repairs,
Maximum capacity is a maximum number of shipments/repair travels that can be conducted from a DC by a vehicle. It is calculated as a relation of average travel time (including preparations) to the work hours available in a DC. To simplify the calculation, we assume the travel time as a demanded time of delivery (three and six hours accordingly).
pt - time needed to prepare a shipment, in minutes, presumed twenty minutes,
$fleet$ - total number of vehicles at company's disposal,
$budget$ - budget for maintenance of the entire designed DC network,

decision variables

XD_{ij} - O - 1 variable that will be 1 if a DC i serves the deliveries to customer zone j, and 0 otherwise,
XR_{ij} - O - 1 variable that will be 1 if a DC i serves the repairs to customer zone j, and 0 otherwise,
NDV_i - number of vehicles in DC i serving the deliveries of products,
NRV_i - number of vehicles in DC i serving the deliveries of repairs,

goal variables
$d_1^+, d_1^-, d_2^+, d_2^-, d_3^+, d_3^-, d_4^+, d_4^-, d_{k1}^+, d_{k1}^-, d_{k2}^+, d_{k2}^-$ - goal constraint deviational variables, defined as
d_k^+ = overachievement of goal k,
d_k^- = underachievement of goal k, where:
$d_k^+, d_k^- \geq 0$, for all k.

3.2 Model constraints

A. System constraints

(1) $\Sigma_i XD_{ij} = 1,$ for all j

Each customer zone j is served by one transsshipment site.

(2) $\Sigma_i XR_{ij} = 1,$ for all j

Each customer zone j is served by one repair station.

(3) $\Sigma_j dd_j XD_{ij} / NDV_i \leq msc,$ for all i

Product demands covered by a DC does not exceed its shipment capacity

(4) $\Sigma_j dr_j XR_{ij} / NRV_i \leq mrc,$ for all i

Repair demands covered by a DC does not exceed its service capacity.

(5) $\Sigma_i NDV_i + \Sigma_i NRV_i \leq fleet,$

Total number of utilized vehicles in DC network is at most equal to the number of vehicles at company's disposal. (This constraint is optionally considered as a system one or as the goal constraint in other version of the model.)

B. Goal constraints

(6) $\Sigma_i \max(XD_{ij}, XR_{ij})\, fcp_i + \max(XD_{ij})\, fcdu + \max(XR_{ij})\, fcru + (NDV_i + NRV_i) fcvu + d_1^- - d_1^+ = budget,$

Fixed location costs of establishing the network of DC's are limited by budget constraint. While costs of possession vary for different sites location, the other fixed costs (*fcdu, fcru* and *fcvu*) are standard for all sites. Function max(*spec.*) sets the value of the variable on 1 if there is at least one customer zone served in the specified way, and 0 otherwise.

(7) $\Sigma_i \max (XD_{ij})\, ptcs_i + d_2^- - d_2^+ = 0,$

Transportation cost of providing all transshipment DCs with goods from origin factory is to be minimized to possible extent.

(8) $\Sigma_i \Sigma_j XD_{ij}\, stcd_{ij}\, dd_j + d_3^- - d_3^+ = 0,$

Transportation cost for shipments from all DCs to assigned customer zones is to be minimized.

(9) $\Sigma_i \Sigma_j XR_{ij}\ stcr_{ij}\ dr_l + d_4^- - d_4^+ = 0,$

Transportation cost for repairs from all DCs to assigned customer zones is to be minimized.

(10) $XD_{ij}\,(tt_{ij}+pt) + d_{k1}^- - d_{k1}^+ \le 180,$ for all i, j

It is required for each DC that the time of tools delivery does not exceed three hours.

(11) $XR_{ij}\,(tt_{ij}+pt) + d_{k2}^- - d_{k2}^+ \le 360,$ for all i, j

It is required for each DC that the time of repair expedition does not exceed six hours.

3.3 Objective function

The designation of each goal has been indicated with the constraints formulation. The list of prioritized goals is as follows:

Minimize: $P1\ d_1^+, P2\ d_2^+, P3\ d_3^+, P4\ d_4^+, P5\ d_{k1}^+, P6\ d_{k2}^+,$

4 Model properties discussion

Let us examine some model assumptions to define its properties and to determine its position among the other location models.

(1) DC location problems may be classified by the way in which each demand point is allocated to a single distribution center, i.e., each demand point can send and receive via only single DC or not. Formulation presented in this paper involvs a single allocation of homogenous demands.

(2) In the presented model, a DC can be located in any existing location and theoretically in all points of customer demand. It results in not planar location model but in a discrete one.

(3) Since the fixed location cost of establishing the DC varies across the sites the preliminated budget does not define the feasible number of centers. The problem can be considered as location covering problem rather than the maximum covering problem.
(4) Most location models make the assumption that we are dealing with a single product or service and that all demands are identical. In our case the single set of DCs will serve two different demands: tools deliveries and repair services. In model building we use two patterns of demands to find the optimal compromise for DC location.
(5) The model that we have developed is a static model, i.e., the inputs do not depend on time. Inputs that may depend on time include demands, costs and available and preexisting candidate DC locations. In a dynamic problem formulation--being under our study--we are concerned not only with the question where to locate the distribution centers but also with the question of when to invest in a new center or to close the existing one. Since the ARCHIMEDES company plans to expand the list of its customers, the most appropriate dynamic approach seems to be that of identifying the optimal evolution of distribution centers over the time.

5 Conclusions

The paper considers the real-life discrete intermediate distribution centers location problem in which the following items are given:

- single origin manufacturing plant
- n demand locations
- m existing DC sites
- l new candidate DC sites; $l \leq n$
- two patterns of demand at each demand location: for products and for services
- fixed location cost for each DC location
- distance-related shipment operation cost for each DC location
- distance-related service operation cost for each DC location
- fixed cost associated with using a vehicle
- products delivery time standard
- service delivery time standard
- budget limitations
- total vehicles number

The integer goal programming formulation is proposed to achieve several conflicting goals with the relaxation (deviations) being under control. The following goals have been specified:

(1) Establishing DC network covering all demands (products and services) respecting the budget as much as possible.
(2) Minimizing the distribution system operating costs to possible extent.
(3) Improving the quality of system services by satisfying the products and repair demands within the target time standards as much as possible.

The numerical problems of the model are not considered.

References

Campbell J.F.(1994), "Integer Programming Formulations of Discrete Hub Location Problems", *Europ. Journal of Operation Research,* 72 (2), 387-405

Daskin M.S. (1995), *Network and discrete location. Models, algorithms and applications,* Wiley, New York

Erlenkotter D. (1977), "Facility Location with Price Sensitive Demands: Private, Public, and Quasi-public" *Management Sci.,* Vol. 24, No. 4, 378-386

Geoffrion A. M. and Graves G.W. (1971), "Multicommodity Distribution System Design by Benders Decomposition", *Management Sci.,* Vol. 20, No. 5, 822-844

Mirchandani P. B. and Francis R. L. (eds) (1990), *Discrete location theory,* Wiley, New York

Taylor B. W., Moore L. J., Clayton E. R., Davies K. R. and Rakes T. R. (1985), "An Integer Nonlinear Goal Programming Model for the Deployment of State Highway Patrol Units", *Management Sci.,* Vol. 20, No. 5, 1335-1347

Witkowski J., (1995), *Logistic strategy of industrial enterprises,* Scientific Papers of Wroclaw University of Economics, nr 702, Wroclaw *(in Polish)*

Distance Games and Goal Programming Models of Voting Behaviour[1]

František Turnovec

Center for Economic Research and Graduate Education, Charles University,
& Economics Institute of the Academy of Sciences of the Czech Republic,
Politických vězňů 7, 111 21 Prague 1, Czech Republic

Abstract: In this paper we focus attention on multiple-vote procedures that provide the possibility to use multi-criteria optimization and game-theoretical models of voting behaviour. So called distance games are introduced in which players select a location in some feasible allocation space defined as a convex hull of a finite number of points by the following process: each player is selecting a weighting vector of generic points to minimize a distance between the point that is expected to be selected and his preferred location, and then aggregated weights are used to determine the selected point as a convex combination of generic points. It is shown how to use distance games for an analysis of a rational behaviour of sophisticated voters in so called portfolio voting procedures.

Keywords: Distance games, Kuhn-Tucker conditions, Nash equilibrium, portfolio voting, voting procedure

1 Distance Games

Let

$$\boldsymbol{a}_i = (a_{i1}, a_{i2}, \ldots, a_{ik})^T \, , \; i = 1,2,\ldots,m$$

be generic points defining a feasible allocation space as the convex hull (a set of all convex combinations) of $\mathbf{a}_i$:

$$\wp = \{\boldsymbol{p} \in \mathbb{R}^k \colon \boldsymbol{p} = \sum_{i=1}^{m} \boldsymbol{a}_i \alpha_i, \; \sum_{i=1}^{m} \alpha_i = 1, \; \alpha_i \geq 0\}$$

Let

[1] This research was undertaken with support from the European Commission's Phare ACE Programme 1994 No. 94-0666-R.

$$b_j = (b_{j1}, a_{j2}, \ldots, a_{jk})^T, \; j = 1,2,\ldots,n$$

be some points of individually preferred locations of n individuals (one point for each individual), in general case different from $\mathbf{a}_i$. The problem is to select a single location $\mathbf{p}^0 \in P$ such that the distance of all individually preferred points $\mathbf{b}_j$ to $\mathbf{p}^0$ is "as small as possible". Let us assume the following method of selection:

Each individual j presents his "weighting vector"

$$w_j = (w_{j1}, w_{j2}, \ldots, w_{jk})$$

such that

$$\sum_{i=1}^{m} w_{ji} = 1, \; w_{ji} \geq 0$$

Then the individual weights are aggregated into "social" weights

$$\alpha_i^0 = \frac{\sum_{j=1}^{n} w_{ji}}{n}$$

and location

$$p^0 = \sum_{i=1}^{m} a_{is} \alpha_s^0$$

is selected.

Using a distance function d we can define the distance of the r-th individual's preferred location from selected point

$$d(p_0, b_r) = d(\sum_{i=1}^{m} a_i \alpha_i^0, b_r) = d(\frac{1}{n} \sum_{i=1}^{m} a_i \sum_{j=1}^{n} w_{ji}, b_r)$$

Assuming that each individual is making his decision independently on all other individuals we can model the problem of individual weights selection by the minimization of each r-th individual distance from selected point

$$d(\frac{1}{n}(\sum_{i=1}^{m} a_i w_{ri} + \sum_{j \neq r} \sum_{i=1}^{m} a_i w_{ji}), b_r) \tag{1}$$

where the r-th individual controls only his individual weights w_{ri}.

Assuming Euclidean distance d in (1) we receive

$$\sqrt{\sum_{s=1}^{m}\Big[\frac{1}{n}\Big(\sum_{i=1}^{m} a_{is}w_{ri}+\sum_{j\neq r}\sum_{i=1}^{m} a_{is}w_{ji}\Big) - b_{rs}\Big]^2} \qquad (2)$$

Clearly the root of the sum of square deviations has the minimum in the same point as the sum of square deviations, so minimization of (2) can be substituted by minimization of the sum of squares function

$$\sum_{s=1}^{m}\Big[\frac{1}{n}\Big(\sum_{i=1}^{m} a_{is}w_{ri}+\sum_{j\neq r}\sum_{i=1}^{m} a_{is}w_{ji}\Big) - b_{rs}\Big]^2$$

By a distance game we mean here the non-cooperative game of n players with pay-off (minimizing) functions

$$M_r(\mathbf{w}_1,\dots,\mathbf{w}_r,\dots,\mathbf{w}_n)=\sum_{s=1}^{k}\Big[\frac{1}{n}\Big(\sum_{i=1}^{m} a_{is}w_{ri}+\sum_{j\neq r}\sum_{i=1}^{m} a_{is}w_{ji}\Big)-b_{rs}\Big]^2 \qquad (3)$$

and strategy sets

$$W^{(r)} = \left\{\mathbf{w}_r \in \mathbb{R}^m : \sum_{i=1}^{m} w_{ri} = 1,\ w_{ri} \geq 0\right\} \qquad (4)$$

2 Kuhn-Tucker Conditions for Equilibrium Points

Nash equilibrium in the game (3)-(4) with respect to the minimizing behaviour of the players can be defined as a collection of weight vectors

$$\mathbf{w}^0 = (\mathbf{w}_1^0,\dots,\mathbf{w}_r^0,\dots,\mathbf{w}_n^0)$$

such that

$$F_r(\mathbf{w}_1^0,\dots,\mathbf{w}_r^0,\dots,\mathbf{w}_n^0) = \min_{\mathbf{w}_r\in W^{(r)}} F_r(\mathbf{w}_1^0,\dots,\mathbf{w}_{r-1}^0,\mathbf{w}_r,\mathbf{w}_{r+1}^0,\dots,\mathbf{w}_n^0) \qquad (5)$$

for all $r = 1, 2, \dots, n$.

Pay-off functions (3) are continuously differentiable and convex in $\mathbf{w}_r$ and the strategy sets (4) are convex and closed. Given the weight vectors $\mathbf{w}_1^0, \dots, \mathbf{w}_{r-1}^0, \mathbf{w}_{r+1}^0, \dots, \mathbf{w}_n^0$ let us define Lagrangian for the minimization problem (5) of the r-th player:

$$L_r(w_r,\lambda_r) = M_r(w_1^0,\ldots,w_{r-1}^0,w_r,w_{r+1}^0,\ldots,w_n^0) - \lambda_r(\sum_{i=1}^{m} w_{ri}-1)$$

For a set of weights

$$w^0 = (w_1^0,\ldots,w_r^0,\ldots,w_n^0)$$

to be a Nash equilibrium in the game (3) - (4) it is necessary and sufficient (Kuhn-Tucker) that there exists a vector

$$\boldsymbol{\lambda}^0 = (\lambda_1^0,\ldots,\lambda_r^0,\ldots,\lambda_n^0)$$

such that $\mathbf{w}^0$, λ^0 is a solution of the following system:

$$\frac{\partial L_r(w_r,\lambda_r)}{\partial w_{ri}} \geq 0 \tag{6}$$

$$w_{ri}\frac{\partial L_r(w_r,\lambda_r)}{\partial w_{ri}} = 0 \tag{7}$$

$$\frac{\partial L_r(w_r,\lambda_r)}{\partial \lambda_r} = 0 \tag{8}$$

$$w_{ri} \geq 0 \tag{9}$$

($i = 1, 2, \ldots, m$, and $r = 1, 2, \ldots, n$). The system (6) - (9) can be written as

$$\frac{2}{n}\sum_{s=1}^{k}(\sum_{i=1}^{m} a_{is}w_{ri}+\sum_{j\neq r}\sum_{i=1}^{m} a_{is}w_{ji})a_{ts} - \lambda_r \geq 2\sum_{s=1}^{k} b_{rs}a_{ts}$$

$$w_{ri}[\frac{2}{n}\sum_{s=1}^{k}(\sum_{i=1}^{m} a_{is}w_{ri}+\sum_{j\neq r}\sum_{i=1}^{m} a_{is}w_{ji})a_{ts} - \lambda_r - 2\sum_{s=1}^{k} b_{rs}a_{ts}] = 0$$

$$\sum_{i=1}^{m} w_{ri} = 1$$

$$w_{ri} \geq 0$$

(t = 1, 2, ..., m and r = 1, 2, ..., n). This system can be solved by techniques used for linear complementarity problems (see e.g. LEMKE 1965, enumerating all solutions TURNOVEC 1970).

3 An Example

To illustrate the concept of a distance game we shall use a simple and rather transparent example.

Let us have two generic points

$$\boldsymbol{a}_1 = (10,\ 0)^T, \quad \boldsymbol{a}_2 = (0,\ 10)^T$$

and two individuals with preferred location points

$$\boldsymbol{b}_1 = (6,\ 4)^T, \quad \boldsymbol{b}_2 = (7,\ 3)^T$$

Then we can construct the two person distance game with minimizing pay-off functions

$$M_1(w_1,w_2) = \left(\frac{1}{2}[10w_{11}+10w_{21}]-6\right)^2 + \left(\frac{1}{2}[10w_{12}+10w_{22}]-4\right)^2$$

$$M_2(w_1,w_2) = \left(\frac{1}{2}[10w_{11}+10w_{21}]-7\right)^2 + \left(\frac{1}{2}[10w_{12}+10w_{22}]-3\right)^2$$

and the strategy sets

$$W^{(1)} = \{w_{11}+w_{12} = 1,\ w_{11}\geq 0,\ w_{12}\geq 0\}$$

$$W^{(2)} = \{w_{21}+w_{22} = 1,\ w_{11}\geq 0,\ w_{12}\geq 0\}$$

The Kuhn-Tucker conditions for the Nash equilibrium points:

$$100w_{11}+100w_{21} - \lambda_1 \geq 120$$

$$100w_{12}+100w_{22} - \lambda_1 \geq 80$$

$$100w_{11}+100w_{21} - \lambda_2 \geq 140$$

$$100w_{12}+100w_{22} - \lambda_2 \geq 70$$

$$w_{11}(100w_{11}+100w_{21} - \lambda_1 - 120) = 0$$

$$w_{12}(100w_{12}+100w_{22} - \lambda_1 - 80) = 0$$

$$w_{21}(100w_{11}+100w_{21} - \lambda_2 - 140) = 0$$

$$w_{22}(100w_{12}+100w_{22} - \lambda_2 - 70) = 0$$

$$w_{11}+w_{12} = 1$$

$$w_{21}+w_{22} = 1$$

$$w_{11},\ w_{12},\ w_{21},\ w_{22} \geq 0$$

In this case we receive a unique solution

$$w_1^0 = (0.6, 0.4),\ \lambda_1^0 = 10,\ \ w_2^0 = (0.7, 0.3),\ \lambda_2^0 = 0$$

and the aggregated social weights

$$\alpha_1^0 = \frac{w_{11}^0 + w_{21}^0}{2} = \frac{0.6+0.7}{2} = 0.65$$

$$\alpha_2^0 = \frac{w_{12}^0 + w_{22}^0}{2} = \frac{0.4+0.3}{2} = 0.35$$

Compromise location selected by this process

$$p_0 = \alpha_1^0 a_1 + \alpha_2^0 a_2 = 0.65\begin{pmatrix}10\\ \\0\end{pmatrix} + 0.35\begin{pmatrix}0\\ \\10\end{pmatrix} = \begin{pmatrix}6.5\\ \\3.5\end{pmatrix}$$

minimizes the distance of the each individual preferred location from the selected location given the weights of other individual.

4 Portfolio Voting

By voting we usually mean the following pattern of collective choice: There is a set of alternatives and a group of individuals. Individual preferences over the alternatives are exogenously specified and are supposed to be orderings. The group is required to choose an alternative (or a subset of the set of alternatives) on the basis of the stating and aggregating of all individual preferences, or to produce a ranking of alternatives from the most preferred to the least preferred.

We should distinguish between a voting method and a counting method. While a voting method says how an individual choice is presented by the voter, a counting method says how the indiviual choices are to be aggregated. The selected voting method together with a counting method define a voting procedure.

There are basically two groups of voting methods: a) Ranking methods, when voting is done by an explicit or implicit presentation of a ranking of alternatives. b) Multiple-vote methods, when voting is done by distributing more votes among the candidates. The most commonly used single-vote method, by which voting is done by presenting a single vote for one and only one selected alternative, may be considered as a special case of a multiple-vote method.[2] While many different sophisticated counting methods can be employed together with ranking voting systems (Condorcet's voting procedures, Hare, Coombs, Dodgson, Copeland etc.), the counting method for various multiple-vote systems (such as approval voting, cumulative voting, Borda's voting, interval voting) is simple: The votes for different alternatives are summed up and a resulting ranking is given by the total number of votes for each alternative.

In this section we focus attention on multiple-vote procedures that provide the possibility to use optimization models of voting behaviour and to consider a multi-attribute character of alternatives leading to multi-criterial decision making of a rational voter.

In stating an individual voter's optimal choice problem as a multi-criteria optimization problem, we shall start with several trivial and, in a sense, simplifying assumptions:

a) *Multiple issues*. We shall suppose that there is a list of major political issues $\boldsymbol{\pi} = (\pi_1, \pi_2, \ldots, \pi_k)$ where $k > 1$ and that an individual voter is able to rank the issues by their importance to him. Let us denote the voter's ranking over this list by $\mathbf{b}_j = (b_{j1}, b_{j2}, \ldots, b_{jk})$, where b_{js} is an integer between 0 and t, expressing the position of the s-th issue in the j-th voter's ordering, or a cardinal measure (say, between 0 and 100) expressing the relative importance of the issue to the voter.

b) *Observability of candidates' position*. We suppose that each candidate or party publicly states his own ranking over the list of major political issues. Let us denote the i-th candidate ranking by $\mathbf{a}_i = (a_{i1}, a_{i2}, \ldots, a_{ik})$.

c) *Trustworthiness* of the candidates. We suppose that there is no uncertainty about the future position of the candidates (publicly declared rankings express the true position of the candidates).

Let us suppose the following proportional voting rule for electing a committee,

---

[2] Properties of voting procedures based on ranking have been presented extensively in many books and papers on public choice theory, (see e.g. DUMMET 1984, NURMI 1987, STRAFFIN 1980, in a framework of multi-criteria decision making HWANG and LIN 1987).

called a *portfolio voting rule* (TURNOVEC, 1995): Each voter chooses among party lists. Let n be the number of voters, m be the number of parties, and v be the number of seats in a committee. Each voter has k votes (as many votes as the number of seats). By x_{ji} let us denote a number of seats assigned by the j-th voter to the i-th party. For simplicity, we suppose that all voters take part in the election and use all their votes, then

$$\sum_{i=1}^{m} x_{ji} = v, \; w_{ji} = x_{ji}/v$$

(for all $i = 1,2,\ldots,m$ and $j = 1,2,\ldots,n$) and $\mathbf{w}_j = (w_{j1}, w_{j2}, \ldots, w_{jm})$ is an j-th *individual voting portfolio*, or an j-th voter feasible voting strategy. Then $\boldsymbol{\alpha} = (\alpha_1, \alpha_2, \ldots, \alpha_m)$ such that

$$\alpha_i = \frac{\sum_{j=1}^{n} w_{ji}}{n}$$

is a *social voting portfolio* according to which the seats are to be distributed in a committee. As before we denote the i-th voter ranking over a list of major political issues by $\mathbf{b}_i = (b_{i1}, b_{i2}, \ldots, b_{ik})$. Each party declares its own ranking over the issues $\mathbf{a}_i = (a_{i1}, a_{i2}, \ldots, a_{ik})$.

The individual voter's rational behaviour can then be defined by an optimal solution to the following goal programming problem:

minimize

$$d\left(\sum_{i=1}^{m} w_{ji}\boldsymbol{a}_i, \boldsymbol{b}_j\right) \tag{10}$$

subject to

$$\sum_{i=1}^{m} w_{ji} = 1, \quad w_{ji} \geq 0 \tag{11}$$

where d is a distance between the j-th individual ranking and the aggregate committee ranking generated by the j-th voter's voting portfolio. An optimal solution $\mathbf{w}_j^0 = (w_{j1}^0, w_{j2}^0, \ldots, w_{jm}^0)$ to the problem (10) - (11) we can call an *j-th individual optimal voting portfolio*.

An obvious criticism of this model of rational voting behaviour in portfolio voting follows from the fact that the level of satisfaction of an individual voter with the results of voting depends not only on his decision, but also (and to a very great extent) on decisions of many other voters. It ignores strategic aspects of voting. A sophisticated voter is aware of the fact that his individual voting portfolio will certainly differ from a social voting portfolio and his satisfaction should be measured by a distance between his individual ranking and the aggregate committee ranking generated by the social voting portfolio, rather than by the individual voter's portfolio.

The distance between the r-th individual ranking and the aggregate committee ranking generated by a social voting portfolio α can be measured by the r-th voter's distance function

$$\delta_r(\boldsymbol{w}_r, \boldsymbol{w}_1, \ldots, \boldsymbol{w}_{r-1}, \boldsymbol{w}_{r+1}, \ldots, \boldsymbol{w}_n) = d\left(\frac{1}{n}\sum_{i=1}^{m}(w_{kj} + \sum_{j \neq r} w_{ji})\boldsymbol{a}_i, \boldsymbol{b}_r\right) \tag{12}$$

where only the individual portfolio variables w_{ri} are under the control of the r-th voter. Therefore, we can formulate a game of the n voters with the pay-off functions (12) and the strategy sets

$$W^{(r)} = \{\boldsymbol{w}_r \in \mathbf{R}_m, \sum_{i=1}^{m} w_{ri} = 1, w_{ri} \geq 0\} \tag{13}$$

The game (12) - (13) is nothing else but a distance game (3) - (4), and a Nash equilibrium analysis can provide a starting point for an investigation of a sophisticated voting behaviour in "portfolio type" voting procedures. We can also consider other possibilities: different levels of information of the voters about the other voters' preferences and use approaches of games against nature, games with sophisticated and indifferent players, a coalitional cooperation etc.

References

Dummet, M., 1984, *Voting Procedures*, Clarendon Press & Oxford University Press, New York.

Hwang, Ching-Lai and Lin Ming-Jeng, 1987, *Group Decision Making under Multiple Criteria*, Springer-Verlag, Berlin.

Lemke, C., 1965: *Bimatrix Equilibrium Points and Mathematical programming*. Management Science, **11**, 681-689.

Nurmi, H., 1987, *Comparing Voting Systems*, D. Reidel Publ. Comp., Dordrecht.

Straffin, P. D., 1980, *Topics in the Theory of Voting*, Birkhäuser, Boston.

Turnovec, F., 1970, *The Double Description Method for Basic Solutions with Orthogonality Property*, CORE Discussion Paper No. 7004, Université Catholique de Louvain.

Turnovec, F., 1995, *Voting as an Optimal Choice*, Prague Economic Papers, **3**, 277-284.

Dynamic Goal Programmıng Models

Tadeusz Trzaskalik

Department of Econometrics, Academy of Economics in Katowice, Poland
Department of Operations Research, University of Łódź, Poland

1. Introduction

Multiobjective dynamic programming deals with multi-period decision processes. There are two main approaches to multi- objective dynamic problems: vector approach and scalarization approach. Vector dynamic approach was first developed by Brown and Strauch (1965). The aim of solving vector dynamic programming problem is to find a set of efficient solutions and Pareto-optimal vectors in the criterion space (Klotzler 1978). In scalarization approach dynamic problem may be transformed into a corresponding single objective dynamic programming problem. This transformation is based on decision maker's (DM) preference. One of the scalarization methods is goal programming approach, first described by Levary (1984). A detailed review of the state of the art can be found in Li and Haimes (1989) and Trzaskalik (1995).

The aim of the paper is to discuss some aspects of goal programming approach to multiple objective dynamic programming. In the first approach, called dynamic goal approach values of multiperiod criteria and multiperiod target goal values given by the DM are considered. Weight coefficients are used to measure different importance of deviations from given multiperiod values. The "nearest" solution is looked for. In the second approach, called dynamic hierarchical goal approach weight coefficients are not introduced. We will treat such a problem as a hierarchical one setting hierarchy for deviations from values proposed by the DM. In the next two models values of period criteria are considered. We assume that a set of those values is given by the DM. In dynamic period goal approach weight coefficients are used to measure deviations from given target period values. The "nearest" solution is looked for again. In dynamic hierarchical period approach hierarchy of deviations from period target values is set by the DM.

The paper is organized as follows. In Section 2 multiple criteria discrete decision process is described. In Section 3 dynamic goal programming model is described. In Section 4 dynamic hierarchical goal programming model is considered. In Section 5 dynamic period goal programming model is introduced. In Section 6 dynamic hierarchical period goal programming model is described. Illustrative numerical examples are given in Section 7. Final remarks in Section 8 end the paper.

The earlier versions of period goal programming algorithms are described in Trzaskalik (1990). Modified hierarchical approach applied in Algorithm 2 is described in Trzaskalik (1992).

2. Multiple Criteria Discrete Decision Process

We consider a discrete decision process which consists of T periods. Let Y_t be the set of all feasible state variables for period t and $X_t(y_t)$ - the set of all feasible decision variables for period t and state $y_t \in Y_t$ (t=1,...,T). We assume that all these sets are finite. Let D_t be the set of all period realizations d_t in period t defined as follows:

$$D_t = \{d_t = (y_t, x_t) : y_t \in Y_t,\ x_t \in X_t(y_t)\} \tag{1}$$

and $D_t(y_t)$ - the set of all period realizations in period t starting at y_t, defined in the following way:

$$D_t(y_t) = \{d_t = (y_t, x_t) : x_t \in X_t(y_t)\} \tag{2}$$

We assume that for t=1,...,T there are given transformations $\Omega_t : D_t \to Y_{t+1}$.
A sequence of period realizations

$$d = (d_1, d_2, \ldots, d_T) = (y_1, x_1, y_2, x_2, \ldots, y_T, x_T) \tag{3}$$

is said to be a process realization if:

$$\underset{t=1,\ldots,T}{\forall} \quad y_{t+1} = \Omega_t(y_t, x_t) \tag{4}$$

Let D be the set of all process realizations.

We assume that in each period t there are defined K period criteria functions $F_t^k : D_t \to R$ (t=1,...,T, k=1,...,K). F is vector- valued criterion function for the whole process and its components $F^1, \ldots, F^K$ are functions of F_t^k and are understood as follows:

$$F^k = \Phi^k(F_1^k, \ldots, F_T^k) \tag{5}$$

Components F^k are said to be multi-period criteria functions. The simplest and the most important form for applications is an additive one:

$$F^k(d) = \sum_{t=1}^{T} F_t^k(d_t) \tag{6}$$

which will be applied in further considerations.

3. Dynamic Goal Programming Model

Let us consider target goal vector

$$\bar{F} = [\bar{F}^1, F^2, \dots, \bar{F}^K] \tag{7}$$

and weight coefficients vectors $\dot{c}$ and $\bar{c}$ containing penalty coefficients for multiperiod deviations from target goals for k = 1,...,K:

$$\dot{c} = [\dot{c}^1, \dot{c}^2, \dots, \dot{c}^K] \tag{8}$$

$$\bar{c} = [\bar{c}^1, \bar{c}^2, \dots, \bar{c}^K] \tag{9}$$

We define auxiliary variables

$$\dot{h}^k(d) = \begin{cases} (F^k(d) - \bar{F}^k) & \text{if } F^k(d) - \bar{F}^k \geq 0 \\ 0 & \text{if } F^k(d) - \bar{F}^k < 0 \end{cases} \tag{10}$$

$$\bar{h}^k(d) = \begin{cases} 0 & \text{if } F_t^k(d) - \bar{F}_t^k \geq 0 \\ (\bar{F}^k - F^k(d)) & \text{if } F_t^k(d) - \bar{F}_t^k < 0 \end{cases} \tag{11}$$

and formulate dynamic goal programming model:

$$Min\{\sum_{k=1}^{K} (\dot{c}^k \dot{h}^k(d) + \bar{c}^k \bar{h}^k(d)) : d \in D\} \tag{12}$$

Proposed interactive algorithm solving the problem is described below.

Algorithm 1

1. DM sets target goal vector F and weight vectors $\dot{c}$ and $\bar{c}$.
2. We find $\dot{h}^k$ (d) and $\bar{h}^k$ (d) according to formulas (10) and (11).
3. Using a modified dynamic programming approach solve problem (12). Let D be the set of solutions of that problem.
4. Ask the DM to choose realization $\bar{d} \in \bar{D}$ which has not been considered before.
5. The DM makes one of the following decisions:
   a) accepts realization $\bar{d}$ as the final one,
   b) points to another solution from $\bar{D}$,
   c) repeats the whole procedure with changed target goal vector or changed weight vectors,
   d) breaks the procedure.

4. Dynamic Hierarchical Goal Programming Model

If it is not possible to define weight deviation vectors $\dot{c}$ and $\bar{c}$ we can apply modified hierarchical approach for solving goal programming problem. We assume that criteria are numerated in such a way that deviations from target values of criterion number 1 are more important than deviations from target values of period criterion number 2, deviations from target values of period criterion number 2 are more important than deviations from target values of criterion number 3 (which are also less important than deviations from target values of criterion number 1) and so on. Solution of the dynamic hierarchical goal programming model is described in Algorithm 2.

Algorithm 2

1. DM defines single hierarchy of criteria.
2. Set k=1 and $D^1 = D$.
3. DM sets target goal value $\bar{F}^k$ and tolerance coefficients $\dot{\mu}^k$ and $\bar{\mu}^k$ ($\dot{\mu}^k, \bar{\mu}^k \in [0,1)$) or tolerance limit vectors $\dot{l}^k$ and $\bar{l}^k$.

These values allow the creation of relaxed sets D^{k+1} such that

a) for the tolerance coefficients

$$D^{k+1} = \{d \in D^k \;:\; \dot{h}^k(d) \le \dot{\mu}^k \bar{F}^k \;\wedge\; \bar{h}^k(d) \le \bar{\mu}^k \bar{F}^k\} \tag{13}$$

b) for the tolerance limits

$$D^{k+1} = \{d \in D^k \;:\; \dot{h}^k(d) \le \dot{l}^k \;\wedge\; \bar{h}^k(d) \le \bar{l}^k\} \tag{14}$$

4. If $k \le K-1$, set $k = k+1$ and proceed to step 3.
5. Solve problem:

$$\text{Min } \{ \left(\dot{h}^K(d) + \bar{h}^K(d)\right) : d \in D^{K+1} \} \tag{15}$$

Let $\bar{D}$ be the set of solutions of that problem.

6. The DM makes one of the following decisions:
   a) accepts one of the realizations $\bar{d} \in \bar{D}$ as the final one,
   b) points to a realization $d^{K+1} \in D^{K+1}$ as a final one,
   c) repeats the whole procedure with changed target goal values or changed parameters,
   d) breaks the procedure.

5. Dynamic Period Goal Programming Model

Let us consider period target goal matrix

$$F = \begin{bmatrix} \bar{F}_1^1 & \bar{F}_1^1 & \cdots & \bar{F}_T^1 \\ \cdot & \cdot & \cdot & \cdot \\ \bar{F}_1^K & \bar{F}_2^K & \cdots & \bar{F}_T^K \end{bmatrix} \tag{16}$$

and period weight coefficients matrices $\dot{C}$ and $\bar{C}$ containing penalty coefficients for period deviations for t=1,...,T from target goals for k=1,...,K:

$$\dot{C} = \begin{bmatrix} \dot{c}_1^1 & \dot{c}_2^1 & \cdots & \dot{c}_T^1 \\ \cdot & \cdot & \cdot & \cdot \\ \dot{c}_1^K & \dot{c}_2^K & \cdots & \dot{c}_T^K \end{bmatrix} \qquad C = \begin{bmatrix} \bar{c}_1^1 & \bar{c}_2^1 & \cdots & \bar{c}_T^1 \\ \cdot & \cdot & \cdot & \cdot \\ \bar{c}_1^K & \bar{c}_2^K & \cdots & \bar{c}_T^K \end{bmatrix} \tag{17}$$

We define auxiliary variables

$$\dot{h}^k_t(y_t, x_t) = \begin{cases} (F_t^k(y_t, x_t) - \bar{F}_t^k) & \text{if} \quad F_t^k(y_t, x_t) - \bar{F}_t^k \geq 0 \\ 0 & \text{if} \quad F_t^k(y_t, x_t) - \bar{F}_t^k < 0 \end{cases} \tag{18}$$

$$\bar{h}^k_t(y_t, x_t) = \begin{cases} 0 & \text{if} \quad F_t^k(y_t, x_t) - \bar{F}_t^k \geq 0 \\ (\bar{F}_t^k - F_t^k(y_t, x_t)) & \text{if} \quad F_t^k(y_t, x_t) - \bar{F}_t^k < 0 \end{cases} \tag{19}$$

and formulate dynamic period goal programming model:

$$\text{Min} \ \{\sum_{k=1}^{K} \sum_{t=1}^{T} (\dot{c}_t^k \dot{h}_t^k(y_t, x_t) + \bar{c}_t^k \bar{h}_t^k(y_t, x_t)) : (y_1, x_1, \ldots, y_T, x_T) \subset D\} \tag{20}$$

Proposed interactive algorithm solving the problem is described below.

Algorithm 3

1. DM sets target goal matrix $\bar{F}$ and weight matrices $\dot{C}$ and $\bar{C}$.
2. We find $\dot{h}_t^k(y_t, x_t)$ and $\bar{h}_t^k(y_t, x_t)$ according to formulas (18) and (19).
3. Using a standard dynamic programming approach solve problem (20). Let $\bar{D}$ be the set of solutions of that problem.
4. Ask the DM to choose realization $\bar{d} \in \bar{D}$ which has not been considered before.
5. The DM makes one of the following decisions:
   a) accepts realization $\bar{d}$ as the final one,
   b) points to another solution from $\bar{D}$ and goes back to step 5,
   c) repeats the whole procedure with changed target goal matrix or changed weight matrices,
   d) breaks the procedure.

6. Dynamic Hierarchical Period Goal Programming Model

If it is not possible to define weight deviation matrices $\dot{C}$ and $\bar{C}$ we can apply modified hierarchical approach for period goal programming problem . We assume the same numeration of criteria as before. Solution of the dynamic period hierarchical goal programming model is described in Algorithm 4.

Algorithm 4

1. DM defines single hierarchy of criteria.
2. Set k=1 and $D^1 = D$.
3. DM sets target goal vector $\bar{F}^k = [\bar{F}_1^k, \ldots, \bar{F}_T^k]$ and period tolerance coefficient vectors $\dot{\mu}^k = [\dot{\mu}_1^k, \ldots, \dot{\mu}_T^k]$, $\bar{\mu}^k = [\bar{\mu}_1^k, \ldots, \bar{\mu}_T^k]$, $(\dot{\mu}_t^k, \ \dot{\mu}_t^k \in [0,1))$ or period tolerance limit vectors $\dot{l}^k = [\dot{l}_t^k, \ldots, \dot{l}_t^k]$, $\bar{l}^k = [\bar{l}_t^k, \ldots, \bar{l}_t^k]$. These values and procedure of generating feasible states and decisions allow the creation of relaxed sets D^{k+1} such that
   a) for the tolerance coefficients

$$D^{k+1} = \{d \in D^k : \forall_{t=1,\dots,T} \; \dot{h}_t^k(y_t, x_t) \le \dot{\mu}_t^k \dot{F}_t^k \wedge \bar{h}_t^k(y_t, x_t) \le \bar{\mu}_t^k \bar{F}_t^k\} \tag{21}$$

b) for the tolerance limits

$$D^{k+1} = \{d \in D^k : \forall_{t=1,\dots,T} \; \dot{h}_t^k(y_t, x_t) \le \dot{l}_t^k \wedge \bar{h}_t^k(y_t, x_t) \le \bar{l}_t^k\} \tag{22}$$

4. If $k \le K-1$, set $k = k+1$ and proceed to step 3.
5. Using a standard dynamic programming approach solve problem:

$$\text{Min } \{\sum_{t=1}^{T}(\dot{h}_t^K(y_t, x_t) + \bar{h}_t^K(y_t, x_t)) : d \in D^{K+1}\} \tag{23}$$

Let $\bar{D}$ be the set of solutions of that problem.

6. The DM makes one of the following decisions:

a) accepts one of the realizations $\bar{d} \in \bar{D}$ as the final one,

b) points to a realization $d^{K+1} \in D^{K+1}$ as a final one,

c) repeats all the procedure with changed target goal vectors or changed parameters,

e) breaks the procedure.

7. Illustrative Numerical Examples

We consider a process which consists of 3 periods, in which:

$Y_t = \{0,1\}$ for t=1,2,3,4

$X_t(0) = \{0,1\}$, $X_t(1) = \{0,1\}$ for t=1,2,3

Values of period criteria are given in Table 1.

Period realization	F_1^1	F_1^2	F_2^1	F_2^2	F_3^1	F_3^2
(0,0)	7	5	6	8	4	9
(0,1)	6	6	3	9	3	8
(1,0)	3	9	7	8	10	2
(1,1)	4	10	7	4	3	3

Table 1

Example 1

Dynamic goal programming approach

Assume that there are given vectors:

$F = [18, 18]$, $c = [1, 1]$, $\bar{c} = [1, 1]$

Applying modified dynamic programming procedure we obtain:

$$D = \{\overset{1}{d}, \overset{2}{d}\}$$

and

$d^1 = (0,0,\ 0,1,\ 1,0)$ $F(d^1) = [20, 16]$,

$\dot{h}^1(d^1) = 2$, $\bar{h}^1(d^1) = 0$, $\dot{h}^2(d^1) = 0$, $\bar{h}^2(d^1) = 2$

$d^2 = (1,0,\ 0,1,\ 1,0)$ $F(d^2) = [16, 20]$,

$h^1(d^2) = 0$, $h^1(d^2) = 2$, $h^2(d^2) = 2$, $h^2(d^2) = 0$

Example 2

Dynamic hierarchical goal programming model .

Assume that the DM defines the following hierarchy of criteria: the most important is criterion 1, the less importatnt - criterion 2. The DM sets:

$$\bar{F}^1 = 18 \quad \dot{l}^1 = 3 \quad \bar{l}^1 = 3$$

We consider such period realizations, for which

$$\dot{h}^1(d) \leq 3 \ \wedge \ \bar{h}^1(d) \leq 3$$

Set D^2 contains 10 realizations.

The DM sets:

$$\bar{F}^2 = 18 \quad \dot{l}^2 = 3 \quad \bar{l}^2 = 3$$

We consider such period realizations, for which

$$\dot{h}^2(d) \leq 3 \ \wedge \ \bar{h}^2(d) \leq 3$$

We obtain

$$D^3 = \{ d^1, d^2, d^3 \}$$

$d^1 = (0,0, 0,1, 1,0)$ $F(d^1) = [20, 16]$
$d^2 = (1,0, 0,1, 1,0)$ $F(d^2) = [16, 20]$
$d^3 = (1,1, 1,1, 1,0)$ $F(d^3) = [21, 16]$

The DM makes the final choice among these realizations.

Example 3

Period goal programming approach .

Assume that there are given matrices $\bar{F}$, $\dot{C}$ and $\bar{C}$:

$$\bar{F} = \begin{bmatrix} 5 & 7 & 9 \\ 8 & 7 & 8 \end{bmatrix} \qquad \dot{C} = \begin{bmatrix} 1 & 1 & 1 \\ 1 & 1 & 1 \end{bmatrix} \qquad \bar{C} = \begin{bmatrix} 1 & 1 & 1 \\ 1 & 1 & 1 \end{bmatrix}$$

Values of realizations are given in Table 2.

Period realization	h_1^1	h_1^2	h_2^1	h_2^2	h_3^1	h_3^2	h_1^1	h_1^2	h_2^1	h_2^2	h_3^1	h_3^2
(0,0)	2	0	0	1	0	1	0	3	1	0	5	0
(0,1)	1	0	0	2	0	0	0	2	4	0	6	0
(1,0)	0	1	0	1	1	0	2	0	0	0	0	6
(1,1)	0	2	0	0	0	0	1	0	0	3	6	5

Table 2

Applying single criterion dynamic programming procedure we obtain

$\bar{D} = \{ d^1, d^2 \}$ where $d^1 = (0,1, 1,0, 0,1)$ $F(d^1) = [16,22]$
$d^2 = (1,1, 1,0, 0,1)$ $F(d^2) = [14,26]$

The DM makes one of the decisions described in step 5 of Algorithm 3.

Example 4

Hierarchical period goal programming approach.

Assume that the DM defines the following hierarchy of criteria: the most important is criterion 1, the less important - criterion 2. The DM sets target goal vector and period tolerance limit vectors for criterion 1 as follows:

$\bar{F}^1 = [\,5\ 7\ 9\,]$, $\dot{l}^1 = [\,2\ 2\ 0\,]$, $\bar{l}^1 = [\,2\ 2\ 6\,]$

We consider such period realizations, for which

$$\dot{h}_1^{\,1}(y_1, x_1) \le 2 \quad \wedge \quad \bar{h}_1^{\,1}(y_1, x_1) \le 2$$
$$\dot{h}_2^{\,1}(y_2, x_2) \le 2 \quad \wedge \quad \bar{h}_1^{\,1}(y_2, x_2) \le 2$$
$$\dot{h}_1^{\,1}(y_1, x_1) \le 0 \quad \wedge \quad \bar{h}_1^{\,1}(y_1, x_1) \le 6$$

The DM sets target goal vector and period tolerance limit vectors for criterion 2 as follows:

$\bar{F}^2 = [\,8\ 7\ 8\,]$, $\dot{l}^2 = [\,1\ 1\ 1\,]$, $\bar{l}^2 = [\,2\ 2\ 2\,]$

We consider such period realizations, for which

$$\dot{h}_1^{\,2}(y_1, x_1) \le 1 \quad \wedge \quad \bar{h}_1^{\,2}(y_1, x_1) \le 2$$
$$\dot{h}_2^{\,2}(y_2, x_2) \le 1 \quad \wedge \quad \bar{h}_1^{\,2}(y_2, x_2) \le 2$$
$$\dot{h}_1^{\,2}(y_1, x_1) \le 1 \quad \wedge \quad \bar{h}_1^{\,2}(y_1, x_1) \le 2$$

We have:

$$D^2 = \{\, d^1, d^2 \,\}$$

where

$$d^1 = (0{,}1,\ 1{,}0,\ 0{,}0)$$
$$d^2 = (1{,}0,\ 0{,}0,\ 0{,}0)$$

The DM makes the final choice among these realizations.

8. Final Remarks

There are some possible extensions for goal programming approaches discussed above.

1. Multiperiod goal programming model can be defined by means of family of functions L^p and have the form :

$$\text{Min} \ \left\{ \left(\sum_{k=1}^{K} \left(\dot{c}^k \dot{h}^k(d) + \bar{c}^k \bar{h}^k(d) \right)^p \right)^{1/p} : d \in D \right\}$$

2. Period goal programming model could be defined by means of family of functions L^p and have the form:

$$\text{Min} \ \left\{ \left(\sum_{k=1}^{K} \sum_{t=1}^{T} \left(\dot{c}_t^k \dot{h}_t^k(y_t, x_t) + \bar{c}_t^k \bar{h}_t^k(y_t, x_t) \right)^p \right)^{1/p} : d \in D \right\}$$

If p=1, we obtain models discussed in the paper.

3. Formulating dynamic hierarchical period approach we can also apply group hierarchy for positive and negative deviations from the given targe goal vectors and combine goal programming approach with modified dynamic hierarchical approach and modified group hierarchy approach, discribed by Trzaskalik (1996).

4. Capital expenditure problem where all projects under consideration are NPV-positive and the firm wishes to evaluate other financial goals during its financial planning horizon, described in Jog, Michalowski and Trzaskalik (1995) can be reformulated as a period goal programming hierarchical model.

REFERENCES

Brown,T.A and R.R.Strauch (1965) "Dynamic Programming in Multiplicative Lattices". Journal of Mathematical Analysis and Applications, 12,364-370.

Jog,V.,Michalowski,W. and T.Trzaskalik (1995) "An Interactive Multi -objective Programming Approach to the Capital Expenditure Planning Problems". Research in Finance, 13,261-274.

Klotzler, R.(1978) "Multiobjective Dynamic Programming". Math. Operationsforsch.Statist.Ser.Optimization 9,423-426.

Levary, R.R.(1984) "Dynamic Programming Models with Goal Objectives" Int.J.Systems Sci. 15,3,309-314.

Li,D. and Y.Y.Haimes (1989) "Multiobjective Dynamic Programming. The State of the Art". Control Theory and Advanced Technology 5,471- 483.

Trzaskalik, T.(1990) "Wielokryterialne dyskretne programowanie dynamiczne. Teoria i zastosowania w praktyce gospodarczej (Multi-criteria Dynamic Programming. Theory and Economic Applications). Wydawnictwa Uczelniane Akademii Ekonomicznej w Katowicach (in Polish).

Trzaskalik, T.(1992) "Hierarchical Approach to Multi-criteria Dynamic Programming". INFOR 30,138-147.

Trzaskalik, T.(1994) "Multiple Criteria Discrete Dynamic Programming. Mathematics Today vol.XII-A, 173-199.

Trzaskalik,T.(1996) "Hierarchical Problems in Multiple Criteria Dynamic Programming. Submitted for IFORS, Vancouver (July 8-12, 1996).

Estimation Through the Imprecise Goal Programming Model

Belaïd Aouni, Ossama Kettani and Jean-Marc Martel

Faculté des sciences de l'administration, Université Laval, Sainte-Foy (Québec). G1K 7P4. Canada.

Abstract. The recent studies suggest that mathematical programming could be a good alternative to the conventional statistical analysis methods as the least squares method and the least absolute method. In fact, the mathematical programming models provide more flexibility for modelling the estimation context. This flexibility gives to the analyst a platform where his knowlege and experience can be an integral part of the parameters estimation.

Moreover, the mathematical programming gives the possibility to take in account the imprecision associated with some variable values. This paper suggests an estimation model which enables the analyst to integrate his experience and judgement in a context where the values of the dependant variable are imprecise and expressed by an interval.

Keywords. Goal Programming model, Imprecise Goals, Statistical Estimation.

1 Introduction

With the intention of understanding, explaining or predicting a phenomenon behaviour (dependant variable) resorting to an estimation process consists in a) identifying the factors (independant variables) that play a significant role in the phenomenon behaviour, b) specifying the general function form that enables to put the phenomenon behaviour in relation with the identified factors and c) estimating the parameters of this response function from a sample analysis that describes the observable behaviour of the studied phenomenon. This estimation is led within the framework of an optimization process, through which the gap or distance between the observed behaviour and the one predicted of the studied phenomenon is minimized.

Statistical methods are generally used as a tool of data analysis and parameter estimation. In general, these methods depend on the deviation minimization between the observed data and the estimated values by the mathematical model. Among the main statistical techniques used for the response function parameters estimation, we find a) the least squares method and b) the least absolute value method. These two methods are essentially based on minimizing estimation errors. These errors are, in fact, a distance function between the observed values and the estimated values.

The multiple linear regression is one of the most popular approaches in the estimation process. This approach can be distinguished by a) a response function of an additive form, b) an estimation procedure of parameters based on the least squares method where the distance is expressed in an Euclidean way and, above all, by c) a great facility (in algorithmic terms) of use. This facility is due to the quadratic convex form of distance functions as well as the fact that these functions are to be optimized (minimized) without additionnal constraints. Thus, these properties enable to transform the quadratic optimization problem in a linear equation set whose resolution technique efficiency is well known.

Recent works suggest that mathematical programming constitutes an interesting alternative to some conventional statistical methods such as that of the least squares. Resorting to mathematical programming in the statistical estimation process developped further to the works of Charnes *et al.* (1986), Sueyoshi (1986) and Freed and Glover (1981-b) who introduced the Goal Programming (GP) techniques as an estimation tool, Kettani (1988) and Oral and Kettani (1989) in the multicriteria choice analysis and Kettani *et al.* (1996) in the field of real estate evaluation.

Besides, the mathematical programming tools have greatly been used in discriminant analysis. For example, Rao (1971), Zeleny (1973) and Pekelman and Subarta (1974) used mathematical integer programming to determine discriminant functions in order to obtain a better classification rate compared to the standard statistical methods. Several linear and non linear mathematical formulations of the discriminant analysis problem were proposed by Freed and Glover (1981-a; 1982; 1986), Bajgier and Hill (1982), Glorfeld and Gaither (1982), Stam and Joachimsthaler (1989), Koehler (1990), Koehler and Erenguc (1990), Glover (1990) and Rubin (1990).

In a statistical estimation process, the sample data relative to the dependant variable as well as to the independant variables are generally considered as precise values, it is to say that they are not marred by uncertainty nor imprecision. Now, it is possible that the values of these variables may be imprecise. It is quite possible for example that the variable values can be imprecise and that the best we can do is to express them through an interval.

The GP model is a particular case of the distance function model that consists in minimizing the deviation sum between the obtained solutions and the fixed goals for the considered objectives. Charnes *et al.* (1986) introduced the possibility of using the GP as a parametric estimation tool where the deviations or the distance between the observed values and those estimated are to be minimized.

Resorting to mathematical programming to estimate the response function parameters offers several advantages of which: a) there are less assumptions to be satisfied on the nature of historical data, b) a slightest need as regards the number of observed data and c) a greater flexibility in the estimation context modelization in the sense that the analyst may frame the estimation process by turning to account some a priori informations at his disposal due to notably his knowlege and experience.

The first objective of this paper is to generalize penalty function forms related to deviations by introducing the concept of satisfaction functions developped in Martel and Aouni (1990). This is to be situated in the context where some of the observable values y_i, used for estimation purposes, will be considered as imprecise. In a such context, the analyst can only specify an inferior limit and a superior limit for these values. The model formulation of the GP with imprecise goals developped by Martel and Aouni (1996) will be used to integrate the analyst judgement and experience in the estimation process.

The remainder of this paper will be divided into three sections. Section 2 will deal with the general form of the optimization problem subjacent to the statistical estimation process, section 3 with the suggested method to treat contexts where the observable values of the dependant variable are expressed within an interval of possible values and finally, in section 4 we will draw some conclusions.

2 Statistical estimation

The statistical estimation is mainly based on the estimation of the response function parameters whose generic form is generally *a priori* specified. This function is meant to explain in what way the behaviour of the dependant variable results of behaviour conjunction of independant variables. The sample analysis, made in an appropriate way, enables to test the validity of the generic form of the response function and, if need be, to estimate its parameters.

Let us note by A a sample of n observations, by y_i the observed value of the dependant variable in the context of the observation i and by x_{ij} the observed value of the independant variable j in the context of the same observation and let us note by f the response function. The statistical function aims to estimate the parameters of the function f so that: the distance between $f(x_{ij})$ and y_i will be minimal, for i=1, 2,...,n.

Thus, the degree of confidence in the obtained estimations for the estimated parameters is higher that the distance is small between the observed values of the dependant variable and its estimated values obtained through the response function. In fact, the statistical estimation methods can be mainly distinguished by the form of the response function, the definition adopted from the concept of distance and its treatment in the estimation process. Among the parametric estimation methods, we will focus on the two following ones: the very popular method of the least squares and the least absolute values method which in fact opened the way to the mathematical programming as an estimation tool.

2.1 The least squares method

In this method, the concept of distance is defined in Euclidean terms, it is to say that it is measured by the square of the difference between the observed value and the estimated value of the dependant variable. Formally, the estimation process in the method of the least squares depends on the following optimization problem:

Program 2.1: $$\underset{\text{on the parameters of } f}{\text{Minimize}} \quad \sum_{i=1}^{n} \left(y_i - f(x_{ij})\right)^2 .$$

Thus this program resolution enables to determine the parameters of f so that the Euclidean distance be minimal. When f has the following additive form: $f(x_{ij}) = \beta_0 + \sum_{j=1}^{m} \beta_j x_{ij}$, whereas β_0 and β_j (for j=1, 2, ..., m) are the parameters to be estimated, the method of the least squares becomes what we call the multiple linear regression and the resolution of the program 2.1, thanks to its convexity properties, is simply reduced to the resolution of a linear equation system.

2.2 The least absolute value method

Here, the concept of distance is rather defined in linear terms. Thus, the distance is measured by the difference, in an absolute value, between the observed value and the estimated value of the dependant variable. So, the estimation process in the least absolute value method may be described by means of the following optimization problem:

Program 2.2: $$\underset{\text{on the parameters of } f}{\text{Minimize}} \quad \sum_{i=1}^{n} |y_i - f(x_{ij})| .$$

Thus the parameters of f are determined so that the linear distance be minimal. When f has the additive form described in (2.1), the optimization program 2.2 becomes: $\underset{\beta_0, \beta_1, \beta_2, \dots, \beta_m}{\text{Minimize}} \sum_{i=1}^{n} |y_i - (\beta_0 + \sum_{j=1}^{m} x_{ij}\beta_j)|$.

In fact, the program (2.2) corresponds to the GP model whose the following linear version may be obtained by means of the transformation procedure proposed by Charnes and Cooper (1977):

Program 2.3: Minimize $Z = \sum_{i=1}^{n} \left(\delta_i^+ + \delta_i^-\right)$

Subject to:

$$\beta_0 + \sum_{j=1}^{m} x_{ij}\beta_j - \delta_i^+ + \delta_i^- = y_i; \qquad i=1, 2, \ldots, n;$$

$$\beta_0,\ \beta_j,\ \delta_i^- \text{ and } \delta_i^+ \geq 0;\ i=1, 2, \ldots, n \text{ and } j=1, 2, \ldots, m.$$

These two methods of parametric estimation make a few naive and simplistic treatment of the data with a context where it is possible that some particular observations may cause a certain fuzziness in the estimation. Moreover, there may be more reliable observations compared to others but these estimation methods do not enable to take in account such situations.

As discussed in Kettani *et al.* (1996), the last formulation present an advantage comparatively to the least squares method. So, contrary to the least squares method, the least absolute value method is less affected by extreme values.

Generally, the least squares method and that of the least absolute value method consider the observed values y_i of the dependant variable as precise values. It is possible that the analyst may collect only one imprecise information on the value of y_i. The next section will be devoted to the estimation processus in this context.

3 A model of estimation with imprecise observations

In this section, we will examinate how we can deal with the imprecision related to the observed values y_i of the dependant variable. The real estate evaluation is one of many contexts where the analyst may estimate the market value of property by using imprecise information. For example, the selling price of a property does not necessarily represent exactly its market value. In the case of a repossessed asset, the selling price of the property is often inferior to its market value. In such situation, the observed data as regards the selling price of the properties might well lead to non adequate results for the parameter estimation that will be used for the real estate taxation. Then to remedy to that, the analyst will establish an interval within which the market value of this property is probably located. The values of y_i^l and y_i^u correspond to the inferior and superior limit of this interval. We consider that the dependant variable varies, for each observation y_i, between an inferior limit y_i^l and a superior limit y_i^u. In other terms, the values y_i of the dependant variable are defined on the following interval: $[y_i^l, y_i^u]$.

The objective of the estimation process with imprecise data is to establish values $\hat{y}_i$ that belong as much as possible to the interval $[y_i^l, y_i^u]$. In other words, we consider that there are no estimation errors when $\hat{y}_i$ is within this interval. In the opposite case, the distance between $\hat{y}_i$ and the interval $[y_i^l, y_i^u]$ represents an estimation error whose scope is function of several factors, such that the analyst value system, the caracteristics of the estimation context and stakes related to this estimation. To deal with this context, we will examinate how the interval GP formulation will be used to obtain the estimation parameters.

In a previous paper, Martel and Aouni (1996) propose a formulation of the GP model in an imprecise environment. In this formulation, the authors suggest the concept of the satisfaction function in the GP model with imprecise goal values. This concept is inspired by the works of Martel and Aouni (1990).

For each objective (or observation), the satisfaction functions are independantly defined for negative and positive deviations. For each observation or group of observations, the analyst may define a satisfaction function relative to distances that we can qualify as negative or positive deviations according to the fact that we are respectively below y_i^l or beyond y_i^u.

The general form of the satisfaction function is as followed:

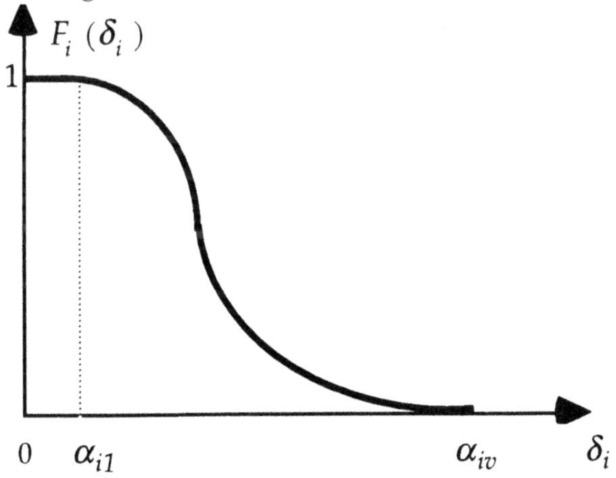

Fig. 3. Satisfaction function type.

The analyst is entirely satisfied when the deviations δ_i from the goals are located within the interval $[0, \alpha_{i1}]$ (α_{i1} is a threshold of indifference). So, within the area of indifference $[0, \alpha_{i1}]$ the deviation of an observation is not penalized and the degree of satisfaction takes its maximal value which is 1. Outside this interval, the satisfaction function is decreasing monotonous. Besides, any other estimation leading to a deviation that exceeds the threshold α_{iv} (it's a veto threshold) is inadmissible. The relative importance of different observations is represented in part by the vector W_i and partially by the thresholds of the satisfaction function.

The analyst experience and judgement are taken into consideration in the estimation model by the satisfaction functions $F_i(\delta_i)$. In that sense, the reformulation of the program 2.3 takes the following form:

$$\text{Maximize } Z=\sum_{i=1}^{n}\left(W_i^{+}F_i^{+}(\delta_i^{+})+W_i^{-}F_i^{-}(\delta_i^{-})\right)$$

Subject to:

$$\beta_0+\sum_{j=1}^{m}x_{ij}\beta_j-\delta_i^{+}+\delta_i^{-}=\zeta_i \quad i=1,2,\ldots,n;$$

$$Cx\leq c \quad \text{(the system constraints);}$$

$$\delta_i^{+} \text{ and } \delta_i^{-}\leq\alpha_{iv}; \ \delta_i^{+} \text{ and } \delta_i^{-}\geq 0 \ ;i=1,2,\ldots,n.$$

In this formulation, goals ζ_i are any point in the interval $[y_i^l, y_i^u]$ ($\zeta_i \in [y_i^l, y_i^u]$) and the thresholds of indifference α_{i1} of the satisfaction functions enable to characterize the imprecision expressed by the interval where : $\alpha_{i1}^{+}=y_i^u-\zeta_i$ and $\alpha_{i1}^{-}=\zeta_i-y_i^l$ (α_{i1}^{+} and α_{i1}^{-} are the thresholds of indifference associated to the positive and the negative deviations respectively). This new formulation of the estimation model enables the treatment of imprecision relative to the dependant variable and the analyst to introduce explicitly his knowledge and experience with the aid of various types of satisfaction functions. Moreover, by this estimation model formulation, the analyst can introduce other constraints as the system constraints.

Besides, the choice of satisfaction functions is not limitative. This tasks is due to the analyst from his judgement and his experience as to the reliability of the observed values. In order to help him to make such choice, he may be inspired by six types of satisfaction function proposed by Martel and Aouni (1990). It is with the aid of an interactive process that the analyst is called to specify his satisfaction functions.

The linear representation of these different types of satisfaction function needs the introduction of binary variables. Thus, the equivalent linear mathematical model will consist of quadratic expressions with integer numbers 0-1. The modified linearization procedure of Oral and Kettani (Oral and Kettani, 1992; Aouni, 1996) may be used to find an equivalent linear formulation.

4 Conclusion

The parametric estimation model proposed in this paper enables the analyst to integrate his experience and judgement in a context where the values of the dependant variable are imprecise and expressed by an interval. The values of the estimation parameters are obtained by a process where the analyst is very present. Moreover, the satisfaction functions enable him to express his confidence in the observed values. Besides, he may carry out an analysis of sensitivity in order to see the impact of the variation of the discrimination thresholds, relative to satisfaction functions, on the estimation parameters.

This model enables a more enlightened (less naive) treatment of the historical data where the analyst may, for example, attach more importance to some observations which he judges more reliable.

References

Aouni, B., "Linéarisation des expressions quadratiques en programmation mathématique : des bornes plus efficaces," In *Proceedings of the annual conference of the Administrative Sciences Association of Canada, Management Science Division*, Vol. 17, 1996 (38-46).

Bajgier, S. M. and A. V. Hill, "An Experimental Comparison of Statistical and Linear Programming Approaches to the Discriminant Problem," *Decision Sciences*, Vol. 13, 1982 (604-618).

Charnes, A., W. W. Cooper and T. Sueyoshi, "Least Squares/Ridge Regression and Goal Programming/Constrained Regression Alternatives," *European Journal of Operational Research*, Vol. 27, No. 1, 1986 (146-157).

Freed, N. and F. Glover, "A Linear Programming Approach to the Discriminant Problem," *Decision Sciences*, Vol. 12, 1981-a (68-74).

Freed, N. and F. Glover, "Evaluating Alternative Linear Porgramming Models to Solve the Two-Group Discriminant Problem," *Decision Sciences*, Vol. 17, 1986 (151-162).

Freed, N. and F. Glover, "Linear Programming and Statistical Discriminant: the LP Side," *Decision Sciences*, Vol. 13, 1982 (172-175).

Freed, N. and F. Glover, "Simple but Powerful Goal Programming Models for Discriminant Problems," *European Journal of Operational Research*, No. 7, 1981-b (44-60).

Glorfeld, L. W. and N. Gaither, "On Using Linear Programming in Discriminant Problems," *Decision Sciences*, Vol. 13, 1982 (167-171).

Glover, F., "Improved Linear Programming Models for Discriminant Analysis," *Decision Sciences*, Vol. 21, 1990 (771-785).

Kettani, O., *Nouvelles formulations de choix multicritères basées sur la programmation mathématique*. Non published Ph.D. Thesis, Faculté des Sciences de l'Administration, Université Laval, Ste-Foy (Québec), Canada, 1988.

Kettani, O., M. Oral and Y. Siskos (1996), "A Multiple Criteria Analysis Model for Real Estate Evaluation," *Journal of Global Optimisation*, 1996 (to appear).

Koehler, G. J. and S. Erenguc, "Minimizing Misclassification in Linear Discriminant Analysis," *Decision Sciences*, Vol. 21, 1990 (63-85).

Koehler, G. J., "Considerations for Mathematical Programming Models in Discriminant Analysis," *Managerial and Decision Economics*, Vol. 11, No. 4, 1990 (227-234).

Martel, J.-M. and B. Aouni, "Incorporating the Decision Maker's Preferences in the Goal Programming Model," *Journal of Operational Research Society*, Vol. 41, No. 12, 1990 (1121-1132).

Martel, J.-M. and B. Aouni, "Incorporating the Decision Maker's Preferences in the Goal Programming Model with Fuzzy Goal Values: A New Formulation," In Tamiz, M. (ed.), *Multi-Objective Programming and Goal Programming*, Lecture Notes in Economics and Mathematical Systems, Springer-Verlag, 1996.

Oral, M. et O. Kettani , "A Mathematical Programming Model for Market Share Prediction," *International Journal of Forecasting*, Vol.5, No.1, 1989 (59-68).

Oral, M. and O. Kettani, "A linearization Procedure for Quadratic and Cubic Mixed-Integer Problems," *Operations Research*, Vol. 40, Supp. No. 1, 1992 (S109-S116).

Pekelman, D. et K.S. Subarta, "Mathematical Programming Models for Determination of the Attribute Weights," *Management Science*, Vol. 20, 1974 (1217-1229).

Rao, M. R., "Cluster Analysis and Mathematical Programming," *Journal of the American Statistical Association*, Vol. 66, 1971 (622-626).

Rubin, P. A., "A Comparison of Linear Programming And Parametric Approaches to the Two-Group Discriminant Problem," *Decision Sciences*, Vol. 21, 1990 (373-386).

Stam, A. and E. A. Joachimsthaler, "Solving the Classification Problem in Discriminant Analysis Via Linear and Nonlinear Programming Methods," *Decision Sciences*, Vol. 20, 1989 (285-293).

Sueyoshi, T., *Goal Programming/Constrained Regression and Alternative Approaches to Statistical Estimation*. Non published Ph.D. Thesis, The University of Texas at Austin, 1986.

Zeleny, M., "Compromise Programming," *Multiple Criteria Decision Making*, Cochrane and Zeleny (Eds.), Columbia, University of South Carolina Press, 1973 (263-301).

Part 2

Multiple Objective Theory

Characteristics of the Efficient Solutions of Bicriteria and Tricriteria Network Flow Problems

Adli Mustafa[1,2] and Mark Goh[2]

[1] School of Mathematical Sciences, Universiti Sains Malaysia, 11800 USM Pulau Pinang, Malaysia
[2] Faculty of Business Administration, National University of Singapore, 10 Kent Ridge Crescent, Singapore 119260

Abstract. The ease of understanding their pictorial representation and the better computational performances of algorithms that have been developed to solve them, are among some of the advantages of single objective network flow problems. Even though these advantages have been extensively explored for the single objective case, there seems to be a lack of progress in exploring these advantages for the multiple objective case. In this paper, some interesting characteristics of the efficient solutions of bicriteria and tricriteria network flow problems are described. The descriptions are based on our observations of the compromise solutions obtained from the use of a readily available user-friendly software package called DINAS (Dynamic Interactive Network Analysis System).

Keywords. Network flow optimization, multiobjective programming

1 Introduction

Let $G = (N, A)$ be a network with a node set N and an arc set A. A general multiple objective network flow (MONF) problem can be stated as follows:

$$\text{Minimize} \quad f_m(\boldsymbol{x}) = \sum_{(i,j)\in A} c_{ij}^m x_{ij}, \qquad m = 1, \cdots, p \tag{1}$$

$$\text{subject to} \quad \sum_{j\in N} x_{ij} - \sum_{j\in N} x_{ji} = b_i \qquad \forall i \in N,\ i \neq j \tag{2}$$

$$l_{ij} \le x_{ij} \le u_{ij} \qquad \forall (i,j) \in A \tag{3}$$

$$x_{ij} \ge 0 \qquad \forall (i,j) \in A \tag{4}$$

(MONF-F)

where

x_{ij} = flow through arc (i,j)
u_{ij} = upper bound on flow through arc (i,j)
l_{ij} = lower bound on flow through arc (i,j)
b_i = supply-demand balance at node i
c_{ij}^m = cost of shipping one unit of flow from node i to node j in the m-th objective function.

When $p = 2$, the problem is called a bicriteria network flow (BCNF) problem and when $p = 3$, it is called a tricriteria network flow (TCNF) problem. Depending on the problem that was formulated, the cost associated with each arc may represent distance, delivery time, shipping cost, reliability of delivery, or deterioration of commodities.

Let X and Y denote the set of feasible solutions to the MONF problem in the decision space and in the objective space respectively. Since X is a compact convex set, Y is also a compact convex set. The function $F = (f_1, f_2, \cdots, f_p)$ maps the points $x \in X$ onto $y = (f_1(x), f_2(x), \cdots, f_p(x)) \in Y$. An *efficient solution* to the MONF problem is defined as a solution $x^0 \in X$ in which there does not exist $x \in X$ such that $f_i(x) \leq f_i(x^0)$ for all i and $f_i(x) < f_i(x^0)$ for at least one i.

MONF problems are special cases of multiple objective linear programming (MOLP) problems. As such, they can be solved using any existing method for solving MOLP problems (Shin and Ravindran 1991, and Steuer 1986). For the single objective case, the special topological and combinatorial structure of network flow problem has extensively been explored to allow the development of a number of efficient algorithms as solution tools (Golden and Magnanti 1977, and Ahuja *et al.* 1993). However, for the multiple objective case, the exploration is still limited. This is evident from the surveys of Current and Min (1986), Current and Marsh (1993), and Ulungu and Teghem (1994).

This paper describes some interesting characteristics of the efficient solutions of the BCNF and the TCNF problems. These characteristics are as observed from the use of a readily available user-friendly package called DINAS (Dynamic Interactive Network Analysis System) (Ogryczak *et al.* 1992). Even though DINAS is meant for multiobjective transshipment problems involving facility location, omitting the facility location part of the formulation makes it a solution tool for the multiobjective transshipment problem or MONF problem.

The rest of the paper is organized as follows. Section two describes the characteristics of the efficient solutions of the BCNF problem. Section three describes the characteristics of the efficient solutions of the TCNF problem. Some discussion on intercepting cycles is given in Section four. Section five concludes the paper.

2 Bicriteria case

Figure 2.1 shows a TCNF problem of shipping a total of 14 units of flow from nodes A, C and H to nodes I, E and G in a way which minimizes $f_1(x)$, $f_2(x)$ and $f_3(x)$ as defined in (MONF-F). For our discussion on BCNF problem, we consider only the minimization of $f_1(x)$ and $f_2(x)$.

The interactive analysis with DINAS is controlled by specifying the acceptable values (aspiration levels) and the worst acceptable values (reservation levels) for several objectives. This scheme of interaction is known as the reference point approach (Wierbicki 1982). As an initial guide in the interactive analysis, a pay-off matrix and two reference vectors (the utopia vector and the nadir vector) will be computed and displayed.

The summarization of our observation of the compromise solutions to the BCNF problem obtained from DINAS is as follows:

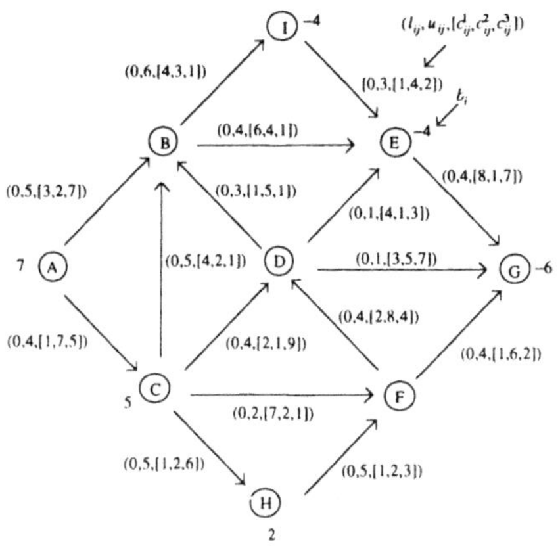

Fig. 2.1. An example network

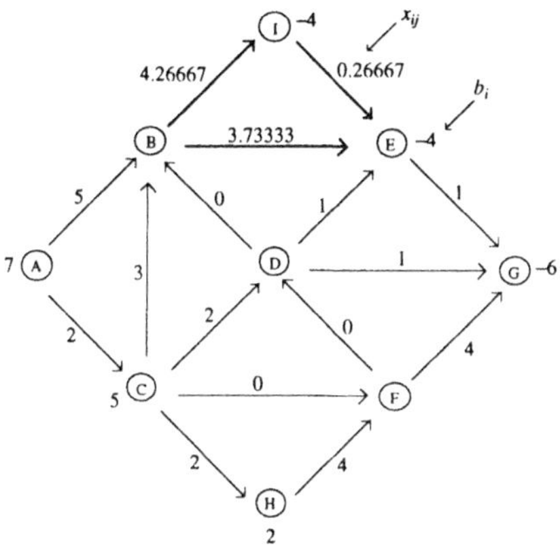

Fig. 2.2. A compromise solution (BCNF)

Except for the solutions associated with the minimization of each objective function separately which are always integer-valued, the compromise solutions need not be integer-valued. In the noninteger compromise solutions, arcs with noninteger flow form a unique cycle and the deviation of the flow from the nearest integer value on all these arcs is a constant.

In Figure 2.2, one of the noninteger compromise solution obtained from DINAS for the example network in Figure 2.1 is shown. The noninteger flow arcs form cycle B-I-E-B, and the deviation of flow from the nearest integer value on all these arcs is 0.26667. An explanation to the above observation will next be provided.

Figure 2.3 illustrates a typical feasible region and an efficient frontier of a BCNF problem in the objective space. The efficient frontier is the piecewise linear curve joining y^1, y^2, y^3, y^4, y^5 and y^6.

X_E, the set of efficient feasible solution to the BCNF problem, consists of a set of efficient basic feasible solutions (denoted by X_{BE}) as well as a set of efficient

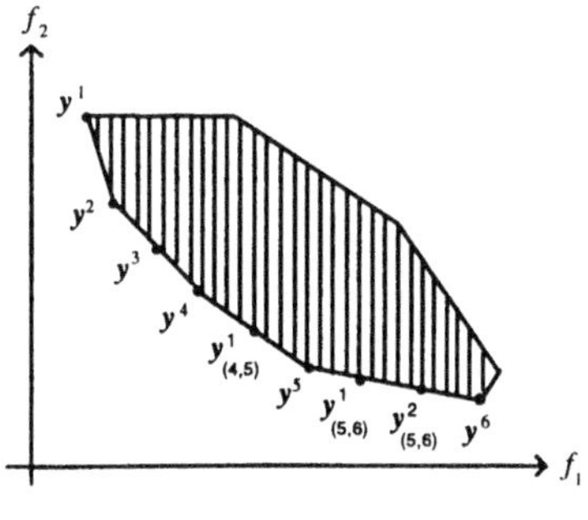

Fig. 2.3. A feasible region and an efficient frontier

nonbasic feasible solutions (denoted by X_{NE}). X_{BE}, which is also known as the set of efficient extreme point solutions, corresponds to extreme points in the decision space. X_{BE} need not correspond to extreme points in the objective space. X_{NE}, on the other hand, are the points on the boundary line of the efficient frontier. It is well known that every efficient extreme point solution of the BCNF problem is an integer-valued solution and can be represented as a *rooted basis tree*. Since the solutions associated with the minimization of each objective function separately are efficient extreme point solutions, they are always integer-valued. The characteristics of X_{NE} will be discussed next.

Let $\boldsymbol{x}^i$ and $\boldsymbol{x}^j$ be two adjacent efficient extreme point solutions. Suppose that a nonbasic arc (*k*, *l*) is the *entering arc* in moving from $\boldsymbol{x}^i$ to $\boldsymbol{x}^j$. When arc (*k*, *l*) is entered into the basis tree of $\boldsymbol{x}^i$, a *unique cycle* is formed. In moving from $\boldsymbol{x}^i$ to $\boldsymbol{x}^j$, only the flows along the arcs forming the unique cycle will change. The change of flow in these arcs is given by (an integer value) $\Delta \geq 0$, which is equal to the maximum allowable change of flow while maintaining feasibility. The arc in the unique cycle which restricts the amount of flow change will be the *leaving arc*. When there is no change of flow in the unique cycle, we are at $\boldsymbol{x}^i$ and when the change of flow is Δ, we are at $\boldsymbol{x}^j$. For the change of flow taking on any value between 0 and Δ, we obtain efficient nonbasic feasible solutions which are represented as points on the boundary line of the efficient frontier in the objective space joining point $\boldsymbol{y}^i$ and point $\boldsymbol{y}^j$, where $\boldsymbol{y}^i = (f_1(\boldsymbol{x}^i), f_2(\boldsymbol{x}^i))$ and $\boldsymbol{y}^j = (f_1(\boldsymbol{x}^j), f_2(\boldsymbol{x}^j))$.

Pulat *et al.* (1992) have indicated that there will be $\Delta - 1$ number of integer-valued efficient nonbasic feasible solutions in between point $\boldsymbol{y}^i$ and $\boldsymbol{y}^j$. Each of these solutions corresponds to a different integral flow in the unique cycle associated with moving from $\boldsymbol{x}^i$ to $\boldsymbol{x}^j$. On the same line in *Y*, these integer solution points are equidistant from each other (Lee and Pulat 1993). In Figure 2.3, these solutions are points $\boldsymbol{y}^1_{(4,5)}$, $\boldsymbol{y}^1_{(5,6)}$ and $\boldsymbol{y}^2_{(5,6)}$. $\boldsymbol{y}^k_{(i,j)}$ represents the *k*-th efficient integer but nonbasic feasible point $\boldsymbol{x}^k_{(i,j)}$ between two adjacent extreme point solutions $\boldsymbol{x}^i$ and $\boldsymbol{x}^j$, where $\boldsymbol{y}^k_{(i,j)} = (f_1(\boldsymbol{x}^k_{(i,j)}), f_2(\boldsymbol{x}^k_{(i,j)}))$. All other

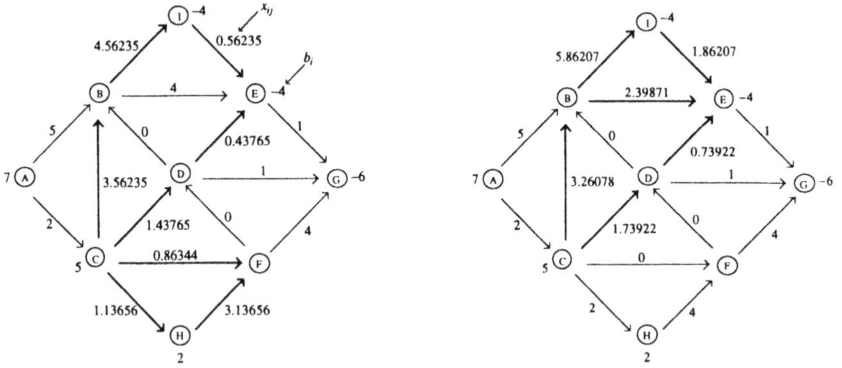

(a) Two distinct cycles

(b) Two intercepting cycles

Fig. 3.1. Compromise solutions (TCNF)

solutions are noninteger. These noninteger solutions correspond to nonintegral flow in the unique cycle associated with moving from x^i to x^j. Thus, this explains the formation of a unique cycle with a constant flow deviation from the nearest integer values on all the arcs forming it.

3 Tricriteria case

We again consider the network in Figure 2.1. This time around all three objective functions are considered. The summarization of our observation of the compromise solutions to the TCNF problem obtained from DINAS is as follows:

> *Except for the solutions associated with the minimization of each objective function separately which are always integer-valued, the compromise solutions need not be integer-valued. In the noninteger compromise solutions, arcs with noninteger flow form a unique cycle, two distinct cycles or two intercepting cycles. In each cycle, the deviation of flow from the nearest integer value on all arcs forming it is a constant. However, for intercepting cycles, the deviation of flow on arcs which are common to both cycles is a net deviation of both cycles.*

In Figure 3.1, two of the noninteger compromise solutions obtained from DINAS for the example network under consideration are shown. Figure 3.1(a) shows the formation of two distinct cycles while Figure 3.1(b) shows two intercepting cycles. We next provide an explanation to the above observation.

In the objective space, the efficient frontier or the set of efficient solution to the TCNF problem is a subset of the boundary of a three-dimensional convex polyhedron. Fruhwirth and Mekelburg (1994) associate a connected graph called the *efficient point graph* to the efficient frontier of the tricriteria linear program problem. An example of an efficient point graph is depicted in Figure 3.2. Since TCNF model is actually a special tricriteria linear program model, the description of the efficient point graph is also valid for the TCNF model.

From the description of the structure of the efficient point graph, an efficient point of a TCNF problem could be located on a vertex, on an edge or on the interior of a face of Y. For points on the efficient vertices and on the efficient edges, their characteristics have been described earlier in our discussion on the BCNF case. We now look at the characteristics of points on the interior of the efficient faces.

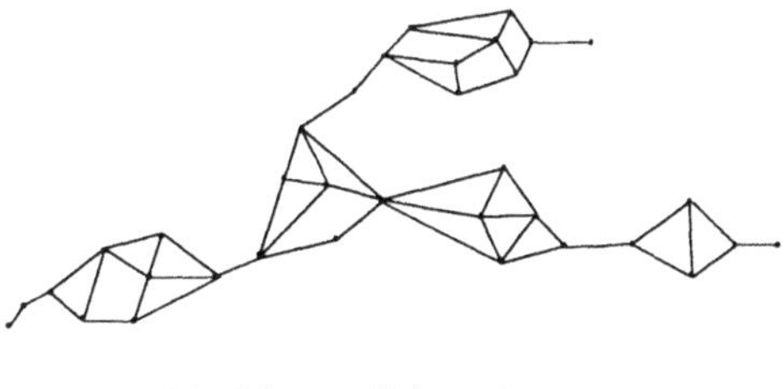

Fig. 3.2. An efficient point graph

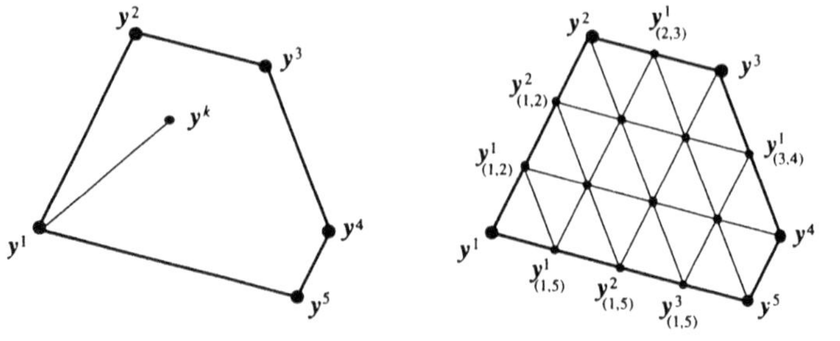

(a) An efficient face

(b) An efficient face with gridlines

Fig. 3.3. Efficient faces

It is well known that the relative interior of a face of Y is efficient if and only if all incident edges are efficient. If an edge is efficient, both incident vertices are efficient. Suppose we have an efficient face of Y with five incident edges as depicted in Figure 3.3(a). Let $\boldsymbol{y}^i$, $i=1,\cdots,5$ be extreme points respectively representing extreme point solutions $\boldsymbol{x}^i$, $i=1,\cdots,5$ in the decision space X, where $\boldsymbol{y}^i = (f_1(\boldsymbol{x}^i), f_2(\boldsymbol{x}^i), f_3(\boldsymbol{x}^i))$. The adjacency of the extreme points are indicated by the edges connecting them.

Let $\boldsymbol{y}^k$ be an interior point in an efficient face. Since an efficient face lies on a plane, we require only two independent vectors as a basis for the plane and also as a basis for the efficient face. Referring to Figure 3.3(a), we treat point $\boldsymbol{y}^1$ as an origin and the adjacent efficient edges $(\boldsymbol{y}^1, \boldsymbol{y}^2)$ and $(\boldsymbol{y}^1, \boldsymbol{y}^5)$ as the basis. From our discussion of the BCNF case, we observe that each efficient edge relates to flows along a unique cycle. Since the line $(\boldsymbol{y}^1, \boldsymbol{y}^k)$ can be represented as a linear combination of the two adjacent efficient edges $(\boldsymbol{y}^1, \boldsymbol{y}^2)$ and $(\boldsymbol{y}^1, \boldsymbol{y}^5)$, reaching $\boldsymbol{y}^k$ from $\boldsymbol{y}^1$ can be done by simultaneously changing the flows along the two cycles related to the two adjacent efficient edges. Any of the other extreme points in Figure 3.3(a) and their respective adjacent efficient edges could be chosen as the origin and the basis instead. The effect would be the same. An alternative way of visualizing the formation of the two cycles associated with $\boldsymbol{y}^k$ is to consider the effect of simultaneously entering two arcs (*i.e.*, moving from $\boldsymbol{x}^1$ to $\boldsymbol{x}^2$ and also from $\boldsymbol{x}^1$ to $\boldsymbol{x}^5$) to a rooted basis tree associated with $\boldsymbol{x}^1$.

Also, let $\boldsymbol{y}^k_{(i,j)}$ represent the k-th efficient integer but nonbasic feasible point $\boldsymbol{x}^k_{(i,j)}$ between two adjacent extreme point solutions $\boldsymbol{x}^i$ and $\boldsymbol{x}^j$, where $\boldsymbol{y}^k_{(i,j)} = (f_1(\boldsymbol{x}^k_{(i,j)}), f_2(\boldsymbol{x}^k_{(i,j)}), f_3(\boldsymbol{x}^k_{(i,j)}))$. As mentioned earlier, on the same line in Y, these integer solution points are equidistant from each other (Lee and Pulat 1993). In Figure 3.3(b), a detailed diagram of the efficient face in Figure 3.3(a) is shown. *Gridlines* which are parallel to the efficient edges and passing through points $\boldsymbol{y}^k_{(i,j)}$ are drawn. A gridline is basically a line where the movement of points along or parallel to it represents a flow adjustment in the cycle associated with the efficient edge which it is parallel with. From the diagram, it is clear that there are integer nonbasic efficient point solutions in the interior of an efficient face. These integer point solutions are the interceptions of the gridlines. It is also

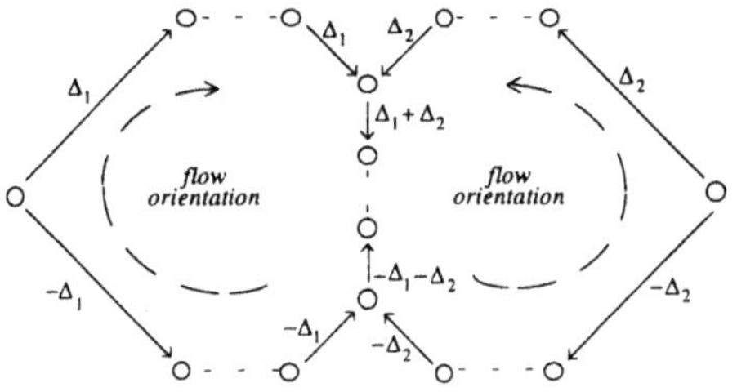

Fig. 3.4. Interception of two cycles

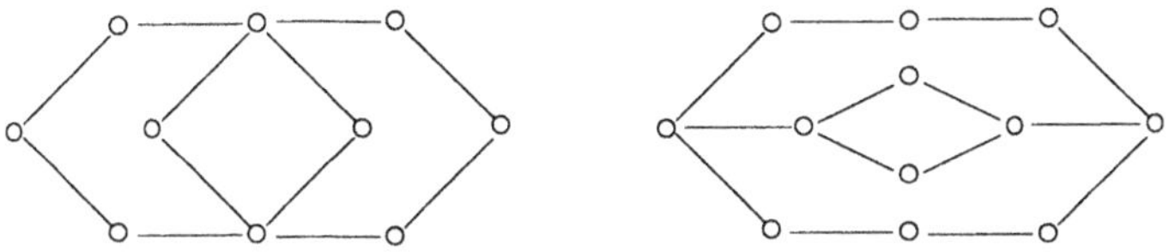

Fig. 3.5. Possible structures for the interception of two cycles

clear that there will not be any integer point solution in area between the gridlines. An area bounded by gridlines will be referred to as a *cell of an efficient face*. More discussion about gridlines will be given in Section 4.

When the change of flows along the two cycles associated with a basis are integers, y^k becomes an integer-valued efficient solution (these are points at the interceptions of the gridlines). When the change of flows along one cycle is integer-valued and the other is not, y^k would be a noninteger efficient solution and the arcs with noninteger flows will form a cycle (cycle associated with an efficient edge) with deviation of flows from the nearest integer value on all these arcs a constant (these are points on the gridlines but not at the interception points). When the flow changes along both cycles are noninteger, y^k would be a noninteger efficient solution with the noninteger flow arcs forming exactly two cycles (cycles associated with two adjacent efficient edges) (these are points not on the gridlines). The cycles could be distinct or intercepting. For intercepting cycles, the place of interception can be at only one position (*i.e.*, at a sequence of adjacent arcs) (refer to Figure 3.4 for an example). This is due to the fact that the formation of the two cycles can be visualized as a result of simultaneously entering two arcs to a rooted basis tree. Structures such as in Figure 3.5 will not be obtained since these structures require the removal of more than two arcs to turn them back into trees.

4 Discussion

We next look at some of the important and interesting observations about the relationship between intercepting cycles and cells of an efficient face.

In Figure 4.1(a), it is interesting to note that when two cycles intercept, a third cycle comprising the distinct arcs of the two intercepting cycles will be formed.

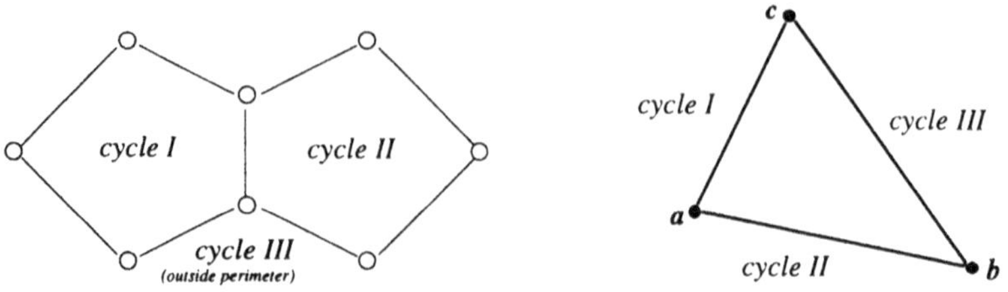

(a) Intercepting cycles (b) A cell of an efficient face

Fig. 4.1. Relationship between intercepting cycles and a cell of an efficient face

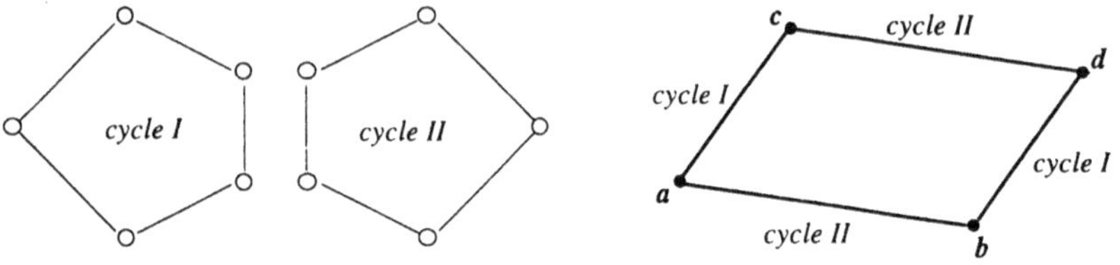

(a) Non-intercepting cycles (b) A cell of an efficient face

Fig. 4.2. Relationship between non-intercepting cycles and a cell of an efficient face

An n unit of flow increase in any of the cycle can always be represented as a simultaneous n unit of flow increase in the other two cycles. For example, a unit flow increase in the clockwise direction of cycle I is equivalent to a simultaneous one unit flow increase in the anti-clockwise direction of cycle II and in the clockwise direction of cycle III. This flow increase relationship among the three cycles can accurately be represented as a triangle called a cell of an efficient face as shown in Figure 4.1(b). In the diagram, edges (a,c) and (b,c) represent a unit flow increase in the clockwise direction of cycle I and cycle III, respectively, while edge (a,b) represents a unit flow increase in the anti-clockwise direction of cycle II. This observation about intercepting cycles and their representation using a triangle, in a way, helps in explaining the formation of an efficient face with gridlines as shown in Figure 3.3(b) and also in explaining the relationship between the cycles representing each efficient edges as shown in Figure 3.3(a). Since in Figure 4.1(a) we observed that there can only be 3 cycles associated with the interception of a pair of cycles that constitute a basis of an efficient face, then, all gridlines and efficient edges on that efficient face can only represent the 3 cycles.

When the two cycles associated with an efficient face of a TCNF problem are distinct (as shown in Figure 4.2(a)), a cell of an efficient face will be a parallelogram as shown in Figure 4.2(b). There will be no diagonal line in the parallelogram as there is no cycle which could be represented as a simultaneous flow increase in cycle I and cycle II.

5 Conclusion

Some interesting characteristics of the efficient solutions of bicriteria and tricriteria network flow problems have thus been presented. The descriptions should offer some insight into the structure of BCNF and TCNF solutions and hopefully would lead to further exploration. With computer graphics becoming popular in multiple objective programming (see for example Korhonen 1990 and Antunes *et al.* 1992), the descriptions could help in the development of decision support tools which could approximate and visualize the efficient solution set of BCNF and TCNF problems.

For certain BCNF and TCNF problems, the nature of the actual situation that was modeled requires the flows along the arcs to be integer-valued. The descriptions about the characteristics of the efficient solutions of BCNF and TCNF problems, among others, should also provide us with simple means of obtaining integer-valued efficient solutions to these problems once we are on the efficient frontier.

References

Ahuja, R.K., Magnanti, T.L. and Orlin, J.B., *Network Flows: Theory, Algorithms and Applications*, Prentice-Hall, Englewoods Cliffs, NJ (1993).

Antunes, C.H., Alves, M.J., Silva, A.L. and Climaco, J.N., An integrated MOLP method base package - A guided tour of TOMMIX. *Computers Ops Res.* **19**, 7, 609-625 (1992).

Current, J. and Min, H., Multiobjective design of transportation networks: taxonomy and annotation. *Eur. J. Opl Res.* **26**, 187-201 (1986).

Current, J. and Marsh, M., Multiobjective transportation network design and routing problems: taxanomy and annotation. *Eur. J. Opl Res.* **65**, 4-19 (1993).

Fruhwirth, B. and Mekelburg, K., On the efficient point set of tricriteria linear programs. *Eur. J. Opl Res.* **72**, 192-199 (1994).

Golden, B.L. and Magnanti, T.L., Deterministic network optimization: a bibliography. *Network* **7**, 149-183 (1977).

Korhonen, P., The multiple objective linear programming decision support system VIG and its applications. *Readings in Multiple Criteria Decision Aid* (Edited by C.a. Bana e Costa), Springer-Verlag (1990).

Lee, H. and Pulat, P.S., Bicriteria network flow problems: integer case. *Eur. J. Opl Res.* **66**, 148-157 (1993).

Ogryczak, W., Studzinski, K. and Zorychta, K., DINAS: A computer-assisted analysis system for multiobjective transshipment problems with facility location. *Computers Ops Res.* **19**, 7, 637-647 (1992).

Pulat, P.S., Huarng, F. and Lee, H., Efficient solutions for the bicriteria network flow problem. *Computers Ops Res.*. **19**, 7, 649-655 (1992).

Shin, W.S and Ravindran, A., Interactive multiple objective optimization: survey I - continuous case. *Computers Ops Res.* **18**, 97-114 (1991).

Steuer, R.E., *Multiple Criteria Optimization: Theory, Computation, and Application*, John Wiley & Sons, New York (1986).

Ulungu, E.L. and Teghem, J., Multi-objective combinatorial optimization problems: a survey. *J. Multi-Criteria Decision Analysis* **3**, 83-104 (1994).

Wierzbicki, A.P., A mathematical basis for satisficing decison making. *Mathl Modeling* **3**, 391-405 (1982).

Non-connected Efficiency Graphs in Multiple Criteria Combinatorial Optimization

Matthias Ehrgott[1] and Kathrin Klamroth[2]

[1] Department of Mathematics, University of Kaiserslautern, PO Box 3049, 67653 Kaiserslautern, Germany; partially supported by the Deutsche Forschungsgemeinschaft (DFG) and grant ERBCHRXCT930087 of the European HC&M Programme

[2] Department of Mathematics, University of Kaiserslautern, PO Box 3049, 67653 Kaiserslautern, Germany; supported by the Deutsche Forschungsgemeinschaft (DFG)

Abstract. In multiple criteria optimization an important research topic is the topological structure of the set X_e of efficient solutions. Of major interest is the connectedness of X_e, since it would allow the determination of X_e without considering non-efficient solutions in the process. We review general results on the subject, including the connectedness result for efficient solutions in multiple criteria linear programming. This result can be used to derive a definition of connectedness for discrete optimization problems. We present a counterexample to the connectivity conjecture, namely that the sets of efficient solutions of the shortest path problem and the minimal spanning tree problem are connected. We will also present a general method to construct non-connected efficiency graphs for these problems.

Keywords. Multiple criteria, efficient solutions, combinatorial optimization, connectedness

1 Introduction

In this paper we consider multiple criteria optimization problems of the form

$$\min \ f(x)$$

$$\text{s.t.} \quad x \in X$$

where $f : I\!R^n \to I\!R^Q$ and $X \subseteq I\!R^n$. Although we assume familiarity with the basic concepts of multicriteria optimization we will briefly give the most important definitions.

In the general case $I\!R^Q$ is ordered by a cone K (see [16] for general results on orders defined by cones). In multiple criteria optimization the notion of optimality is usually replaced by efficiency, since in general different solution values in $\mathbb{R}^Q$ exist which can be considered as "best" solutions of the problem in the sense that they cannot be improved. $x_e \in X$ is called efficient solution if $(\{f(x_e)\} - K) \cap f(X) = \{f(x_e)\}$. Note that for $Y, K \subset \mathbb{R}^Q$ the set $Y - K$ is defined to be $\{y - k \mid y \in Y, k \in K\}$. The set of all efficient solutions is denoted by X_e. Most of the research in this area has been devoted to the case where $K = \mathbb{R}^Q_+$. Then the ordering defined by K is the component-wise order and $x_e \in X$ is efficient if there is no $x \in X$ dominating x_e, i.e. there is no $x \in X$ such that $f_q(x) \leq f_q(x_e)$, $q = 1, \ldots, Q$ and strict inequality holds in at least one case.

2 General Results on Connectedness

The topic of this paper will be connectedness of X_e. Although of great importance for the construction of algorithms there is not much work done in that field. For general "continuous" problems we are dealing with the question: Is X_e connected in the topological sense? In order to state a general result providing the answer to this question, we define a set $Y \subset \mathbb{R}^Q$ to be K–compact if $(\{y\} - K) \cap Y$ is compact for all $y \in Y$.

Theorem 1 ([14]) *If K is a closed, convex, pointed (i.e. $x \in K \Rightarrow -x \notin K$) cone such that $int(K) \neq \emptyset$ and $Y = f(X)$ is closed, convex and K–compact, then X_e is connected.*

Algorithmically this result can be applied to determine the set X_e by local search methods, i.e. given any efficient solution $x \in X_e$ we can search locally to find efficient solutions in the neighbourhood of x until eventually the whole set X_e is determined. The result of Theorem 1 has been generalized in [9] to the case where $\mathbb{R}^n$ and $\mathbb{R}^Q$ are replaced by locally convex spaces. Several authors proved connectedness of X_e for special types of functions [15, 2]. Also several results on the connectedness of the set of weakly efficient solutions are known [15], where $x_{we} \in X$ is said to be weakly efficient if $(f(x_{we}) - int(K)) \cap f(X) = \emptyset$. In the following we will only consider the case $K = \mathbb{R}^Q_+$. In this case efficient solutions are often called Pareto optimal solutions.

For combinatorial optimization problems, however, this connectedness result is not applicable. In this case we will use the concept of connectedness of a graph rather than the topological connectedness of Theorem 1. We define a graph, the nodes of which represent the efficient solutions of the combinatorial problem. Edges are introduced between all pairs of nodes which are "adjacent" in some sense. If the resulting graph is connected this fact will allow the development of algorithms to find X_e by neighbourhood search among adjacent solutions. The most promising definition of adjacency seems to be by using the link between these two concepts of connectedness, which is provided by linear programming. A multiple criteria linear program (MCLP) is defined as

$$\min(c^1 x, \ldots, c^Q x) \text{ where } c^q \in I\!R^n, q = 1, \ldots, Q$$

$$\text{s.t. } x \in X = \{x \in I\!R^n \mid x \geq 0, a^k x \leq b_k, k = 1, \ldots, m\}$$

Obviously Theorem 1 immediately implies that X_e is connected in this special case. Before the general result of Theorem 1 was known the connectedness result for MCLP had been proved by various authors [3, 6, 17]. The most important solutions in linear programming are basic solutions which correspond to extreme points of the polyhedral feasible set X, and fundamental solutions which correspond to extreme rays of X, if X is unbounded. Let B and F denote the sets of basic feasible and fundamental solutions, respectively. Then $x_e^1, x_e^2 \in X_e \cap B$ are said to be adjacent if they have $m-1$ basic variables in common and $\alpha x_e^1 + (1-\alpha)x_e^2$ is efficient for all $\alpha \in [0,1]$. Furthermore $x_e \in B \cap X$ and $x_f \in F \cap X$ are said to be adjacent if $x_e + \beta x_f$ is efficient for all $\beta \geq 0$. Now let $\mathcal{B}$ be the index set of all efficient basic feasible solutions and let $\mathcal{F}$ be the index set of all fundamental solutions which are adjacent to an efficient basic feasible solution. The main result in [10] is the following:

Theorem 2 **([10])** *Define a graph $G = (V, E)$ by introducing a node for each index in $\mathcal{B} \cup \mathcal{F}$ and an edge between two nodes if the corresponding solutions are adjacent. Then G is connected.*

Theorem 2 is used in MCLP simplex-algorithms: The set of all efficient extreme points of X is determined by pivoting among efficient bases only, i.e. by moving from efficient extreme point to adjacent efficient extreme point.

Thus in this paper we will restrict ourselves to combinatorial optimization problems which have a linear programming formulation, i.e. feasible solutions of the combinatorial problem correspond to basic feasible solutions of the associated linear program. Two feasible solutions of the combinatorial problem are then called adjacent if the corresponding basic feasible solutions of the associated LP are adjacent. The relations between connectedness in nonlinear (continuous), linear and combinatorial multicriteria problems are illustrated in Figure 1.

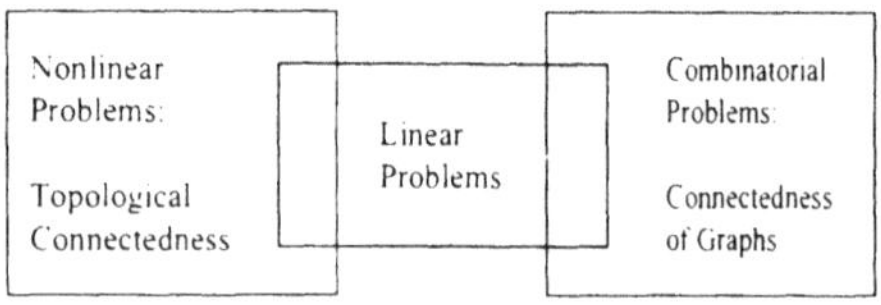

Figure 1: Connectedness in Nonlinear, Linear and Combinatorial Problems

3 Combinatorial Problems

3.1 The Shortest Path and the Spanning Tree Problem

In [13] a multiple criteria dynamic programming algorithm for the shortest path problem is derived. This algorithm allows finding all efficient paths from node s to node t in a given directed graph $G = (V, A)$. The multiple criteria shortest path problem can be formulated as a linear program by

$$\min(c^1 x, \ldots, c^Q x)$$

$$\text{s.t.} \sum_j x_{ij} - \sum_j x_{ji} = \begin{cases} 1 & i = s \\ 0 & i \notin \{s, t\} \\ -1 & i = t \end{cases}$$

According to the definitions in Section 2 two paths from s to t are adjacent if they correspond to two adjacent basic feasible solutions of the above LP. These basic feasible solutions represent spanning trees of the underlying directed graph G, i.e. the two paths are adjacent if the corresponding spanning trees have $n - 2$ edges in common.

Another important discrete optimization problem is the spanning tree problem: Given an undirected graph $G = (V, E)$, find $\min(c^1(T), \ldots, c^Q(T))$ such that T is a spanning tree of G. The linear programming formulation of this problem is:

$$\min(c^1 x, \ldots, c^Q x)$$

$$\begin{aligned} \text{s.t.} \sum_{e \in E} x_e &= n - 1 \\ \sum_{e \in E(S)} x_e &\leq |S| - 1 \qquad \forall S \subset V \\ &\quad \text{where } E(S) = \{e = [i, j] \in E \mid i, j \in S\} \\ x_e &\geq 0 \end{aligned}$$

Again according to Section 2 two (efficient) spanning trees are adjacent if they have $n - 2$ edges in common. We will now formally introduce the efficiency graph corresponding to a spanning tree problem and a shortest path problem, which has been introduced as Pareto graph in [4].

Definition 1 *Let $G = (V, E)$ with edge costs $c^1, \ldots, c^Q : E \to I\!R$ be a given graph. The efficiency graph $\mathcal{EG}^T(G)$ for the spanning tree problem on G is defined as follows: Its node set is the set of efficient spanning trees of G. Two nodes are joined by an edge if the corresponding spanning trees are adjacent. Analogously the efficiency graph $\mathcal{EG}^{P(s,t)}(G)$ for the shortest path problem on G with end nodes s and t is defined: Its node set is the set of efficient paths from s to t. Two nodes are joined by an edge if the corresponding paths are adjacent, where adjacency is defined as above.*

Connectivity Conjecture: $\mathcal{EG}^T(G)$ and $\mathcal{EG}^{P(s,t)}(G)$ are connected.

The connectivity conjecture has been stated in [4] and [7] for the general matroid optimization problem and the matroid intersection problem, respectively. Also a theorem in [13] is a reformulation of the connectivity conjecture for the shortest path problem. The important implication is that if the conjecture were true it would be possible to find all efficient solutions of the spanning tree and shortest path problem by neighbourhood search, i.e. by exchanges of one edge in the trees which correspond to efficient basic feasible solutions of the linear programming formulations of the two problems. In particular the approximation algorithms stated in [8] or [12] would find all efficient spanning trees and thus be exact.

Theorem 3 *$\mathcal{EG}^T(G)$ and $\mathcal{EG}^{P(s,t)}(G)$ are not connected in general.*

The proof is provided by Example 1 and Lemma 1.

Example 1 *Consider the graph $G_1 = (V, E)$ given in Figure 2. There are 12 efficient spanning trees of G_1, listed in Table 1. Obviously each efficient spanning tree contains all edges with cost $(0,0)$. Therefore in Table 1 we only list edges with positive costs.*

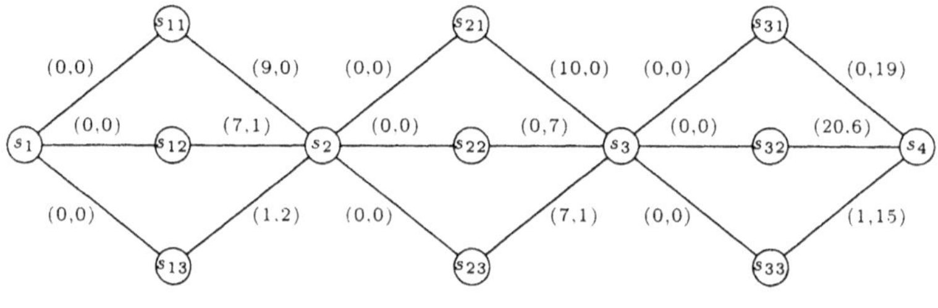

Figure 2: Graph G_1 has Nonadjacent Efficient Spanning Trees

It is easy to see that T_8 is not adjacent to any other efficient spanning tree.

We will now look at the problem of finding efficient paths from s_1 to s_4 in the same graph of Figure 2. Clearly if $\mathcal{EG}^{P(s_1,s_4)}(G_1)$ is not connected, the same holds in the directed case: G_1 can be made directed by just orienting each arc from left to right in Figure 2. Lemma 1 then provides the counterexample to Theorem 2.

Lemma 1 *In Example 1, $\mathcal{EG}^{P(s_1,s_4)}(G)$ and $\mathcal{EG}^T(G)$ are isomorphic.*

The proof is omitted, we refer to [5].

Table 1: Efficient Spanning Trees of G_1

Efficient Tree	Edges	Cost
T_1	$[s_{13}, s_2][s_{22}, s_3][s_{31}, s_4]$	(1,28)
T_2	$[s_{13}, s_2][s_{22}, s_3][s_{33}, s_4]$	(2,24)
T_3	$[s_{13}, s_2][s_{23}, s_3][s_{31}, s_4]$	(8,22)
T_4	$[s_{13}, s_2][s_{23}, s_3][s_{33}, s_4]$	(9,18)
T_5	$[s_{13}, s_2][s_{21}, s_3][s_{33}, s_4]$	(12,17)
T_6	$[s_{11}, s_2][s_{23}, s_3][s_{33}, s_4]$	(17,16)
T_7	$[s_{11}, s_2][s_{21}, s_3][s_{33}, s_4]$	(20,15)
T_8	$[s_{12}, s_2][s_{22}, s_3][s_{32}, s_4]$	(27,14)
T_9	$[s_{13}, s_2][s_{23}, s_3][s_{31}, s_4]$	(28,9)
T_{10}	$[s_{13}, s_2][s_{21}, s_3][s_{31}, s_4]$	(31,8)
T_{11}	$[s_{11}, s_2][s_{23}, s_3][s_{31}, s_4]$	(36,7)
T_{12}	$[s_{11}, s_2][s_{21}, s_3][s_{31}, s_4]$	(39,6)

3.2 Generalization

A disconnected efficiency graph may also occur in more general situations and not only in the example presented above. We will show that it is possible to extend any graph in such a way that the efficiency graph for the problem on the extended graph is not connected. This holds for both the spanning tree and the shortest path problem.

Let $G = (V, E)$ be an arbitrary given (connected) graph. We construct a graph $\tilde{G}$ containing G as a subgraph such that $\mathcal{EG}^T(\tilde{G})$ is not connected. First we will explain the procedure by an example and outline the general case later. Consider the graph G and its efficiency graph $\mathcal{EG}^T(G)$ in Figure 3.

Figure 3: The Graph G and its Efficiency Graph $\mathcal{EG}^T(G)$

Obviously G has three spanning trees which are all efficient. They have the weights $c(T_1) = (2, 13)$, $c(T_2) = (8, 10)$, $c(T_3) = (10, 3)$, and each is adjacent to the two other efficient spanning trees. We now construct a graph G' for which one efficient tree is adjacent to only one other efficient tree in $\mathcal{EG}^T(G')$. Therefore we introduce new nodes v', v_0, v_1, v_2 and new edges $[v_i, v']$, $i = 0, 1, 2$ and $[v, v_i]$, $i = 0, 1, 2$, where v is an arbitrary node of G. G' is shown in Figure 4.

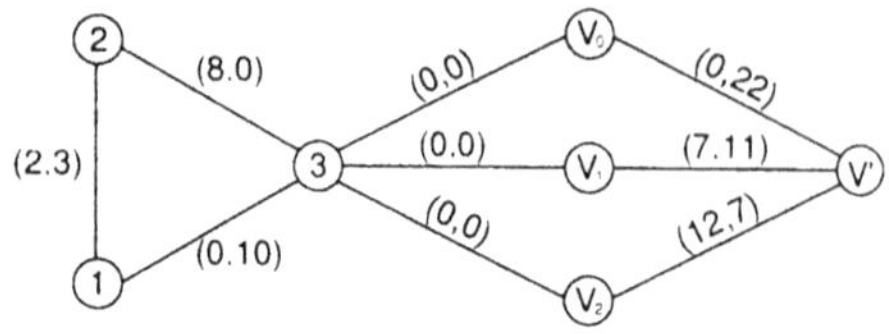

Figure 4: The Graph G'

Table 2: Weights of Spanning Trees T_{ij} of G'

$c(T_{ij})$	j=1	j=2	j=3
i=0	(2,35)	(8,32)	(10,25)
i=1	(9,24)	(15,21)	(17,14)
i=2	(14,20)	(20,17)	(22,10)

The spanning trees T_{ij} of G' consist of a spanning tree T_j of G, all edges $[v, v_k]$, $k = 0, 1, 2$, and exactly one edge $[v_i, v']$, $i \in \{0, 1, 2\}$. The weights of T_{ij} are given in Table 2.

Concerning adjacency of spanning trees of G' we observe that $T_{k,l}$ and $T_{i,j}$ are adjacent if and only if $k \neq i$ and $j = l$ or $k = i$ and T_l, T_j are adjacent spanning trees of G. Furthermore from Table 2 we see that T_{22} is dominated by T_{13}, T_{03} is dominated by T_{11}, and T_{12} is dominated by T_{21}. Consequently spanning tree T_{02} is adjacent to only one other efficient spanning tree, namely T_{01}. We can now apply an analogous construction for graph G' and will end up with a graph G'' which has one efficient spanning tree which is not connected to any other tree in $\mathcal{EG}^T(G'')$.

We will now briefly outline this procedure for the general case. For the detailed method, including proofs, we refer to the full length paper [5]. Let $c : E \to I\!R^Q_+$ be the cost-function on the edges of G. For simplicity we restrict ourselves to the case $Q = 2$. Note that the results also hold for $Q > 2$.

We will assume that G has at least one efficient spanning tree which is not lexicographically minimal with respect to costs (c_1, c_2) and (c_2, c_1). Let $T_1, \ldots, T_m$ be the set of efficient spanning trees of G. Furthermore let $T_k \in \{T_1, \ldots, T_m\}$ be such that T_k is not lexicographically minimal and such that it has as few incident edges in $\mathcal{EG}^T(G)$ as possible. We denote the costs of T_i by (a_i, b_i) and will use $(x, y) = (a_k, b_k)$ for easier distinction. We assume that the numbering of efficient trees is such that the costs are ordered lexicographically, i.e. $a_1 \leq \ldots \leq x \leq \ldots \leq a_l \leq \ldots a_m$ and $b_1 \geq \ldots \geq y \geq \ldots \geq b_l \geq \ldots \geq b_m$. Furthermore w.l.o.g let $T_l \in \{T_{k+1}, \ldots, T_m\}$ be such that T_k and T_l are connected by an edge in $\mathcal{EG}^T(G)$.

In the following we will construct a graph G' containing G as a subgraph such that there exists at least one efficient spanning tree of G' which has fewer incident edges in $\mathcal{EG}^T(G')$ than T_k in $\mathcal{EG}^T(G)$. Thus applying this procedure iteratively we will be able to find a graph $\tilde{G}$ containing G such that $\mathcal{EG}^T(\tilde{G})$ has an isolated node, i.e. the efficiency graph of $\tilde{G}$ is disconnected. We distinguish two cases and extend G in two different ways:

Extension 1. First let us assume that there exists $n \in I\!N$ and $0 < \epsilon < \min\{x - a_1, a_l - x\}$ such that

$$x > \frac{1}{n}a_l + \frac{n-1}{n}(a_1 + \epsilon) \quad (1)$$

$$y > \frac{1}{n}b_l + \frac{n-1}{n}(b_1 + \epsilon) \quad (2)$$

Then we proceed as in the example above and define $G' := (V(G'), E(G'))$ by $V(G') = (V(G) \cup \{v'\} \cup \{v_0, \ldots, v_n\}$ and $E(G') = E(G) \cup \{[v, v_0], \ldots, [v, v_n]\} \cup \{[v_0, v'], \ldots, [v_n, v'],$ where v is an arbitrary node of $V(G)$. Let $C := -(n-1)(y - b_1 - \epsilon)$ and assign the following costs to the additional edges:

$$\begin{aligned} c(v, v_i) &= (0, 0); \qquad i \in \{0, \ldots, n\} \\ c(v_0, v') &= (a_1 - x - \epsilon, C + b_1 - b_l + \epsilon) \\ c(v_i, v') &= ((i-1)(x - a_1 - \epsilon), C + (i-1)(y - b_1 - \epsilon)); \quad i \in \{1, \ldots, n\} \end{aligned}$$

Lemma 2 *If conditions (1) and (2) hold, then there exists a spanning tree T' of G' such that the number of edges incident to T' in $\mathcal{EG}^T(G')$ is strictly smaller than the number of edges incident to T_k in $\mathcal{EG}^T(G)$.*

The proof is along the lines of the arguments in the example above. The observations concerning adjacency and dominance of spanning trees are carried over to the general construction, where we make use of the defined costs of the additional edges.

Extension 2. In the second case we consider the situation that (1) or (2) do not hold. Then let $G^* := (V(G^*), E(G^*))$ where $V(G^*) = V(G) \cup \{v^*, v_1, v_2\}$, $E(G^*) = E(G) \cup \{[v, v_1], [v, v_2], [v_1, v^*], [v_2, v^*]\})$ and v is an arbitrary node of $V(G)$. We assign the following costs to the additional edges:

$$\begin{aligned} c(v, v_i) &= (0, 0), \; i = 1, 2 \\ c(v_1, v^*) &= (0, \beta) \\ c(v_2, v^*) &= (a_l - a_1 - \delta, 0) \end{aligned}$$

where $\beta \geq \max\{\frac{x + a_l - 2a_1 - \delta}{x - a_1}, b_1 - b_l\}$ and $\delta > 0$ is sufficiently small. If we now choose $n = \lfloor \frac{x + a_l - 2a_1 - \delta - \epsilon}{x - a_1 - \epsilon} \rfloor + 1$ and $\epsilon > 0$ small enough, we see that conditions (1) and (2) hold for the graph G^*. Hence after appropriate renumbering of the efficient trees we have exactly the situation of the first case with T_{1k} in the place of T_k and T_{2k} in the place of T_l. Moreover T_{1k} is not adjacent to more spanning trees of G^* than T_k was for G.

Analogously to Example 1 Extension 1 and Extension 2 can be easily transferred to the shortest path problem by replacing "spanning tree" by "$s-v-$path" respectively "$s-v'-$path" and T by P in Extensions 1 and 2. Then Lemma 2 can be reformulated as follows:

Lemma 3 *If conditions (1) and (2) hold, then there exists an efficient path from s to v' in G' which is incident to fewer edges in $\mathcal{EG}^{P(s,v')}(G')$ than the (not lexicographically minimal) efficient path from s to v in G which has the fewest incident edges in $\mathcal{EG}^{P(s,v)}(G)$.*

Theorem 4 *For a given graph $G = (V, E)$ and costs $c^1, \ldots, c^Q : E \to I\!R_+$ there exists a graph $\tilde{G}_T$ and costs $\tilde{c}^1, \ldots, \tilde{c}^Q : E(\tilde{G}_T) \to I\!R_+$ containing G as a subgraph such that $\mathcal{EG}^T(\tilde{G}_T)$ is not connected.*

Analogously, for a given graph $G = (V, E)$, vertices $s, v \in V(G)$ and costs $c^1, \ldots, c^Q : E \to I\!R_+$ there exists a graph $\tilde{G}_P$ and costs $\tilde{c}^1, \ldots, \tilde{c}^Q : E(\tilde{G}_P) \to I\!R_+$ containing G as a subgraph such that $\mathcal{EG}^{P(s,\tilde{v})}(\tilde{G}_P)$ is not connected.

Despite the negative result of Theorem 4 we note the following result. We consider a subset of the efficient spanning trees of G, the so called extremal efficient spanning trees, which are optimal for some convex combination of the Q costs. It is known that the set of extremal efficient spanning trees *is connected*, see [4].

4 Conclusions and Future Research

First let us note that shortest path and spanning tree problems are not the only discrete multiple criteria problems for which the set of efficient solutions is not connected in general. The method described in Section 3.2 can also be applied to construct examples of non-connected efficiency graphs for multiple criteria matroid optimization problems, where the matroid is either a partition or a transversal matroid.

Despite the negative results of Theorem 3 and Theorem 4 we remark that according to our experience a disconnected graph $\mathcal{EG}^T(G)$ appears only very rarely. We carried out computational tests together with M. Lind from Aarhus University, Denmark [11]. He implemented a program for finding efficient spanning trees based on the connectedness hypothesis. The approach is as follows: First all extremal efficient spanning trees are found. Then a neighbourhood search is used to find non-extremal efficient spanning trees. A total of 50 randomly generated graphs with 10 to 50 nodes was tested and no example of a disconnected efficiency graph was found. In these tests we compared the efficient solutions found under the hypothesis of connectedness with all efficient solutions calculated by an enumeration approach.

Therefore we conclude that, although the efficiency graph is not connected in general, a procedure based on the connectedness hypothesis, as proposed in [8] and [12, 1], yields a very good approximation of the set of efficient spanning trees. In many cases all efficient spanning trees will be found and in many others only few will be missing. On the other hand the approach implemented in [11] is much faster than an enumeration approach to find all efficient solutions. Running times were within minutes of CPU-time even for larger graphs of 50 nodes, whereas for some graphs with 50 nodes and even for dense graphs with 20 nodes we were not able to find the set of all efficient spanning trees using the enumeration method within 10 hours of computing time.

With respect to future research topics, we shall focus on two main directions. The first concerns the concept of connectivity. By now we do not know any combinatorial problem for which the connectivity conjecture is true, if the current definition is used. Hence it should be investigated if there exist other concepts of connectivity, by introducing other definitions of adjacency of efficient solutions, e.g. of the spanning tree or shortest path problem, such that the corresponding efficiency graph is connected. Such a concept would then have all the advantages pointed out in Section 1. The second direction of research is related to the question: Do there exist combinatorial optimization problems such that their efficient solutions are "connected"? This holds e.g. for special cases of the spanning tree problem, namely if G contains only one cycle. Then any two spanning trees differ by only one edge and hence $\mathcal{EG}^T(G)$ is connected. But obviously we are interested in problems where the connectivity conjecture is true for every instance.

References

[1] K.A. Andersen, K. Joernsten, and M. Lind. On bicriterion minimal spanning trees: An approximation. Technical Report 95/13, Department of Operations Research, University of Aarhus, 1995.

[2] E.U. Choo and A.D. Atkins. Connectedness in multiple linear fractional programming. *Management Science*, 29:250–255, 1983.

[3] J.G. Ecker and I.A. Kouada. Finding efficient points for linear multiple objective programs. *Mathematical Programming*, 8:375–377, 1975.

[4] M. Ehrgott. On matroids with multiple objectives. *To appear in Opimization*, 1996.

[5] M. Ehrgott and K. Klamroth. Connectedness of efficient solutions in multiple criteria combinatorial optimization. *European Journal of Operational Research*, 97, 1997.

[6] J.P. Evans and R.E. Steuer. A revised simplex method for linear multiple objective programs. *Mathematical Programming*, 5:375–377, 1973.

[7] H.W. Hamacher. Matroid intersections with several objectives. Presented at ECCO VII, Milan, February 21.-23., 1994, 1994.

[8] H.W. Hamacher and G. Ruhe. On spanning tree problems with multiple objectives. *Annals of Operations Research*, 52:209–230, 1994.

[9] S. Helbig. On the connectedness of the set of weakly efficient points of a vector optimization problem in locally convex spaces. *Journal of Optimization Theory and Applications*, 65:257–271, 1990.

[10] H. Isermann. The enumeration of the set of all efficient solutions for a linear multiple objective program. *Operational Research Quarterly*, 28(3):711–725, 1977.

[11] M. Lind. Neighbourhood search for the bicriterion spanning tree problem. Program implemented at University of Aarhus and Universität Kaiserslautern, February 1995, 1995.

[12] M. Lind. On bicriterion spanning tree problems. Presented at the Twelfth International Conference on MCDM, Hagen, June 18.-23., 1995, 1995.

[13] E.Q.V. Martins. On a multicriteria shortest path problem. *European Journal of Operational Research*, 16:236–245, 1984.

[14] P.H. Naccache. Connectedness of the set of nondominated outcomes in multicriteria optimization. *Journal of Optimization Theory and Applications*, 25:459–467, 1978.

[15] A.R. Warburton. Quasiconcave vector maximization: Connectedness of the sets of Pareto-optimal and weak Pareto-optimal alternatives. *Journal of Optimization Theory and Applications*, 40:537–557, 1983.

[16] P.L. Yu. Cone convexity, cone extreme points and nondominated solutions in decision problems with multiobjectives. *Journal of Optimization Theory and Applications*, 14:319–377, 1974.

[17] P.L. Yu and M. Zeleny. The set of all nondominated solutions in linear cases and a multicriteria simplex method. *Journal of Mathematical Analysis and Applications*, 49:430–468, 1975.

An Augmented Lagrangian Scalarization for Multiple Objective Programming

Matthew L. TenHuisen[1] and Malgorzata M. Wiecek[2]

[1] Mathematical Sciences Department, University of North Carolina at Wilmington, Wilmington, NC 28403-3297, USA.

[2] Fachbereich Mathematik, Universitaet Kaiserslautern, D-67653 Kaiserslautern, Germany (on sabbatical leave from the Department of Mathematical Sciences, Clemson University, Clemson, SC 29634-1907, USA)

Abstract. The multiple objective program (MOP) is related to its ε-constraint single-objective counterpart for which an augmented Lagrangian is developed. The resulting scalarization generates q_k-approachable points that are identified as locally efficient and/or efficient points of the MOP. An illustrative example is enclosed.

Keywords. Multiple objective programming, locally efficient points, augmented Lagrangian, approachable points

1 Introduction

Multiple objective programs (MOPs) encountered in the field of engineering and management often involve nonconvex objective and/or constraint functions. In structural design, nonconvex sets and functions were used by Osyczka (1984). Design problems in chemical engineering studied by Jahn et al. (1992) as well as table olive preparations systems studied by Kopsidas (1995) led to nonconvex bicriteria problems. The lack of convexity profoundly affects the structure of the efficient (and nondominated) set as well as causes substantial difficulties in finding its elements. Many methods, when applied to nonconvex MOPs, may fail to generate all nondominated solutions or locally nondominated solutions that are of particular interest in engineering applications.

Some recent research papers, however, have been specially oriented toward the characterization of the efficient set for nonconvex problems. Bernau (1990) used penalty functions in order to determine efficient solutions of nonconvex MOPs. Kostreva et al. (1992) developed a method for generating both locally and globally efficient solutions of polynomial, so possibly strongly nonconvex, MOPs. Kostreva and Wiecek (1993) and Isac et al. (1995) studied nonconvex MOPs emerging from linear complementarity problems.

TenHuisen (1993) and TenHuisen and Wiecek (1994) initiated a direction of solving nonconvex MOPs using generalized Lagrangian duality theory and proposed an approach to bicriteria programs based on specific quadratic Lagrangian functions (1997). This paper follows upon the recent study and presents an augmented Lagrangian scalarization for MOPs. The scalarization generates *locally* efficient points, that cannot be found by means of other methods, and is also capable of finding efficient solutions.

In Section 2, basic concepts are defined and the notion of approachability and the related augmented Lagrangian are introduced. Approachability and efficiency are studied in Section 3. They are illustrated with an example in Section 4. Section 5 concludes the paper.

2 Problem formulation

Consider the MOP formulated as

MOP	minimize	$f(x)$
	subject to	$x \in X$,

where $f(x) = [f_1(x), f_2(x), ..., f_m(x)]$ and each $f_i(x)$, $i = 1, ..., m$, is a real-valued function defined on $X \subset R^n$. We assume that each $f_i(x)$, $i = 1, ..., m$, is continuous and the feasible region X is a nonempty and compact set in R^n.

A point $x^0 \in X$ is said to be an efficient solution of MOP if there is no other point $x \in X$ such that $f(x) \leq f(x^0)$, (i.e. $f_i(x) \leq f_i(x^0)$ for all $i = 1, ..., m$, with strict inequality holding for at least one i). The set of all efficient solutions is denoted by X_E. A point $x^0 \in X$ is called a properly efficient solution of MOP (in the sense of Geoffrion) if x^0 is efficient and there exists a scalar $M > 0$ such that for each i, $1 \leq i \leq m$, and each $x \in X$ satisfying $f_i(x) < f_i(x^0)$, there exists at least one j, $j \neq i$, with $f_j(x) > f_j(x^0)$ and $(f_i(x) - f_i(x^0))/(f_j(x^0) - f_j(x)) \leq M$. Properly efficient solutions are denoted X_{PE}.

The image of the feasible region X under the vector-valued mapping $[f_1(x), .., f_m(x)]$ is called the outcome set, and the image of an efficient solution under this same mapping is called a nondominated solution. We denote the outcome set here by Y, the nondominated set by Y_E, and the properly nondominated set by Y_{PE}.

Efficiency, when considered over the feasible set, is referred to as global. However, while dealing with nonconvex MOPs, the notion of locally efficient solutions is necessary. A point $x^0 \in X$ is called a locally efficient solution of MOP if there exists a neighborhood $U(x^0)$ of x^0 such that there is no point $x \in X \cap U(x^0)$ for which $f(x) \leq f(x^0)$. Analogously, locally nondominated (efficient) solutions Y_{LE} (X_{LE}) and locally properly nondominated (efficient) solutions Y_{LPE} (X_{LPE}) can be defined.

Corresponding to the MOP are the ε-constraint problems $P_k(\varepsilon)$, $k = 1, ..., m$, (Haimes and Chankong, 1983),

$\mathbf{P_k(\varepsilon)}$	minimize	$f_k(x)$
	subject to	$f_j(x) \leq \varepsilon_j,\ j = 1, ..., m,\ j \neq k$
		$x \in X$.

Here $\varepsilon = (\varepsilon_1, ..., \varepsilon_{k-1}, \varepsilon_{k+1}, ..., \varepsilon_m)$. Letting the function $f_j(x) - \varepsilon_j$ be denoted by $g_j(x, \varepsilon_j)$, $j = 1, ..., m$, $j \neq k$, and $g_{-k}(x, \varepsilon) = [g_1(x, \varepsilon_1), ..., g_{k-1}(x, \varepsilon_{k-1}), g_{k+1}(x, \varepsilon_{k+1}), ..., g_m(x, \varepsilon_m)]^T$, the inequality constraint in $P_k(\varepsilon)$ can be expressed as $g_{-k}(x, \varepsilon) \leq 0$. Let $E^k = \{ \varepsilon: \min \{f_j(x): x \in X \} \leq \varepsilon_j \leq \max \{f_j(x): x \in X \},\ j = 1, ..., m,\ j \neq k \}$. We assume throughout this paper that $\varepsilon \in E^k$. Related to $P_k(\varepsilon)$ and the outcome set Y are the following concepts defined for k = 1, ..., m. Denote $f_{-k}(x) = [f_1(x), ..., f_{k-1}(x), f_{k+1}(x), ..., f_m(x)]^T$. Define the k-th image set $Y_k = \{ (y, z) \in R^{m-1} x R$: $y = f_{-k}(x), z = f_k(x), x \in X\}$, and, accordingly, its nondominated set Y_{kE}, locally nondominated set Y_{kLE}, and locally properly nondominated set Y_{kLPE}. Define also the k-th lower envelope $Y_k^{env} = \{ (y', z') \in Y_k : z' \leq z$ for all $(y, z) \in Y_k$ for which $y = y' \}$. Let X_k^{env} denote the set of all pre-image points in X of the points in Y_k^{env}. Define now the k-th envelope function $env_k(y) = \min \{ z: (y, z) \in Y_k \}$, and observe that $(y, z) \in Y_k^{env}$ if and only if $env_k(y) = z$. Let $\nabla env_k(y)$ denote the gradient of $env_k(y)$ and $\nabla env_k(y^*) = \nabla env_k(y)|_{y^*}$.

Proposition 1.

a) If $(y, z) \in Y_{kLE}$ for some k, $1 \leq k \leq m$, then $(y, z) \in Y_{jLE}$ for every j = 1, ..., m.
b) If $(y, z) \in Y_{kE}$ for some k, $1 \leq k \leq m$, then $(y, z) \in Y_{jE}$ for every j = 1, ..., m.
c) If $(y, z) \in Y_{kLE}$ for some k, $1 \leq k \leq m$, then $(y, z) \in Y_{LE}$.
d) If $(y, z) \in Y_{kE}$ for some k, $1 \leq k \leq m$, then $(y, z) \in Y_E$.

Proposition 2. A point $(y, z) \in Y_E$ iff $(y, z) \in Y_k^{env}$ for every k = 1, ..., m.

Augmented Lagrangian functions have been studied by many researchers in the context of nonlinear single objective programming. See Rockafellar (1974), Tind and Wolsey (1981), and Minoux (1986) for various forms of those functions, their use and importance. In this paper we are particularly interested in the quadratic Lagrangian related to the quadratic function of form (2.1).

Let $q_k(y)$: $R^{m-1} \to R$ be a quadratic function of the form

$$q_k(y) = -(y - \varepsilon)^T A(y - \varepsilon) - b^T(y - \varepsilon) + c, \quad (2.1)$$

where A is an (m-1)x(m-1) diagonal matrix with all its elements equal to a, a > 0 and finite, $b \in R^{m-1}$, and c is a scalar. Let $\nabla q_k(y)$ denote the gradient of $q_k(y)$ and $\nabla q_k(y^*) = \nabla q_k(y)|_{y^*}$.

Definition 1. Let $(y^0, z^0) \in R^m$. The function $q_k(y)$: $R^{m-1} \to R$ is said to *support* the set $Y_k \subset R^m$ at (y^0, z^0) if $q(y^0) = z^0$ and $q(y) \leq z$ for all $(y, z) \in Y_k$.

Definition 2. A point $x^0 \in X$ is said to be *q_k-approachable* if there exists a quadratic function $q_k(y)$ of the form (2.1) that supports Y_k at (y^0, z^0), where $(y^0, z^0) = (f_{-k}(x^0), f_k(x^0))$.

Define the augmented Lagrangian function as

$$QL_k(x, A, b, \varepsilon) = f_k(x) + g_{-k}^T(x, \varepsilon)Ag_{-k}(x, \varepsilon) + b^T g_{-k}(x, \varepsilon), \quad (2.2)$$

where A and b are defined as in (2.1). Corresponding to $P_k(\varepsilon)$ is the quadratic Lagrangian problem $QL_k(\varepsilon)$:

$QL_k(\varepsilon)$	minimize	$QL_k(x, A, b, \varepsilon)$
	subject to	$x \in X$,

which is the main tool of the proposed scalarization.

3 Approachability and efficiency

In this section we show that $QL_k(\varepsilon)$ produces approachable points that turn out to be locally efficient or efficient solutions of the original MOP. First we relate approachability to efficiency and local efficiency.

Theorem 1. If $x^0 \in X$ is q_k-approachable, then $x^0 \in X_k^{env}$.

Proof: If x^0 is q_k-approachable, then there exists a quadratic function of the form (2.1), with $a > 0$, which supports Y_k at (y^0, z^0), where $y^0 = f_{-k}(x^0)$ and $z = f_k(x^0)$. That is, $q_k(y^0) = z^0$ and $q_k(y) \leq z$ for all $(y, z) \in Y_k$. Therefore, there does not exist $x' \in X$ such that $y' = f_{-k}(x') = y^0$ and $z' = f_k(x') < z^0$. It follows that $x^0 \in X_k^{env}$.

Corollary 1. If $x^0 \in X$ is q_k-approachable for every $k = 1, ..., m$, then $x^0 \in X_E$.

Theorem 2. If for some k $(1 \leq k \leq m)$ there exists a point $(y^0, z^0) \in Y_k^{env}$ such that $env_k(y)$ is continuously differentiable at y^0 and $\nabla env_k(y^0) \leq 0$, then $(y^0, z^0) \in Y_{LE}$. Additionally, if $\nabla env_k(y^0) < 0$ and finite, then $(y^0, z^0) \in Y_{LPE}$.

Proof: Since $env_k(y^0) = z^0$ then

$$z^0 \leq z \text{ for all } (y, z) \in Y_k : \; y = y^0. \tag{3.1}$$

The gradient condition implies that for every $i = 1, ..., m$, $i \neq k$,

$$\lim_{y_i \to y_i^0} \frac{env_k(y) - z^0}{y_i - y_i^0} \leq 0, \tag{3.2}$$

where $y_r = y_r^0$ for $r = 1, ..., m$, $r \neq i$, $y_i = y_i^0 + \Delta y_i$, and that for at least one j, $j \neq k$,

$$\lim_{y_j \to y_j^0} \frac{env_k(y) - z^0}{y_j - y_j^0} < 0. \tag{3.3}$$

Assume now that $(y^0, z^0) \notin Y_{LE}$, that is $(y^0, z^0) \notin Y_{kLE}$, where $(y^0, z^0) = (f_{-k}(x^0), f_k(x^0))$ and $x^0 \in X$. Then, there exists $x' \in X \cap U(x^0)$ for which $(y', z') \leq (y^0, z^0)$, where $(y', z') = (f_{-k}(x'), f_k(x'))$. Due to continuity, y' satisfies (3.2) and (3.3). Consider the following two cases:

i) If $y' = y^0$ then there must be that $z' < z^0$, which contradicts (3.1).

ii) If $z' = z^0$ then there must be that $y' \leq y^0$. Hence, there exists at least one i, $1 \leq i \leq m$, $i \neq k$, for which $y_i' < y_i^0$. Obviously $i=j$, and from (3.3) $env_k(y') > z^0$. Consequently, $env_k(y') > z'$, which contradicts the definition of the k-th envelope function. We conclude that $(y^0, z^0) \in Y_{LE}$.

Now assume that for every $i = 1, ..., m$, $i \neq k$,

$$\lim_{y_i \to y_i^0} \frac{env_k(y) - z^0}{y_i - y_i^0} < 0, \tag{3.4}$$

and finite. We shall show that $(y^0, z^0) \in Y_{kLPE}$. If, for any i, $1 \le i \le m$, $i \ne k$, $f_i(x) < f_i(x^0)$, then $y_i - y_i^0 < 0$ and from (3.4) $env_k(y) - z^0 > 0$ for all y in a neighborhood $V_k(y^0)$ of y^0. Therefore $env_k(y) > f_k(x^0)$. From (3.4) we also conclude that there exists a scalar $\rho_i > 0$ such that $\frac{z^0 - env_k(y)}{y_i - y_i^0} \ge \rho_i$ for every $i = 1, .., m$, $i \ne k$. Therefore, there exists a scalar $M_i = \frac{1}{\rho_i}$ such that $\frac{y_i - y_i^0}{z^0 - env_k(y)} \le M_i$, or equivalently

$$\frac{f_i(x^0) - f_i(x)}{env_k(y) - f_k(x^0)} \le M_i \quad \text{for every } i = 1, .., m, i \ne k. \tag{3.5}$$

Since the limit in (3.4) is finite, there exists a scalar $M_k > 0$ such that

$$\frac{z^0 - env_k(y)}{y_i - y_i^0} \le M_k, \text{ or equivalently}$$

$$\frac{f_k(x^0) - env_k(y)}{f_i(x) - f_i(x^0)} \le M_k. \tag{3.6}$$

(3.5) and (3.6) imply that $(y^0, z^0) \in Y_{kLPE}$, and also $(y^0, z^0) \in Y_{LPE}$.

Theorem 3. If for some k ($1 \le k \le m$) there exists a point $x^0 \in X$ such that x^0 is q_k-approachable, $env_k(y)$ is continuously differentiable at $y^0 = f_{-k}(x^0)$, and $\nabla env_k(y^0) < 0$ and finite, then $x^0 \in X_{LPE}$.

Proof: If $x^0 \in X$ is q_k-approachable, Theorem 1 proves that $x^0 \in X_k^{env}$. Theorem 2, then, implies that $x^0 \in X_{LPE}$.

Lemma 1. Let $q_k(y) = -(y - \varepsilon)^T A(y - \varepsilon) - b^T(y - \varepsilon) + QL_k(x^*, A, b, \varepsilon)$. Then a point x^* minimizes $QL_k(x, A, b, \varepsilon)$ over X if and only if $q_k(y) \le z$ for all $(y, z) \in Y_k$.

Proof: x^* minimizes $QL_k(x, A, b, \varepsilon)$ over X if and only if

$$QL_k(x^*, A, b, \varepsilon) \le QL_k(x, A, b, \varepsilon) \quad \text{for all } x \in X,$$

which is equivalent to

$$QL_k(x^*, A, b, \varepsilon) \le f_k(x) + g_{-k}^T(x, \varepsilon) A g_{-k}(x, \varepsilon) + b^T g_{-k}(x, \varepsilon) \text{ for all } x \in X.$$

Letting $y = f_{-k}(x)$ and $z = f_k(x)$, this becomes

$$QL_k(x^*, A, b, \varepsilon) \le z + (y - \varepsilon)^T A(y - \varepsilon) + b^T(y - \varepsilon) \text{ for all } (y, z) \in Y_k.$$

Consequently

$$-(y - \varepsilon)^T A(y - \varepsilon) - b^T(y - \varepsilon) + QL_k(x^*, A, b, \varepsilon) \le z \text{ for all } (y, z) \in Y_k$$

and

$q_k(y) \leq z$ for all $(y, z) \in Y_k$.

Theorem 4. A point $x^* \in X$ minimizes $QL_k(x, A, b, \varepsilon)$ over X if and only if x^* is q_k-approachable.

Proof. ($\Rightarrow$) Let x^* minimize $QL_k(x, A, b, \varepsilon)$ over X. Lemma 1 proves that

$$q_k(y) \leq z \text{ for all } (y, z) \in Y_k, \tag{3.7}$$

where

$$q_k(y) = -(y - \varepsilon)^T A(y - \varepsilon) - b^T(y - \varepsilon) + QL_k(x^*, A, b, \varepsilon). \tag{3.8}$$

We shall show that $q_k(y)$ supports the set Y_k at (y^*, z^*), where $y^* = f_{-k}(x^*)$ and $z^* = f_k(x^*)$. From (3.8) and (2.2) we get

$$\begin{aligned} q_k(y^*) &= -(y^* - \varepsilon)^T A(y^* - \varepsilon) - b^T(y^* - \varepsilon) + QL_k(x^*, A, b, \varepsilon) \\ &= -(y^* - \varepsilon)^T A(y^* - \varepsilon) - b^T(y^* - \varepsilon) + f_k(x^*) + \\ &\qquad (f_{-k}(x^*) - \varepsilon)^T A(f_{-k}(x^*) - \varepsilon) + b^T(f_{-k}(x^*) - \varepsilon) \\ &= -(y^* - \varepsilon)^T A(y^* - \varepsilon) - b^T(y^* - \varepsilon) + z^* + (y^* - \varepsilon)^T A(y^* - \varepsilon) + b^T(y^* - \varepsilon) \\ &= z^* \end{aligned} \tag{3.9}$$

From (3.7) and (3.9) we conclude that the quadratic function of form (3.8) supports the set Y_k at (y^*, z^*). Therefore, x^* is q_k-approachable.

($\Leftarrow$) Since x^* is q_k-approachable, there exists a quadratic function $q_k(y)$ of form (2.1) that supports the set Y_k at $(y^*, z^*) = (f_{-k}(x^*), f_k(x^*))$. That is

$$q_k(y) \leq z \text{ for all } (y, z) \in Y_k, \tag{3.10}$$

and

$$q_k(y^*) = z^*. \tag{3.11}$$

From (2.1) and (3.11) we obtain $-(y^* - \varepsilon)^T A(y^* - \varepsilon) - b^T(y^* - \varepsilon) + c = z^*$ which implies

$$-(y^* - \varepsilon)^T A(y^* - \varepsilon) - b^T(y^* - \varepsilon) + c + QL_k(x^*, A, b, \varepsilon) = z^* + QL_k(x^*, A, b, \varepsilon).$$

Consequently

$$-(y^* - \varepsilon)^T A(y^* - \varepsilon) - b^T(y^* - \varepsilon) + c + f_k(x^*) + (f_{-k}(x^*) - \varepsilon)^T A(f_{-k}(x^*) - \varepsilon) + b^T(f_{-k}(x^*) - \varepsilon) = z^* + QL_k(x^*, A, b, \varepsilon).$$

Since $y^* = f_{-k}(x^*)$ and $z^* = f_k(x)$, we get

$$-(y^* - \varepsilon)^T A(y^* - \varepsilon) - b^T(y^* - \varepsilon) + c + z^* + (y^* - \varepsilon)^T A(y^* - \varepsilon) + b^T(y^* - \varepsilon) = z^* + QL_k(x^*, A, b, \varepsilon),$$

which yields $c = QL_k(x^*, A, b, \varepsilon)$. Therefore

$$q_k(y) = -(y - \varepsilon)^T A(y - \varepsilon) - b^T(y - \varepsilon) + QL_k(x^*, A, b, \varepsilon). \tag{3.12}$$

From (3.10) and (3.12), Lemma 1 proves that x^* minimizes $QL_k(x, A, b, \varepsilon)$ over X.

Corollary 2. If $x^0 \in X$ minimizes $QL_k(x, A, b, \varepsilon)$ over X for every $k = 1, ..., m$, then $x^0 \in X_E$.

Theorem 5 below introduces a methodology for finding locally properly efficient solutions of MOPs. The following assumption is helpful in establishing this result.

Assumption 1. Let $H_k(y) = q_k(y^0) + \nabla q_k(y^0)(y - y^0)$ be the hyperplane tangent to the quadratic function $q_k(y)$ at y^0, where $y^0 = f_{-k}(x^0)$. If $x^0 \in X$

minimizes $QL_k(x, A, b, \varepsilon)$ over X, then $H_k(y)$ is a good local approximation of $env_k(y)$ in a neighborhood $V(y^0)$ of y^0.

Lemma 2. Let $q_k(y)$ be a quadratic function of form (2.1) and $env_k(y)$ be continuously differentiable. Under Assumption 1, if $q_k(y)$ supports the set Y_k at (y^0, z^0), then $\nabla env_k(y^0) = \nabla q_k(y^0)$.

<u>Proof:</u> Under Assumption 1, for every y in a neighborhood $V(y^0)$ of y^0:

$$env_k(y^0) + \nabla env_k(y^0)(y - y^0) = q_k(y^0) + \nabla q_k(y^0)(y - y^0) \tag{3.13}$$

Since $q_k(y)$ supports the set Y_k at (y^0, z^0), then $q_k(y^0) = env_k(y^0)$ and (3.13) yields $\nabla env_k(y^0)(y - y^0) = \nabla q_k(y^0)(y - y^0)$ for every y in $V(y^0)$. Therefore, $\nabla env_k(y^0) = \nabla q_k(y^0)$.

Theorem 5. If for some k ($1 \le k \le m$) there exists a point $x^0 \in X$ such that x^0 minimizes $QL_k(x, A, b, \varepsilon)$ over X, $env_k(y)$ is continuously differentiable at $y^0 = f_{-k}(x^0)$, and $\nabla q_k(y^0) < 0$, then $x^0 \in X_{LPE}$.

<u>Proof:</u> If x^0 minimizes $QL_k(x, A, b, \varepsilon)$ over X, then by Theorem 4, x^0 is q_k-approachable. Hence, using Theorem 1, $x^0 \in X_k^{env}$ and its image $(y^0, z^0) = (f_{-k}(x^0), f_k(x^0)) \in Y_k^{env}$. If $\nabla q_k(y^0) < 0$, then by Lemma 2, $\nabla env_k(y^0) < 0$. Consequently, Theorem 3 proves that $x^0 \in X_{LPE}$.

Theorem 5 shows that solving $QL_k(\varepsilon)$ for given A, b, and ε so that the gradient condition is satisfied yields a locally nondominated point (y^0, z^0). Application of this theorem is presented in the next section.

4 Example

Consider the following MOP

minimize $\{f_1(x), f_2(x), f_3(x)\}$

subject to $x \in X$,

where $f_1(x) = \frac{1}{2}x_1^4 - \frac{11}{3}x_1^3 + \frac{31}{4}x_1^2 - \frac{21}{4}x_1 + 2x_2 - 4\sqrt{x_2} + \frac{233}{32}$, $f_2(x) = x_1 - \frac{3}{2}$, $f_3(x) = \sqrt{x_2} - 1$, and $X = \{x \in R^2 : 0 \le x_i \le 4, i = 1, 2.\}$. Choosing f_1 as the primary objective, formulate $P_1(\varepsilon)$ and the quadratic Lagrangian problem $QL_1(\varepsilon)$

minimize $f_1(x) + a\,[(f_2(x) - \varepsilon_2)^2 + (f_3(x) - \varepsilon_3)^2] + b_1(f_2(x) - \varepsilon_2) + b_2(f_3(x) - \varepsilon_3)$

subject to $x \in X$.

Let $\varepsilon = (\varepsilon_2, \varepsilon_3) = (\frac{1}{2}, -\frac{1}{2})$, $a = 5$, and $b = (\frac{9}{4}, 2)$. Solving $QL_1(\varepsilon)$ yields the point $x^* = (2, \frac{1}{4})$, for which the optimal value $QL_k(x^*, a, b, \varepsilon) = \frac{475}{96}$. The point x^* produces the point $(y_1^*, y_2^*, z^*) = (f_2(x^*), f_3(x^*), f_1(x^{**})) = (\frac{1}{2}, -\frac{1}{2}, \frac{475}{96})$ in the objective space. By Theorem 5, $x^* \in X_{LPE}$. Observe, however, that x^* is not efficient. Figure 1 shows the image set Y_1 and the quadratic function $q_1(y) = -5y_1^2 - 5y_2^2 + \frac{11}{4}y_1 - 7y_2 + \frac{247}{96}$ supporting Y_1 at the point $(\frac{1}{2}, -\frac{1}{2}, \frac{475}{96})$.

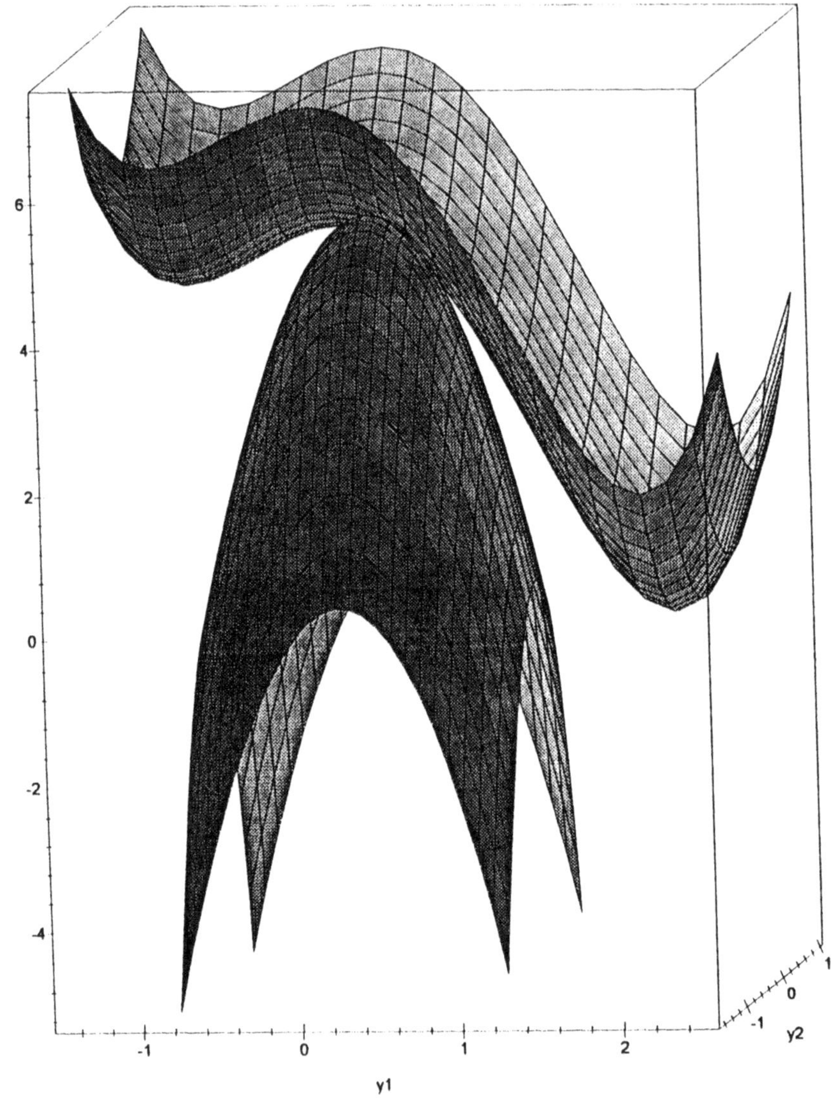

Figure 1. The set Y_1 and the quadratic function $q_1(y)$ supporting it at the locally nondominated point $(\frac{1}{2}, -\frac{1}{2}, \frac{475}{96})$.

5 Conclusions

This paper introduces a nonlinear scalarization specially designed for nonconvex MOPs featuring locally efficient solutions. The scalarization uses a quadratic Lagrangian function related to the ε-constraint formulation of the original MOP, and produces local solutions that are not (globally) efficient, and thus not available by means of other methods. Such solutions, however, are of interest in engineering applications. An example with three objective functions is enclosed.

Future research will include investigation of existence of approachable points as well as application of the methodology to MOPs in engineering.

References

[1] Bernau, H.: Use of exact penalty functions to determine efficient decisions, European Journal of Operational Research, 49 (1990) 348-355

[2] Haimes Y.Y, V. Chankong, *Multiobjective Decision Making - Theory and Methodology*. North Holland, New York 1983

[3] Isac, G., M.M. Kostreva and M.M. Wiecek: Multiple objective approximation of feasible but unsolvable linear complementarity problems, Journal of Optimization Theory and Applications 86 (1995) 389-405

[4] Jahn, J., J. Klose and A. Merkel: On the application of a method of reference point approximation to bicriterial optimization problems in chemical engineering. In: W. Oettli , D. Pallaschke (eds): *Advances in Optimization*. Springer-Verlag, Berlin, 1992, pp. 478-491

[5] Kopsidas, G.C.: Multiobjective optimization of table olive preparation systems. European Journal of Operations Research 85 (1995) 383-398

[6] Kostreva, M.M., T. Ordoyne and M.M. Wiecek: Mulitple-objective programming with polynomial objectives and constraints. European Journal of Operational Research 57 (1992) 381-394

[7] Kostreva, M.M., M.M. Wiecek: Linear complementarity problems and multiple objective programming. Mathematical Programming 60 (1993) 349-359

[8] Minoux, M.: *Mathematical Programming - Theory and Algorithms*. John Wiley, Chichester 1986

[9] Osyczka, A.: *Multicriterion Optimization in Engineering with Fortran Programs*. Ellis Horwood, Chichester 1984

[10] Rockafellar, R.T.: Augmented Lagrange multiplier functions and duality in nonconvex programming. SIAM Journal on Control 12 (1974) 269-285

[11] TenHuisen, M.L.: Generalized Lagrangian Duality in Multiple Objective Programming, Ph.D. thesis, Department of Mathematical Sciences, Clemson University, Clemson, SC 1993

[12] TenHuisen, M.L., M. M. Wiecek: Vector optimization and generalized Lagrangian duality. Annals of Operations Research 51 (1994) 15-32

[13] TenHuisen, M.L., M. M. Wiecek: Efficiency and solution approaches to bicriteria nonconvex programs. Journal of Global Optimization (1997) in print

[14] Tind, J., L.A. Wolsey: An elementary survey of general duality theory in mathematical programming. Mathematical Programming 21 (1981) 241-261

Dominance and Efficiency in Multiobjective Stochastic Linear Programming

F. Ben Abdelaziz [1], P. Lang [2], R. Nadeau [2]

[1]Institut Supérieur de Gestion, Le Bardo 2000, Tunisia
[2]Faculté des Sicences de l'Administration, Université Laval, Ste Foy, Québec, Canada G1K 7P4.

Abstract. In contrast to deterministic multiobjective problems, where the notion of Pareto-efficiency is well accepted, several notions of efficiency are conceivable for the Multiobjective Stochastic Linear Programming problem. Their relevance depends on the available state of information about the decision situation, regarding particularly the Decision Maker's preference structure and probabilistic anticipations. We investigate efficient sets arising naturally from some extreme reference cases of states of information. We study these sets from the point of view of their relative inclusions and provide some indications as to their computability.

Keywords. Efficiency, Multiple Criteria, Stochastic programming.

1 Introduction

The Multiobjective Stochastic Linear Programming (MSLP) problem is often written as:
(1) $"Maximize"\{C(\omega).x | x \in \mathcal{X}(\omega)\}$ where:
(1.1) $X(\omega) = \{x \in \Re^n_+ | A(\omega).x \leq b(\omega)\}$
and C, A and b are random matrices of respective dimensions $m \times n$, $c \times n$ and $c \times 1$ defined on some probability space (Ω, Ξ, P). As in the single objective stochastic case, problem (1) is not well specified, since neither the random constraints nor the maximization of random objectives are mathematically defined. Thus additional specifications are required to make (1) meaningful.

Several approaches have been proposed in the literature to "solve" (1). In one first broad category are "pragmatic" approaches. Some of these popular approaches first transform (1) into a deterministic Multiobjective Program (MP), which is then solved via an interactive method as in PROTRADE of Goicoechea et al. (1979), STRANGE of Teghem et al. (1986) and PROMISE of Urli and Nadeau (1990). These methods are intuitively appealing and have been successfully tested in real-world contexts; however the transformations used at the outset are rather arbitrary and no general criterion is available by which the quality of a solution can be judged.

Formulating criteria to identify "good" solutions is a concern of a second broad category of research. Since problem (1) is in many ways multidimensional, it is natural to view as potentially good a solution that is undominated and thus to seek a counterpart of the deterministic notion of Pareto-efficiency. In that vein, one stream of research dispenses with information on probability measures for problem (1), and views it instead as a parametric problem as in Bitran (1980) and Weber (1985). For these transformations, efficient solutions are defined as Pareto-efficient in the usual sense for all possible realizations of the matrix C. Another stream also concerned with efficiency considers explicitly the probability distributions of the consequences of solutions. For a problem with deterministic constraints, Wolf (1985) uses the marginal distribution of the objectives as a comparison criterion, and identifies particular deterministic multiobjective formulations whose Pareto-efficient solutions are efficient for the stochastic problem according to this criterion. Marti (1990) defines solutions to (1) as solutions to the problem: Maximize $\{E[u(A(\omega).x - b(\omega))] | x \in \mathcal{X} \subseteq \Re^n_+\}$ where u is an unknown utility function.

In sum, the picture emerging from the efficiency-concerned litterature in relation to problem (1) is rather fragmented. In this paper, we attempt to provide a broader perspective on what can be deemed efficient solutions for a problem similar to (1) with a set S of discrete scenarios. This problem is considered here from a decision-making perspective: we assume it has been formulated with a view to helping one or several decision makers arrive at some satisfactory decision. In that setting, what the formulated problem lacks in mathematical specification makes it open to several interpretations; in particular, it is argued here that efficiency is relative, as in Weber (1987), to the DM's *values* and *beliefs*; alternative ***decision contexts*** in that sense will thus lead to different definitions of efficiency.

This paper investigates alternative efficiency definitions that naturally arise under specific decision contexts or states of information (SI) regarding the DM's preference structure and probabilistic beliefs. To provide a broad picture, we concentrate on somewhat extreme SI cases; a large continium of intermediary cases remains to be explored. We investigate relations between the efficient sets thus defined. We also provide some indications as to their computability (this latter aspect, being further elaborated in Ben Abdelaziz et al. 1994, 1995a, 1995b). The problem considered is assumed to have well-defined consequences in terms of its objectives; however, how stochastic constraints are to be satisfied remains to be specified; this is the main concern of Section 2. The rest of this paper is devoted to efficiency notions that emerge under various polar states of information. Fundamental admissible sets are defined in Section 3. Section 4 investigates in more detail admissible solutions "in consequences". The same is done in Section 5 for "distributionally" efficient solutions.

2 Modelling consequences under discrete scenarios

In (1), stochastic constraints (1.1) are not necessarily meant to be satisfied over all states of nature (otherwise they are easily reformulated as deterministic constraints). One has thus to specify what is meant by "satisfying" these constraints. In the litterature, stochastic constraints are usually transformed into some deterministic form (e.g. chance constraints, recourse function). Underlying these transformations are two distinct preoccupations, namely: (i) Measuring the *extent* of violations of these constraints, a modelling concern, and (ii) Establishing limits to or penalties on the violations (and eventually trade-offs on the violation penalties), clearly a much more normative concern.
It seems virtually impossible to obtain a mathematically well-defined formulation of (1) without addressing the first concern. To the extent that a-priori judgments on permissible violations are to be avoided, these violations become additional criteria in their own right. we can propose two approaches to measuring constraint violations that naturally lead to the form:
(2) "Max" $\{\tilde{C}x|Ax \leq b\}$
where C is a random matrix of dimension $(m \times n)$, $\tilde{C} = C^s$ with probability $p_s, s \in \mathcal{S}$, and A and b are deterministic matrices of respective dimensions $(c \times n)$ and $(c \times 1)$.

One approach to measuring violations in (1.1) is to directly consider the vector of these violations as a set of additional criteria to be "maximized". This is the viewpoint taken among others by Wolf (1985) and Marti (1990). A second classical measure of violations is via recourses. For instance, introducing one recourse vector per scenario $s \in \mathcal{S}$. The recourse variables may correspond to concrete corrective actions or merely to a measure of the gap between a desired and a realized state. In both cases, the recourse variables can be translated into a vector of additional random objectives.

In the sequel, we consider a problem of the form (2) with $\mathcal{X}$ compact; we investigate efficient solutions under varying states of information regarding the DM's preference structure and the probability distribution. We use the following notation when comparing any two vectors a and b in $\Re^m$: $a \geq b$ if $a_i \geq b_i \; \forall i \in \mathcal{I}$; $a > b$ if $a \geq b$ and $a \neq b$; and $a \gg b$ if $a_i > b_i \; \forall i \in \mathcal{I}$. Let $\tilde{C}_i$ and $C_i^s \in \Re^n$ denote the ith row respectively of the random matrix $\tilde{C}$ and of the realization C^s.

3 Admissible and unanimous solutions

Incomplete information about the DM's preference structure and subjective probabilistic belief can take several forms; Table 1 summarizes some reference cases. If no utility function can be assumed to exist, little meaning can be ascribed to the probability distribution of outcomes of a decision, one is thus led to compare decisions purely in terms of their consequences. This comparison may further have two forms, depending on how scenarios are viewed; if consequences *per se* are the only aspect that matters to the DM, then to each decision is associated a *set* of consequences, and these sets are to be compared; if however the DM ascribes a meaning to states of the world, it is meaningful to recognize the scenario associated with each consequence, so that the comparison is over *ordered sets* of consequences.

If a utility function u can be hypothesized, it may be possible to ascertain some of its characteristics (e.g. monotony, risk aversion); similarly, it may be possible to circumscribe to some extent the DM's probabilistic beliefs. Our state of information on problem (3) can be thus summarized by a class $\mathcal{U}$ of utility functions and a class $\mathcal{P}$ of probability distributions.

If a utility function cannot be assumed to exist for the DM (SI case 1), a prudent

SI	case1: Existence of a utility function not assumed.
	1.1 State-independent preferences.
	1.2 State-dependent preferences.
SI	case2: A utility function u exists; $u \in \mathcal{U}$, $p \in \mathcal{P}$.
	2.1 Indeterminate probabilistic beliefs.
	2.2 Well-defined probabilistic beliefs.

Table 1: States of information about a decision problem

approach in comparing decisions is to merely consider their consequences and not to attempt any combination or aggregation, as would be done, for instance, by considering the distribution of $\tilde{C}_i \cdot x$. Under state-independent preferences (SI case 1.1):

Definition 1 *Solution $y \in \mathcal{X}$ set-dominates solution $x \in \mathcal{X}$ if*
(i) $\forall i \in \mathcal{I}$, $\forall s \in \mathcal{S}$, $\forall t \in \mathcal{S}$: $C_i^t \cdot y \geq C_i^s \cdot x$, and
(ii) $\exists i_0 \in \mathcal{I}$, $s_0 \in \mathcal{S}$, and $t_0 \in \mathcal{S}$: $C_{i_0}^{t_0} \cdot y > C_{i_0}^{t_0} \cdot x$. The class of set-admissible solutions, noted $\mathcal{A}_s^*$, is the set of solutions $x \in \mathcal{X}$ which are not set-dominated by any $y \in \mathcal{X}$.

Under state-dependent preferences (SI case 1.2), we compare solutions in terms of $m \cdot S$ criteria which leads to a natural counterpart of the notion of Pareto-efficient solution.

Definition 2 *Solution $y \in X$ pointwise dominates solution $x \in X$ if*
(i) $\forall i \in \mathcal{I}$, $\forall s \in \mathcal{S}$: $C_i^s \cdot y \geq C_i^s \cdot x$, and
(ii) $\exists i_0 \in \mathcal{I}$ and $s_0 \in \mathcal{S}$: $C_{i_0}^{s_0} \cdot y > C_{i_0}^{s_0} \cdot x$.

The set of *pointwise admissible* solutions, noted $\mathcal{A}^*$, is the set of solutions $x \in \mathcal{X}$ which are not pointwise dominated by any $y \in \mathcal{X}$.

Let us now consider the second broad case (SI case 2) where a utility function can be assumed, and where our state of information consists of given classes $\mathcal{U}$ and $\mathcal{P}$ of utility functions and of probability distributions. In the sequel we assume that $\mathcal{U} \subseteq \mathcal{U}_1$, where $\mathcal{U}_1$ is the set of non-decreasing functions on $\Re^m$; $E_p[\cdot]$ denotes expectation with respect to probability distribution $p \in \mathcal{P}$.

Let us first consider the following situation (SI case 2.1):

Definition 3 *Solution $y \in \mathcal{X}$ set-($\mathcal{P}$,$\mathcal{U}$) dominates solution $x \in X$ if*
(i) $\forall u \in \mathcal{U}$, $\forall p \in \mathcal{P}$, $\forall q \in \mathcal{P}$: $E_p[u(\tilde{C} \cdot y)] \geq E_q[u(\tilde{C} \cdot x)]$, and
(ii) $\exists u_0 \in \mathcal{U}$, $p_0 \in \mathcal{P}$, $q_0 \in \mathcal{P}$: $E_{q_0}[u_0(\tilde{C} \cdot y)] > E_{p_0}[u_0(\tilde{C} \cdot x)]$.

The set of set-($\mathcal{P}$,$\mathcal{U}$) *admissible* solutions, noted $\mathcal{A}_s(\mathcal{P},\mathcal{U})$, is the set of solutions $x \in \mathcal{X}$ which are not set-($\mathcal{P}$,$\mathcal{U}$) dominated by any $y \in \mathcal{X}$. At another extreme (SI case 2.2), assume the DM has perfectly well formed yet unobserved beliefs as to scenario probabilities. Considering each pair $(u,p) \in \mathcal{U} * \mathcal{P}$ as an evaluation point of view leads to the following alternative definition of admissible solutions.

Definition 4 *Solution $y \in \mathcal{X}$ ($\mathcal{P}$,$\mathcal{U}$)-dominates solution $x \in \mathcal{X}$ if*
(i) $\forall u \in \mathcal{U}$, $\forall p \in P$: $E_p[u(\tilde{C} \cdot y)] \geq E_p[u(\tilde{C} \cdot x)]$, and
(ii) $\exists u_0 \in \mathcal{U}$ and $p_0 \in \mathcal{P}$: $E_{p_0}[u_0(\tilde{C} \cdot y)] > E_{p_0}[u_0(\tilde{C} \cdot x)]$.

The set of ($\mathcal{P}$,$\mathcal{U}$)-admissible solutions, noted $\mathcal{A}(\mathcal{P},\mathcal{U})$, is the set of solutions $x \in \mathcal{X}$ which are not ($\mathcal{P}$,$\mathcal{U}$)-dominated by any $y \in X$.

In the foregoing definitions, dominance is defined in terms of "weak inequalities everywhere with at least one strict inequality", leading to strong admissiblity. Alternative weaker or stricter definitions are conceivable. Weber (1987), for instance, considers weak set-($\mathcal{P}$,$\mathcal{U}$) and weak ($\mathcal{P}$,$\mathcal{U}$) admissible solutions. In a decision setting, however, we think it more useful not to retain solutions which have a disadvantage without any compensating merit (as in weak admissibility), nor to eliminate a solution on grounds of non-uniqueness (as in distinctive admissibility):
Among admissible solutions, one solution case will stand out if it occurs: if there exists a solution which is rated at least as well as any other under all evaluation points of view. Such solutions will be called unanimous.

Definition 5 *(i) The set U_s^* of set-unanimous solutions is the set of solutions $x \in \mathcal{X}$ such that $C_i^s \cdot x \geq C_i^t \cdot y$, $\forall y \in \mathcal{X}$, $\forall i \in \mathcal{I}$, $\forall s \in \mathcal{S}$, $\forall t \in \mathcal{S}$.*
(ii) The set U^* of pointwise unanimous solutions is the set of solutions $x \in \mathcal{X}$ such that $C_i^s \cdot x \geq C_i^s \cdot y$, $\forall y \in \mathcal{X}$, $\forall i \in \mathcal{I}$, $\forall s \in \mathcal{S}$.
(iii) The set $U_s(\mathcal{P}$,$\mathcal{U})$ of set-($\mathcal{P}$,$\mathcal{U}$)-unanimous solutions is the set of solutions $x \in \mathcal{X}$ such that: $E_p[u(\tilde{C} \cdot x)] \geq E_q[u(\tilde{C} \cdot y)]$, $\forall y \in \mathcal{X}$, $\forall p \in \mathcal{P}$, $\forall p \in \mathcal{P}$, $\forall u \in \mathcal{U}$.
(iv) The set $U(\mathcal{P},\mathcal{U})$ of set-($\mathcal{P},\mathcal{U}$)-unanimous solutions is the set of solutions $x \in \mathcal{X}$ such that: $E_p[u(\tilde{C} \cdot x)] \geq E_q[u(\tilde{C} \cdot y)]$, $\forall y \in \mathcal{X}$, $\forall p \in \mathcal{P}$, $\forall p \in \mathcal{P}$, $\forall u \in \mathcal{U}$.

Any such unanimous solution is optimal in the sense of being (weakly) preferred by all points of view, and equivalent in this sense to other unanimous solutions. If such a solution exists, all non-unanimous solutions are dominated and do not deserve consideration.

Theorem 1 *(Ben Abdelaziz et al. 1994):*
(i) $U_s(\mathcal{P}$,$\mathcal{U}) \subseteq \mathcal{U}(\mathcal{P}$,$\mathcal{U})$; $\mathcal{U}_s^* \subseteq \mathcal{U}^*$
(ii) $U_s(\mathcal{P}$,$\mathcal{U}) \neq \emptyset \Rightarrow U_s$ $(\mathcal{P}$,$\mathcal{U}) = \mathcal{A}_s(\mathcal{P}$,$\mathcal{U})$; $U(\mathcal{P}$,$\mathcal{U}) \neq \emptyset \Rightarrow U(\mathcal{P}$,$\mathcal{U}) = \mathcal{A}(\mathcal{P}$,$\mathcal{U})$;

$U_s^* \neq \emptyset \Rightarrow U_s^* = \mathcal{A}_s^*$; $U^* \neq \emptyset \Rightarrow U^* = \mathcal{A}^*$
(iii) $\mathcal{P} \subseteq \mathcal{P}'$, $\mathcal{U} \subseteq \mathcal{U}' \Rightarrow U_s(\mathcal{P}',\mathcal{U}') \subseteq U_s(\mathcal{P},\mathcal{U})$ and $U(\mathcal{P}',\mathcal{U}') \subseteq U(\mathcal{P},\mathcal{U})$.

Thus U_s^* and U^* , and, for large $\mathcal{P}$ and $\mathcal{U}$, $U_s(\mathcal{P},\mathcal{U})$ and $U(\mathcal{P},\mathcal{U})$, are likely to be empty in most practical cases. If there exist no unanimous solutions, the set of admissible solutions is likely to be large, particularly to the extent that the objective $\{C_i^s x|\ s \in \mathcal{S}\ , i \in \mathcal{I}\ , x \in \mathcal{X}\}$ are conflicting.

Definitions 3 and 4 naturally include the special cases where $\mathcal{P} = \{p\}$ or $\mathcal{U} = \{u\}$. We will note for short $E[\cdot]$ for $E_p[\cdot]$ and $\mathcal{A}(\mathcal{U}) = A(\{p\}\ , \mathcal{U})$, $\mathcal{A}(\mathcal{P}) = \mathcal{A}(\mathcal{P},\{u\})$. In the sequel we concentrate on some polar cases where the DM's preference structure cannot be completely specified.

4 Efficiency in consequences

This section's focus is on ill-structured situations where either a utility function cannot be assumed to exist or the available classes of utility functions and probability distributions are quite large. First, we show these two cases to be essentially equivalent. Second, we indicate how the set- and pointwise admissible sets can be operationally identified. Finally, in the case of state-dependent preferences, we investigate subsets of admissible solutions which are of potential interest if not all points of view are of equal importance. Let $\mathcal{P} = \{\pi^s \in \mathcal{P}_1 | \pi_s^s = 1, \pi_t^s = 0\ \forall t \neq s, s \in S\}$ and $\mathcal{U}^* = \{v_i : \Re^m \rightarrow \Re | v_i(z) = z_i, i \in \mathcal{I}\}$ be the classes of degenerate distributions with unit mass on a scenario, and of degenerate utility functions isolating one attribute at a time.

Theorem 2 *(Ben Abdelaziz et al. 1994):*
If $\mathcal{P}^* \subseteq \mathcal{P} \subseteq \mathcal{P}_1$ *and* $\mathcal{U}^* \subseteq \mathcal{U} \subseteq \mathcal{U}_1$ *then:*
(i) $\mathcal{A}_s^* = \mathcal{A}_s(\mathcal{P},\mathcal{U})$; *(ii)* $\mathcal{A}^* = \mathcal{A}(\mathcal{P},\mathcal{U})$; *(iii)* $U_s^* = U_s(\mathcal{P},\mathcal{U})$ *and (iv)* $U^* = (\mathcal{P},\mathcal{U})$.

Thus, according to Theorem 2, under large indeterminacy about the relevant utility function and probability distribution, one is led to examine actions purely in terms of their consequences.

The class of set-admissible solutions, which includes all other types of admissible points, is likely to be very large in most practical situations. Therefore its computability is probably easiest gauged by examining its complement, the class of set-dominated solutions.

The computation of pointwise admissible and unanimous solutions is relatively simple. Letting $\overline{C} = (C_i^s | i \in \mathcal{I}, s \in \mathcal{S})$ denote a matrix of dimension $(S \cdot m) \times n$ formed with the rows C_i^s corresponding to all pairs (scenario, criterion), $\mathcal{A}^*$ is the set of Pareto-efficient solutions to the following multiobjective deterministicblem:

(3) "Maximize" $\{\overline{C} \cdot x | x \in \mathcal{X}\}$.

A solution x is unanimous if for all $(s,i) \in \mathcal{S} \times \mathcal{I}$ it is an optimal solution to the single-objective deterministic problem: Maximize $\{C_i^s \cdot x | x \in \mathcal{X}\}$

In other words, the set U^* is the set of ideal solutions to the multiobjective deterministic problem (3). Thus U^* is likely to be empty in most practical cases. If $U^* = \emptyset$, the set $\mathcal{A}^*$ may be quite large relative to $\mathcal{X}$. Short of a unanimous solution, one may then seek some form of compromise solution. Not all admissible solutions, however, may be of equal interest. In certain decision contexts, a particular criterion may be sufficiently important as to have a sort of veto power; or special attention may be paid to particular scenarios. It may then be important to select solutions that are particularly well-behaved with respect to such "focal" composite viewpoints. Furthermore, if several such composite viewpoints emerge, one would like to obtain, if possible, a solution which is not inferior for any of them.

Definition 6 *(i) Let $\mathcal{C} \subseteq \mathcal{S} \times \mathcal{I}$ be a composite viewpoint. Solution $x \in \mathcal{X}$ is $\mathcal{C}$-advocated if $x \in A$ and there is no $y \in \mathcal{X}$ such that:*
$\forall (s,i) \in \mathcal{C}$, $C_i^s \cdot x \leq C_i^s \cdot y$ *and* $\exists (s,i) \in \mathcal{C}$ *such that* $C_i^s \cdot x < C_i^s \cdot y$.
(ii) Let $\mathcal{F} = \{\mathcal{C}_k \subseteq \mathcal{S} \times \mathcal{I} | k \in \mathcal{K}\}$ be a family of composite viewpoints indexed over the finite set $\mathcal{K}$. A solution $x \in \mathcal{X}$ is $\mathcal{F}$-unanimous if x is $\mathcal{C}_k$-advocated for all $k \in \mathcal{K}$.
(iii) Let $\mathcal{F} = \{\mathcal{C}_k \subseteq \mathcal{S} \times \mathcal{I} | k \in \mathcal{K}\}$ and $F' = \{\mathcal{C}_k' \subseteq \mathcal{S} \times \mathcal{I} | k \in \mathcal{K}'\}$. $\mathcal{F}'$ is finer than $\mathcal{F}$ (noted $\mathcal{F}' \subseteq \mathcal{F}$) if: $\forall k \in \mathcal{K}$, $\exists J_k \subseteq \mathcal{K}'$ such that $\mathcal{C}_k = \bigcup_{j \in J_k} \mathcal{C}_l'$.

Let $\mathcal{P}(\mathcal{C})$ and $U(\mathcal{F})$ denote respectively the sets of $\mathcal{C}$-advocated and $\mathcal{F}$-unanimous solutions. The following inclusions indicate that these sets are intermediate between U and $\mathcal{A}^*$.

Theorem 3 *(Ben Abdelaziz et al. 1994):*
Let $\mathcal{F} = \{\mathcal{C}_k \subseteq \mathcal{S} \times \mathcal{I} | k \in \mathcal{K}\}$. Then:
(i) $\bigcap_{k \in \mathcal{K}} \mathcal{P}(\mathcal{C}_k) \subseteq \mathcal{P}(\bigcup_{k \in \mathcal{K}} \mathcal{C}_k)$; (ii) $\mathcal{F}' \subseteq \mathcal{F} \Rightarrow U(\mathcal{F}') \subseteq U(\mathcal{F})$ and (iii) $U^* \subseteq U(\mathcal{F}) \subseteq \mathcal{P}(\mathcal{C}_k) \subseteq \mathcal{A}\ \forall k \in \mathcal{K}$.

The practical identification of sets $\mathcal{P}(\mathcal{C})$ and $U(\mathcal{F})$ requires applying the multiobjective simplex algorithm over selections of objective rows on problem (4). Particular such sets, and their computation, are illustrated in Ben Abdelaziz et al. (1994).

5 Distributional efficiency

We now consider another polar state of information case where a known probablity $p_s > 0$ is associated with each scenario $s \in S$, and the DM has a utility function u in some class $\mathcal{U} \subseteq \mathcal{U}_1$. In this context, each element $u \in \mathcal{U}$ can be considered as the viewpoint of a potential DM and our concern will be in characterizing admissible solutions in this context: how admissible sets arising from well-known subclasses of $\mathcal{U}_1$ relate to each other; and to what extent admissible solutions can be identified. The answer to the latter question will turn out to be largely negative. We conclude by proposing a subset of admissible solutions which seems more relevant in a single DM context, and which, at least for some classes $\mathcal{U}$, has better computational prospects.

Particular subclasses of $\mathcal{U}_1$ which have elicited considerable interest are the classes of utility function exhibiting *utility independence* (Keeney and Raiffa 1976), which take on one of the following three forms:

$\mathcal{U}_a = \{u \in \mathcal{U}_1 | u(z) = \sum_{i=1}^{m} u_i(z_i),\ u_i \geq 0 \text{ non-decreasing } \forall i \in \mathcal{I}\}$;

$\mathcal{U}_{m1} = \{u \in \mathcal{U}_1 | u(z) = -\prod_{i=1}^{m} -u_i(z_i),\ u_i \leq 0 \text{ non-decreasing } \forall i \in \mathcal{I}\}$;

$\mathcal{U}_{m2} = \{u \in \mathcal{U}_1 | u(z) = \prod_{i=1}^{m} u_i(z_i),\ u_i \geq 0 \text{ non-decreasing } \forall i \in \mathcal{I}\}$;

For the classes $\mathcal{U}_1$, $\mathcal{U}_a$, $\mathcal{U}_{m1}$ and $\mathcal{U}_{m2}$, $\mathcal{U}$-dominance implies a form of stochastic dominance; the implied comparison of distributions, of varying difficulty, is in general quite cumbersome even for the simplest class. Thus, the case $\mathcal{U} = \mathcal{U}_1$ refers to first degree multivariate stochastic dominance; Lehmann (1955), then Levhari et al. (1975) have shown that a solution y $\mathcal{U}_1$-dominates a solution x if and only if $P\{\tilde{C} \cdot x \in B\} \leq P\{\tilde{C} \cdot y \in B\}$ for all $B \in \mathcal{B}$, where $\mathcal{B}$ is the class of decreasing Borel sets (the sets in $\Re^m$ whose characteristic function is non-increasing. Thus the set of $\mathcal{U}_1$-admissible solutions is:
$\mathcal{A}(\mathcal{U}_1) = \{x \in \mathcal{X} | \forall y \in \mathcal{X}\ ,\ [\exists B \in \mathcal{B} : P\{\tilde{C} \cdot x \in B\} < P\{\tilde{C} \cdot y \in B\}]$ or $[P\{\tilde{C} \cdot x \leq a\} \leq P\{\tilde{C} \cdot y \leq a\},\ \forall a \in \Re^m]\}$

For the classes $\mathcal{U}_{m1}$ and $\mathcal{U}_{m2}$, Mosler (1984), showed that $\mathcal{U}$-dominance is equivalent to stochastic dominance in terms of joint distributions, so that:
$\mathcal{A}(\mathcal{U}_{m1}) = \{x \in \mathcal{X} | \forall y \in \mathcal{X},\ [\exists a \in \Re^m : P\{\tilde{C} \cdot x \leq a\} < P\{\tilde{C} \cdot y \leq a\},\ \forall a \in \Re^m]\}$, and $\mathcal{A}(\mathcal{U}_{m2}) = \{x \in \mathcal{X} | \forall y \in \mathcal{X},\ [\exists a \in \Re^m : P\{\tilde{C} \cdot x \gg a\} \geq P\{\tilde{C} \cdot y \gg a\},\ \forall a \in \Re^m]\}$.
Levy and Paroush (1974) showed that $\mathcal{U}_a$-dominance can be verified by comparing marginal distributions, so that:

$\mathcal{A}(\mathcal{U}) = \{x \in \mathcal{X} | \forall y \in \mathcal{X}, [\exists i \in \mathcal{I} \text{ and } a \in \Re : P\{\tilde{C} \cdot x \leq a\} < P\{\tilde{C} \cdot y \leq a\}$ or $[\forall i \in \mathcal{I}, \forall a \in \Re, P\{\tilde{C} \cdot x \leq a\} \leq P\{\tilde{C} \cdot y \leq a\}]\}$. We have the following inclusions:

Theorem 4 *(Ben Abdelaziz 1992):*
(i) $\mathcal{A}(\mathcal{U}_1) \subseteq \mathcal{A}^*$; *(ii)* $\mathcal{A}(\mathcal{U}_{m1}) \subseteq \mathcal{A}(\mathcal{U}_1)$; *(iii)* $\mathcal{A}(\mathcal{U}_{m2}) \subseteq \mathcal{A}(\mathcal{U}_1)$.

As suggested by the implied comparisons of distributions in the four cases above, the sets $\mathcal{A}(\mathcal{U})$ are not generally easy to identify. The following special case of ordered scenarios is the main exception known to us.

Definition 7 *Problem (3) exhibits increasing scenarios if scenario indices can be ordered so that:* $s \in S,\ t \in S,\ s < t \Rightarrow C^s \cdot x \leq C^t \cdot x,\ \forall x \in \mathcal{X}$.

Theorem 5 *(Ben Abdelaziz et al. 1992):*
Under increasing scenarios, $\mathcal{A}(\mathcal{U}_{m1}) = \mathcal{A}(\mathcal{U}_1) = \mathcal{A}^*$.

Thus, in the case of increasing scenarios, the $\mathcal{U}_{m1}$- or $\mathcal{U}_1$-admissible solutions are the Pareto-efficient solutions of (2), independently of any probability distribution, and can be quite easily identified. A better grasp of more general cases is permitted by the notion of "order-preserving zones", namely subsets of $\mathcal{X}$ over which the order of realizations $\{C_i^s x \mid s \in \mathcal{S}\}$ is constant for all criteria $i \in \mathcal{I}$. It turns out (Ben Abdelaziz 1992) that within each order-preserving zone, properties similar to the case of increasing scenarios hold, but that interferences between zones are a major complication.

In a context involving a single DM, however, not all admissible solutions are necessarily of equal relevance: to the extent that some point of view $u \in \mathcal{U}$ does indeed represent the DM's preference structure, only the solutions which are optimal (in an expected utility sense) under some such utility function need to be considered. This motivates the introduction of the following prominent sets.

Definition 8 *(i) The set of distinctly $\mathcal{U}$-advocated solutions is:* $\mathcal{P}^0(\mathcal{U}) = \{x \in \mathcal{X} | \exists u \in \mathcal{U}$ *such that* $\forall y \in \mathcal{X}$, $E[u(\tilde{C}x)) \geq E[u(\tilde{C}y))\}$. *(ii) The set of distinctly $\mathcal{U}$-advocated solutions is:* $\mathcal{P}^0(\mathcal{U}) = \{x \in \mathcal{X} | \exists u \in \mathcal{U}$ *such that* $\forall y \in \mathcal{X}$, $E[u(\tilde{C} \cdot x)) > E[u(\tilde{C} \cdot y))\}$. *(iii) The set of $\mathcal{U}$-advocated solutions is:* $\mathcal{P}(\mathcal{U}) = \mathcal{P}^0(\mathcal{U}) \cup U(\mathcal{U})$.

The set $\mathrm{P}(\mathcal{U})$, however, gathers all solutions of potential interest from the perspective of a single DM. Naturally, these solutions remain admissible.

Theorem 6 *(Ben Abdelaziz et al. 1995a):* $U(\mathcal{U}) \subseteq \mathcal{P}(\mathcal{U}) \subseteq \mathcal{A}(\mathcal{U})$.

Operational methods to generate advocated solutions or to test whether a given solution is advocated or unanimous for the case where $\mathcal{U}$ is the class of non-decreasing additive concave functions are provided in Ben Abdelaziz et al. (1995a,b).

6 Conclusion

In contrast to deterministic multiobjective problems, where the notion of Pareto-efficiency is a well-accepted unifying principle, several notions of efficiency may arise for a MSLP. In a decision context, their appropriateness depends in a large part on our information state about the problem. This paper has considered several polar cases of such information states. Somewhat surprisingly admissible sets under wide indeterminacy are more easily identified. The efficiency concepts proposed here seem relevant not only for the MSLP, but also for multi-criteria problems under uncertainty with more general (e.g. non-polyhedral or discrete) action space, although issues of computatability in such settings remain open. Also remaining to be investigated are the many intermediate forms of incomplete information about the MSLP that can arise in many practical situations.

References

Ben Abdelaziz F. (1992): L'efficacité en programmation multiobjectifs stochastique. Ph.D. Thesis, Université Laval, Québec, Canada.

Ben Abdelaziz F., P. Lang, R. Nadeau (1994): "Pointwise Efficiency in Multiobjective Stochastic Linear Programming". Journal of Operational Research Society 45, 11, 1325-1334.

Ben Abdelaziz F., P. Lang, R. Nadeau (1995a): "Distributional Efficiency in Multiobjective Stochastic Linear Programming". European Journal of Operational Research 85, 399-415.

Ben Abdelaziz F., P. Lang, R. Nadeau (1995b): "Distributional Unanimity in Multiobjective Stochastic Linear Programming". To appear in the Proceedings of the XII Conference on MCDA Coimbra, Portugal.

Bitran G. R. (1980): "Linear Multiple Objective Problems with Interval Coefficients". Management Science 26 no.7, 694-706.

Chankong V., Y.Y. Haimes (1983): *Multiobjective Decision Making, Theory and Methodology*. North Holland.

Goicoechea A., L. Duckstein, M.M. Fagel (1979): "Multiple Objectives Under Uncertainty: An Illustrative Application of PROTRADE". Water Resources Research 15, 203-210.

Keeney R., H. Raiffa (1976): *Decision with Multiple Objectives: Preferences and Value Trade Offs*. Wiley, New York.

Lehmann E.L. (1955): "Ordered Families of Distributions". Annals of Mathematical Statistics 26, 399-419.

Levhari D., J. Paroush, B. Peleg (1975): "Efficiency Analysis for Multivariate Distributions". Review of Economic Studies 42, 87-91.

Levy H., J. Paroush:"Multi-Period Stochastic Dominance", Management Science 21, 428-435

Marti K. (1990): "Computation of Efficient Solutions of Stochastic Optimization Problems with Applications to Regression and Scenario Analysis", pp. 163-188 in R. Slowinski and J. Teghem (eds.).

Mosler K.C. (1984): "Stochastic Dominance Decision Rules when the Attributes are Utility Independent". Management Science 30, 1311-1322.

Teghem J., D. Dufrane, M. Thauvoye (1986): "Strange: An Interactive Method for Multi-Objective Linear Programming Under Uncertainty". European Journal of Operational Research 26, 65-82.

Slowinski R. and J. Teghem (1990), *Stochastic Versus Fuzzy Approaches to Multiobjective Mathematical Programming Under Uncertainty*, Kluwer Academic Publishers.

Urli B., R. Nadeau (1990): "Multiobjective Stochastic Linear Programming With Incomplete Information: A General Methodology", pp. 131-161 in R. Slowinski and J. Teghem (eds.).

Weber R. (1985): "Decision Making Under Uncertainty: A Semi-Infinite Programming Approach". European Journal of Operational Research 19, 104-113.

Weber R. (1987): "Decision Making With Incomplete Information". European Journal of Operational Research 28, 44-57.

Wolf H. (1985): "Die Ermittlung effizienter Lsungen zur stochastischen linearen Optimierungsaufgabe". OR Spektrum 7, 81-90.

A Multi-Criteria Decision Making Problem Associated to Preference Modelling

Jacinto González-Pachón and Mª Isabel Rodríguez-Galiano[1]

[1] Department of Artificial Intelligence, Universidad Politécnica de Madrid. 28660-Boadilla del Monte. Madrid. Spain

Abstract. In this paper, we seek to show the link between inconsistencies in pairwise comparison during a preference modelling process and the conflicting criteria on a Decision Maker's mind when choosing between alternatives. Using a group decision-making technique for solving the multi-criteria decision making problem that arose, we obtain a methodology for addressing intransitivities in preference modelling.

Keywords. Preference Modelling, Pairwise Comparisons, Intransitivities, Borda's Method.

1 Introduction

In Decision Theory, the method of pairwise comparisons is very popular for choosing the optimal alternative when such alternatives are not physically measured. Pairwise comparisons provide a relative evaluation of every pair of alternatives, modelled by a binary relation, S:

$$aSb \text{ if and only if "} a \text{ is at least as good as } b\text{"}$$

If S satisfies certain formal properties, it is possible to use algebraic equations to define it.

For instance, if A is a finite set of alternatives and S is reflexive, transitive and complete -i.e., S is a weak order-, then there is an ordinal value function, $v: A \longrightarrow I\!R$, such that

$$aSb \text{ if and only if } v(a) \geq v(b)$$

If S is reflexive and transitive -i.e., S is a quasi order (q.o.)-, there is a finite family of ordinal value functions over A, $\{v_i\}_{i=1}^{n}$ $v_i : A \longrightarrow I\!R$, such that

$$aSb \text{ if and only if } v_i(a) \geq v_i(b) \; \forall i = 1, \ldots, n$$

For more details about this last result see Ríos Insua (1990). In this paper, a q.o. will be used to model imprecisions in a Decision Maker's (DM) answers;

i.e., in this case, when comparing two alternatives a and b, the DM's standard answer is "I don't know whether I prefer alternative a or alternative b". For a complete study of different kinds of preference structures using other properties for S, see Roubens and Vincke (1985) or, more recently, Abbas et al. (1996).

A partial order -i.e., a non-complete binary relation- often arises if the DM is stating preferences among alternatives using different aspects or dimensions, Roberts (1979). This means that the DM has different criteria on his/her subconscious mind when choosing between two alternatives, and mostly they are in conflict with each other. So, there is a link between incomparabilities in pairwise comparisons and a multi-criteria decision-making problem. This link is shown explicitly in the analytical q.o's representation shown above.

Sometimes, the DM forces him/herself to answer as many questions as he/she can, because people attach more importance to the act of deciding than to the consequence of their decision, French (1986). So, inconsistencies, such as intransitivities, could be more frequent in his/her answers than incomparabilities. Therefore, we consider only a reflexive binary relation S; i.e., S may be non-complete or non-transitive. However, a prespecified model or a prescriptive application requires transitive data. As a result, researchers are often interested in establishing a transitive order.

In this paper, we seek to show a possible link between inconsistencies in DM answers and a multi-criteria decision making problem. In our case, criteria will be defined by ordinal scales.

In Section 2, we present the Mixture of Maximal Quasi Orders (MMQO) as the tool for using intransitivities as new source of information in preference modelling. These maximal q.o.s define a multi-criteria decision making problem associated with the reflexive binary relation S. Additionally, it is possible to define an importance weight for each q.o. in the Mixture from S and its associated MMQO, González-Pachón et al. (1995).

In Section 3, we discuss an ordinal aggregation rule -Borda's Method- for solving the multi-criteria decision-making problem arising from the MMQO, Hwang et al. (1986). Thus, a new methodology for solving intransitivities is obtained, using these intransitivities as source of information.

Finally, we describe some conclusions

2 Mixture of Maximal Quasi Orders (MMQO) and the Approximation Index

Let us give some definitions useful for presenting the MMQO associated with a reflexive binary relation S.

Definition 1 *Let A be a finite set, S a binary reflexive relation on A and $V = \{S_i\}_{i \in I}$ the ordered set of all q.o.s included in S, using inclusion as the order. $S^* \in V$ is a maximal q.o. included in S, if there is no $S' \in V$, such that $S^* \subset S'$.*

Definition 2 *The family of all maximal q.o.s included in a binary relation S will be referred to as* **the Mixture of Maximal Quasi Orders (MMQO) associated with a binary relation** *S and will be denoted by $\mathcal{F}$.*

The following results appear in González-Pachón et al. (1995)

Proposition 1 *If S is a reflexive binary relation on a finite set A, the associated MMQO, $\mathcal{F}$, is not empty and covers S, i.e.,*

$$\mathcal{F} \neq \emptyset \ \ and \ \ S = \bigcup_{Q \in \mathcal{F}} Q$$

The following theorem shows an ordinal value model associated with a general binary reflexive relation. We seek to combine our initial approach to the problem with ordinal values associated with alternatives. Before hand, we firstly recall the following proposition, suggested by Ríos Insua (1990), useful for proving the theorem.

Proposition 2 *Let A be a finite set and S a binary relation on A. S is a q.o. iff there are r ordinal value functions $v_1, ..., v_r$ such that*

$$(a_i, a_j) \in S \Leftrightarrow v_l(a_i) \geq v_l(a_j) \ \ l = 1, .., r$$

for $a_i, a_j \in A$ and some $r \in \mathbf{N}$.

Theorem 1 *Let A be a finite set and S a binary reflexive relation on A. There is a finite set of families of ordinal value functions $\mathcal{V} = \{V_i\}_{j=1}^{q}$ such that*

$$(x, y) \in S \Leftrightarrow \exists\, j \in \{1, ..., q\} \ such\ that \ v(x) \geq v(y) \ \ \forall v \in V_j$$

Example

Suppose we have the binary relation S defined by the graph in Fig. 1 for representing a DM's preferences in a pairwise comparison of the set of alternatives $A = \{a, b, c, d\}$.

The MMQO associated with S is shown in Fig. 2 as Hesse diagrams.

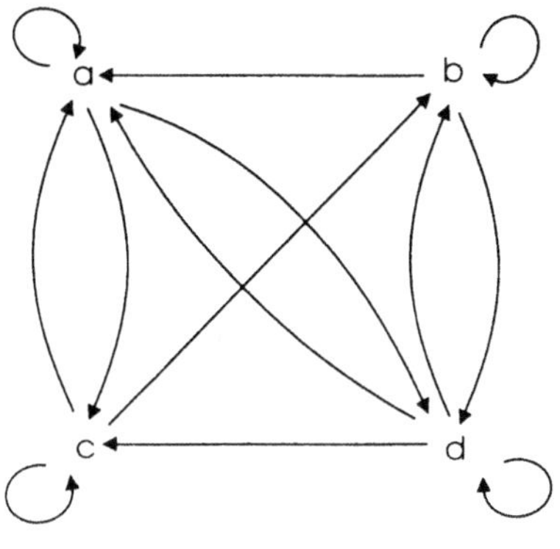

Fig. 1. Binary relation S

We seek therefore is to assign a weight to each q.o. in the MMQO. This weight can be interpreted as the q.o.'s level of importance in the Mixture. The "closer" a q.o. is to S, the higher the value of the above weight.

The concept of "proximity" is given by the cardinal of the q.o.: the higher the cardinal, the greater its proximity to S.

Definition 3 *Let S be a reflexive binary relation on a finite set A and $\mathcal{F} = \{Q_i\}_{i=1}^q$ its associated MMQO. The* **Approximation Index** *associated with a q.o. included in $\mathcal{F}$, Q_i, is defined by*

$$w_i = \frac{c_i}{\sum_{i=1}^{q} c_i}$$

where c_i is the cardinal of Q_i, $i = 1, .., q$.

This definition gives us a hierarchical ordering of q.o.s in the MMQO. This hierarchy is based on q.o.s with the same Approximation Index; i.e., q.o.s with the same cardinal.

Example

In our example, we have four levels, L_1, L_2, L_3 and L_4 with cardinals 10, 9, 8 and 7, respectively (here loops are considered). The level L_1 is formed by the first q.o. Quasi orders 2 and 3 belong to L_2 and 4, 5 and 6 to L_3. Finally, L_4 consists of 7 and 8. Additionally, the cardinal of S is 13.

Therefore, the q.o. of L_1 is defined with 77% $(=100\times\frac{10}{13})$ of the information expressed by the DM using S. So, the q.o.s of L_2, L_3 and L_4 are defined with 69%, 61% and 54% of the above information, respectively.

The weight associated with the q.o. of the first level, L_1, is $w_1 = 10/66$. Weights associated with the q.o. of the second level, L_2, are $w_2 = w_3 = 9/66$. For the third and fourth levels, the weights are $w_4 = w_5 = w_6 = 8/66$ and $w_7 = w_8 = 7/66$, respectively.

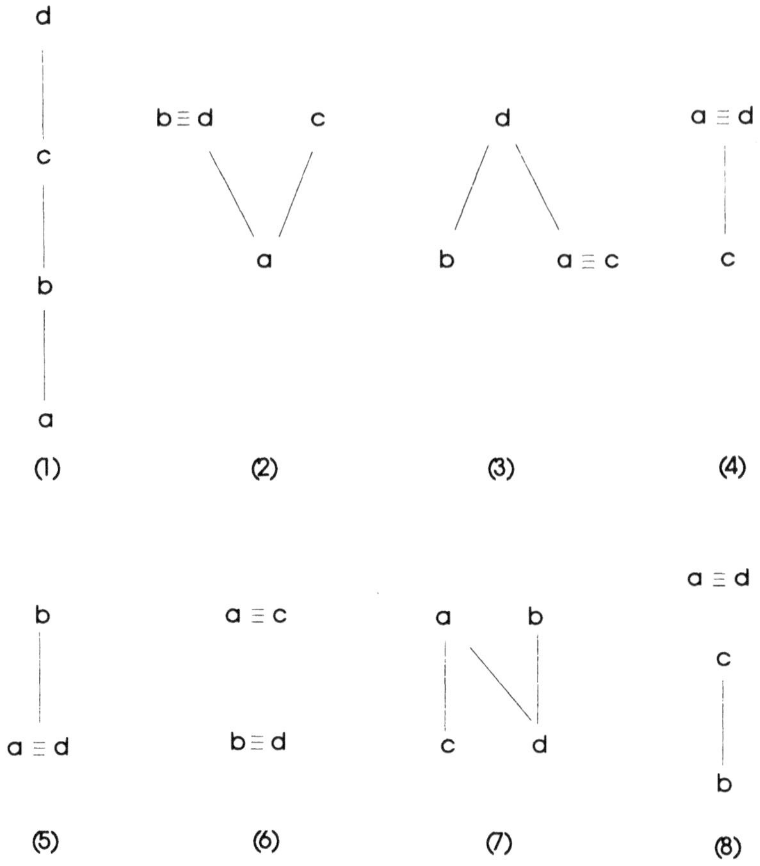

Fig. 2. MMQO associated with S

3 The Associated Multi-Criteria Decision-Making Problem

We are going to describe how to solve the ordinal multi-criteria decision-making problem described in Theorem 1, using Borda's method.

The Borda rule is a well-known method in social choice theory. Its purpose is to aggregate the information contained in a set of binary relations in order to obtain a ranking (weak order), see Hwang et al. (1986).

Consider a finite set I of individuals and a finite set A of alternatives. Let $n = |A|$ be the number of alternatives. Each individual $i \in I$ is associated with a preference Q_i over A, which is a linear ordering of the elements of X, that is, Q_i is complete, transitive and asymmetric binary relation on A.

Given two alternatives a and b, the number of individuals who prefer a to b is $|\{i \in I : aQ_ib\}|$.

The Borda function in a is defined as the number

$$f_B(a) = \sum_{b \in A} |i \in I : aQ_ib|$$

and the alternatives are ranked in the order of the values of f_B.

In this paper we are concerned with aggregation of q.o.s. Therefore, we will use the Borda function in the context of partial orders. Additionally, we seek to identify our ordinal multi-criteria decision-making problem with a group decision-making problem, to solve the former using Borda's method.

Let $\mathcal{F} = \{Q_i\}_{i=1}^q$ be the MMQO associated with a reflexive binary relation, S, on a finite set A and $\{w_i\}_{i=1}^q$ their Approximation Indices. This context may be considered equivalent to: Suppose a set of k individuals, where n_i $(i = 1, \ldots, q)$ have an associated preference Q_i $(i = 1, \ldots, q)$ over A. We may choose $k = \sum_{i=1}^{q} c_i$ and $n_i = c_i$ $i = 1, \ldots, q$, where $c_i = |Q_i|$. So, $k = \sum_{i=1}^{q} n_i$ and $n_i = kw_i$, where w_i is the Approximation Index of Q_i.

In this case, Borda's function gives a ranking over A. Therefore, intransitivities in S are used as information through the MMQO concept for obtaining a linear order over A. This method contrasts with others used for resolving intransitivities, Slater (1961), Mass et al. (1995).

Example

In our example, we consider $k = 66$ individuals $(\sum_{i=1}^{8} c_i = 66)$, where 10 $(= n_1)$ are associated with the first q.o., 9 $(= n_2)$ with the second, 9 $(= n_3)$ with the third, 8 with the fourth, fifth and sixth, and 7 with the seventh and eighth.

Using this context, the Borda score for a is

$$f_B(a) = 10 \times 1 + 9 \times 1 + 9 \times 2 + 8 \times 3 + 8 \times 2 + 8 \times 2 + 7 \times 3 + 7 \times 2 = 128$$

(loops are considered for the score)
The Borda score for b is

$$f_B(b) = 10 \times 2 + 9 \times 3 + 9 \times 1 + 8 \times 3 + 8 \times 2 + 8 \times 1 + 7 \times 2 + 7 \times 1 = 125$$

The Borda score for c is

$$f_B(c) = 10 \times 3 + 9 \times 2 + 9 \times 2 + 8 \times 1 + 8 \times 1 + 8 \times 2 + 7 \times 1 + 7 \times 2 = 119$$

And, finally, the Borda score for d is

$$f_B(d) = 10 \times 4 + 9 \times 3 + 9 \times 4 + 8 \times 3 + 8 \times 2 + 8 \times 2 + 7 \times 1 + 7 \times 2 = 180$$

The alternatives are ranked in the following order using the Borda score

$$d \succ a \succ b \succ c$$

4 Conclusions

DM inconsistencies in pairwise comparison are a potential source of information in preference modelling. This information is obtained using the MMQO concept. The MMQO associated with a reflexive binary relation can be interpreted as criteria hidden in DM's brain and used for choosing between alternatives. This fact led us to consider an ordinal multi-criteria decision-making problem. Addressing this problem as a group decision-making problem, Borda's function gave us a ranking of the alternatives.

Acknowledgment

The first author would like to acknowledge the partial support from the project F.I.S 95/0055-02.

References

Abbas, M., Pirlot, M. and Vincke, P. (1996) 'Preference Structures and Co-comparability Graphs'. *Journal of Multicriteria Decision Analysis*, 5 (2) pp. 81-98.

French (1986) 'Decision Theory, *Ellis Horwood.*

González-Pachón, J. and Ríos-Insua, S. (1995) 'Mixture of Maximal Quasi Orders: A New Approach to Preference Modelling'. (submitted)

Hwang C-L. and Lin, M-J (1986) 'Group Decision Making under Multiple Criteria. (LNEMS 281). *Springer.*

Mass, A., Bezembinder, T. and Wakker, P. (1995) 'On Solving Intransitivities in Repeated Pairwise Choices', *Mathematical Social Sciences*, 29, pp. 83-101.

Ríos-Insua, D. (1990) 'Sensitivity Analysis in Multiobjective Decision Making, (LNMES 347), *New York: Springer.*

Roberts,F. (1972) 'What if Utility Functions do not exist?', *Theory and Decision*, 3, pp. 126-139.

Roberts, F. (1979) 'Measurement Theory', *Addison Wesley.*
Roubens, M. and Vincke, Ph. (1985) 'Preference Modelling', (LNEMS 250), *New York: Springer.*
Slater, P. (1961) 'Inconsistencies in a Schedule of Paired Comparisons', *Biometrica* 48, pp. 303-312.

On Optimality and Duality in Multiobjective Nonsmooth Programming

Vasile PREDA[1] and I.M. STANCU-MINASIAN[2]

[1] Mathematics Faculty, University of Bucharest, 14 Academiei Str., Bucharest, ROMANIA
[2] The Romanian Academy, Centre of Mathematical Statistics, 76100 Bucharest, ROMANIA

Abstract. A multiobjective nonsmooth programming problem is studied. By considering the concept of weak maximum, some general necessary and sufficient optimality conditions are obtained in terms of classical Dini-derivatives. Also, a general Mond-Weir dual is considered and some duality results are given.

Keywords: Nonsmooth programming, weak maximum, optimality conditions, duality theorems.

1. Introduction

In the last time the interest in optimality conditions and duality for nonsmooth single - and multiobjective programming problems is growing. Diewert [4] using classical Dini-derivatives, developed some very simple necessary optimality conditions for nonsmooth single-objective programming problems and showed that these conditions are also sufficient for pseudoconcave objective functions and quasiconvex constraint functions without assuming that one-sided directional derivatives exist.

The results of Diewert [4] are extended to multiobjective nonsmooth programming problems by Bhatia and Aggarwal [3] where the concepts of efficient and proper efficient solutions are studied.

Weir and Mond [10], by considering the concept of weak minima extended different scalar duality results to multiobjective programming problems under a variety of generalized convexity conditions.

On this line and using the approach of Diewert [4], in Bector, Bhatia and Jain [2], by considering the concept of weak maximum, necessary and sufficient conditions, for a nonsmooth multiobjective programming problem, have been developed. Also the duality results have been obtained for Wolfe dual in terms of classical Dini-derivatives.

Preda and Stancu-Minasian [9] introduced a general Mond-Weir dual for nonlinear nonsmooth multiobjective programming problem and duality results are given.

In this paper, by considering the concept of weak maximum, some general necessary and sufficient optimality conditions for multiobjective nonsmooth programming problem are obtained.

Also, a general Mond-Weir dual problem is considered and some duality results are given. Thus we obtain generalizations of the results obtained by Egudo [5], Diewert [4], Bhatia and Aggarwal [3] and Preda and Stancu-Minasian [9].

2. Definitions and Preliminaries

We consider the following nonsmooth multiobjective programming problem.

$$(\mathrm{P}) \qquad \begin{array}{l} \textit{Maximize } f(x) \\ \textit{subject to } x \in X = \{x \in X_0; g(x) \leqq 0\} \end{array}$$

where X_0 is a convex subset of $\mathbf{R}^n$, $f : X_0 \longrightarrow \mathbf{R}^k$, $g : X_0 \longrightarrow \mathbf{R}^m$ with $f(x) = (f_1(x), ..., f_k(x))$, $g(x) = (g_1(x), ..., g_m(x))$, are nondifferentiable proper functions.

Associated to problem (P) is the following single-objective programming problem:

$$(P_\lambda) \qquad \begin{array}{ll} \textit{Maximize} & \lambda^T f(x) \\ \textit{subject to} & x \in X \end{array}$$

where $\lambda \in \mathbf{R}^k_+$ is a preassigned vector.

Definition 2.1 (Diewert [4]) A function $h : \mathbf{R}^n \longrightarrow \mathbf{R}$ is said to be a proper function if $h(x) < +\infty$ for all $x \in \mathbf{R}^n$ and $h(x) > -\infty$ for at least one $x \in \mathbf{R}^n$.

Definition 2.2 (Diewert [4]) The set of feasible directions at a point $x \in \mathbf{R}^n$ for (P) is defined as the set

$F(x) = \{v | v \in \mathbf{R}^n, v^T v = 1, \text{there exists } \delta > 0 \text{ such that } g(x + tv) \leqq 0\ ,$ $0 \leqq t \leqq \delta\}$

Definition 2.3 (Diewert [4]) Let $h : \mathbf{R}^n \longrightarrow \mathbf{R}$. Let $x \in \mathbf{R}^n$, $v \in \mathbf{R}^n$ with $v^T v = \sum_{i=1}^n v_i^2 = 1$. The Dini-derivatives of h at x in the direction v are defined as

$$D_v^{+u} h(x) = \sup_{\{t_n\}} \lim_{n\to\infty} \left\{ \frac{h(x + t_n v) - h(x)}{t_n}; 0 < t_n \leqq \frac{1}{n} \right\}$$

$$D_v^{+l} h(x) = \inf_{\{t_n\}} \lim_{n\to\infty} \left\{ \frac{h(x + t_n v) - h(x)}{t_n}; 0 < t_n \leqq \frac{1}{n} \right\}$$

$$D_v^{-u} h(x) = \sup_{\{t_n\}} \lim_{n\to\infty} \left\{ \frac{h(x - t_n v) - h(x)}{-t_n}; 0 < t_n \leqq \frac{1}{n} \right\}$$

$$D_v^{-l} h(x) = \inf_{\{t_n\}} \lim_{n\to\infty} \left\{ \frac{h(x - t_n v) - h(x)}{-t_n}; 0 < t_n \leqq \frac{1}{n} \right\}$$

Infinite limits are also allowed in the above definitions. Here $D_v^{+u}h(x)$ is the upper right derivative, $D_v^{+l}h(x)$ is the lower right derivative, $D_v^{-u}h(x)$ is the upper left derivative, $D_v^{-l}h(x)$ is the lower left derivative of h evaluated at x in the direction of v.

According to Diewert [4] and Bector et al. [2] we have

$D_v^{+u}h(x) \geqq D_v^{+l}h(x)$; $D_v^{-u}h(x) \geqq D_v^{-l}h(x)$;

$D_v^{-u}h(x) = -D_{-v}^{+l}h(x)$; $D_v^{-l}h(x) = -D_{-v}^{+u}h(x)$;

$D_v^{+u}(-h(x)) = -D_v^{+l}h(x)$; $D_v^{-u}(-h(x)) = -D_v^{-l}h(x)$;

$D_v^{+u}(\alpha_1 h_1(x) + \alpha_2 h_2(x)) \leqq \alpha_1 D_v^{+u}h_1(x) + \alpha_2 D_v^{+u}h_2(x)$,

for $\alpha_1, \alpha_2 \in R$ and $h_1, h_2 : X_0 \longrightarrow R$.

Definition 2.4 (Diewert [4]) The function $h : X_0 \longrightarrow \mathbf{R}$ is said to be a pseudoconcave function on X_0 iff $x^0 \in X_0$, $v \in \mathbf{R}^n$, satisfying $v^T v = 1$, for every $t > 0$, $x^0 + tv \in X_0$, $D_v^{+u}h(x^0) \leqq 0$ implies $h(x^0 + tv) \leqq h(x^0)$, or equivalently $h(x^0 + tv) > h(x^0)$ implies $D_v^{+u}h(x^0) > 0$.

Lemma 2.1 [3] Let x^0 be a feasible solution for problem (P) and let $B(x^0)$ be the set of binding constraints (i.e. $B(x^0) = \{j;\ g_j(x^0) = 0\}$). Then,

$$\left.\begin{array}{ll} D_v^{+u}g_j(x^0) < 0 & \text{for } j \in B(x^0) \\ D_v^{+u}g_j(x^0) < \infty & \text{for } j \notin B(x^0) \end{array}\right\} \text{ implies } v \in F(x^0).$$

3. Optimality Conditions

In this section some necessary and sufficient optimality conditions for the problem (P) are given.

Theorem 3.1. *(Necessary conditions) Let x^0 be a weak maximum solution for (P). Assume that the conditions of Lemma 2.1 are satisfied at x^0. Let L and M be two sets such that $L \cup M = \{1, 2, \ldots m\}$ and $L \cap M = \emptyset$. Then there exist $v \in F(x^0)$, $\lambda \in \mathbf{R}^k$, $y \in \mathbf{R}^m$ such that*

$$D_v^{+u}(\lambda^T f - y_L^T g_L)(x^0) - D_v^{+l}(y_M^T g_M)(x^0) \leqq 0 \tag{3.1}$$

$$y^T g(x^0) = 0 \tag{3.2}$$

$$y \geqq 0 \tag{3.3}$$

$$\lambda \in R^k,\ \lambda \geqq 0,\ \lambda^T e = 1 \tag{3.4}$$

where $e = (1, \ldots 1)^T \in \mathbf{R}^k$, $y_L = (y_i)_{i \in L}$, $g_L = (g_i)_{i \in L}$.

Proof: Using the line of Theorem 1 of [2] we get that there exist $\lambda \in \mathbf{R}_+^k, \lambda^T e = 1$, and $y \in \mathbf{R}_+^m$ such that for all $v \in \mathbf{R}^n$,

$$\sum_{i=1}^{k} \lambda_i D_v^{+u} f_i(x^0) v - \sum_{j=1}^{m} y_j D_v^{+l} g_j(x^0) v = 0 \tag{3.5}$$

Multiplying the equality (3.5) by v, since $v^T v = 1$ and $L \cup M = \{1, 2, \ldots m\}$, $L \cap M = \emptyset$ we obtain

$$\sum_{i=1}^{k} \lambda_i D_v^{+u} f_i(x^0) - \sum_{j \in L} y_j D_v^{+l} g_j(x^0) - \sum_{j \in M} y_j D_v^{+l} g_j(x^0) = 0$$

Now, using the sublinearity for D_v^{+u} and D_v^{+l} we get (3.1).

<u>**Remark:**</u> In the case $L = \emptyset$ we get the necessary optimality conditions given by Theorem 1 from [2].

Theorem 3.2. *(Sufficient Conditions) Let x^0 be a feasible solution for (P). Assume that there exist $y \in \mathbf{R}_+^m$ and $\lambda \in \mathbf{R}_+^k$ with $\lambda^T e = 1$ such that $\lambda^T f(x^0) - y^T g(x^0)$ is finite and the following conditions are satisfied:*

(a1) $\lambda^T f(\cdot) - y^T g(\cdot)$ is proper and pseudoconcave at x^0;

(a2) $D_v^{+u}(\lambda^T f(x^0)) - D_v^{+l}(y_L^T g_L(x^0)) - D_v^{+l}(y_M^T g_M(x^0)) \leqq 0$;

(a3) $y^T g(x^0) = 0$.

Then x^0 is a weak maximum for (P).

Proof: Using (a1), the property $D_v^{+u}(-\varphi(x^0)) = -D_v^{+l}(\varphi(x^0))$ and the sublinearity for D_v^{+u} we obtain :

$$\begin{aligned} 0 &\geqq D_v^{+u}(\lambda^T f(x^0)) - D_v^{+l}(y_L^T g_L(x^0)) - D_v^{+l}(y_M^T g_M(x^0)) = \\ &= D_v^{+u}(\lambda^T f(x^0)) + D_v^{+l}(-y_L^T g_L(x^0)) + D_v^{+l}(-y_M^T g_M(x^0)) \geqq \\ &\geqq D_v^{+u}(\lambda^T f(x^0) - y^T g(x^0)) = D_v^{+u}(\lambda^T f - y^T g)(x^0) \end{aligned}$$

Hence $D_v^{+u}(\lambda^T f - y^T g)(x^0) \leqq 0$. Now, the proof follows along the lines of Theorem 2 of [2].

Theorem 3.3. *If in Theorem 3.2 we replace (a1) by $(a1)'$: $\lambda^T f(\cdot)$ is proper and pseudoconcave at x^0, then the conclusion from Theorem 3.2 still holds.*

Proof: Let v be a direction, $t > 0$ such that $x^0 + tv$ is a feasible point for (P). Hence, for $i \in B(x^0)$, $g_i(x^0 + tv) \leqq g_i(x^0)$. Now, by the definition of D_v^{+u} we get:

$$D_v^{+u}(g_i(x^0)) \leqq 0 \text{ for any } i \in B(x^0) \tag{3.6}$$

Since $y \geqq 0$ and by (a3) we have $y_i = 0$ for $i \notin B(x^0)$. From (3.6) we obtain:

$$\sum_{i \in L} y_i D_v^{+u} g_i(x^0) + \sum_{i \in M} y_i D_v^{+u} g_i(x^0) \leqq 0 \tag{3.7}$$

Using (3.7) and the sublinearity of D_v^{+u} we get:

$$D_v^{+u}(y_L^T g_L(x^0)) + D_v^{+u}(y_M^T g_M(x^0)) \leqq 0 \tag{3.8}$$

Now, by (3.8), (a2) and $D_v^{+u} = -D_v^{+l}$ it results that $D_v^{+u}(\lambda^T f(x^0)) \leqq 0$. Since $\lambda^T f(\cdot)$ is pseudoconcave at x^0, we obtain that $\lambda^T f(x^0+tv) \leqq \lambda^T f(x^0)$ i,e., the relation (17) of [2]. Now, we proceed as in Theorem 2 of [2].

Theorem 3.4. *If in Theorem 3.2 we replace (a1) by:*

(a1)'' $\lambda^T f(\cdot) - y_{I_0}^T g_{I_0}$ is proper and pseudoconcave at $x^0$and for any k, $1 \leqq k \leqq \nu$, $y_{I_k}^T g_{I_k}(\cdot)$ is proper at x^0, where $(I_k)_{0 \leqq k \leqq \nu}$ is a partition of $\{1,2,\ldots,m\}$, i.e., $I_k \cap I_p = \emptyset$ for $k \neq p$ and $\bigcup_{k=0}^{\nu} I_k = \{1,2,\ldots,m\}$, then x^0 is still a weak maximum for (P).

Proof: We consider v and t defined as in Theorem 3.3. Since $y \geqq 0$ and $y_i = 0$ for $i \notin B(x^0)$, by the feasibility of $x^0 + tv$ we get

$$y_{I_k}^T g_{I_k}(x^0+tv) \leqq y_{I_k}^T g_{I_k}(x^0) \quad \text{for } 1 \leqq k \leqq \nu \tag{3.9}$$

Using (3.9) and the definition of D_v^{+u} for $y_{I_k}^T g_{I_k}(\cdot)$ at x^0 it results that:

$$D_v^{+u}(y_{I_k}^T g_{I_k}(x^0)) \leqq 0 \quad \text{for any } 1 \leqq k \leqq \nu \tag{3.10}$$

By $D_v^{+u} = -D_v^{+l}$, the sublinearity of D_v^{+u} and by (3.10) we obtain:

$$D_v^{+u}(\lambda^T f - y_{I_0}^T g_{I_0})(x^0) \leqq 0$$

Now, we proceed as in Theorem 3.3.

<u>**Remark:**</u> For $L = \emptyset$ we get the sufficient optimality conditions given by Bector et al. [2, Theorem 2].

4. Duality

Relative to the problem (P) we consider the following problem (D) which may be regarded as a general Mond-Weir dual for (P) in terms of Dini-derivatives.

(D): *Minimize* $(f(w) - y_{I_0}^T g_{I_0}(w)e)$

subject to

$$D_v^{+u}(\lambda^T f(w) - y_{I_0}^T g_{I_0}(w)) - D_v^{+u}(y_{I_0^C}^T g_{I_0^C}(w)) \leqq 0 \tag{c1}$$

$$y_{I_k}^T g_{I_k}(w) \geqq 0 \quad 1 \leqq k \leqq \nu \tag{c2}$$

$$y \geqq 0,\ y \in \mathbf{R}^m \tag{c3}$$

$$\lambda \in \mathbf{R}^k,\ \lambda \geqq 0,\ \lambda^T e = 1$$

where v is a direction vector and $I_0^C = \{1,2,\ldots,m\} \setminus I_0$.

Theorem 4.1. *(Weak Duality) Let x be feasible for (P) and (w, λ, y) be feasible for (D). We assume that $\lambda^T f(\cdot) - y_{I_0}^T g_{I_0}(\cdot)$ is pseudoconcave. Then $f(x) \leqq f(w) - y_{I_0}^T g_{I_0}(w)e$ for any $v \in \mathbf{R}^n$ with $v^T v = 1$, $0 < t \leqq \delta$ for some $\delta > 0$ such that $x = w + tv \in X$.*

Proof: We proceed by contradicting. Hence, there exists $x, (w, \lambda, y), v, t$ and δ such that $x = w + tv \in X$, $0 < t \leqq \delta$, $v \in \mathbf{R}^n$, $v^T v = 1, (w, \lambda, y)$ is feasible for (D) and

$$f(x) > f(w) - y_{I_0}^T g_{I_0}(w)e \tag{4.1}$$

Since $\lambda \geqq 0$, $\lambda^T e = 1$, by (4.1) we get:

$$\lambda^T f(x) > \lambda^T f(w) - y_{I_0}^T g_{I_0}(w) \tag{4.2}$$

Using $y \geqq 0$, feasibility of x and (4.2) we obtain:

$$\lambda^T f(x) - y_{I_0}^T g_{I_0}(x) > \lambda^T f(w) - y_{I_0}^T g_{I_0}(w)$$

Now, by the pseudoconcavity of $\lambda^T f(\cdot) - y_{I_0}^T g_{I_0}(\cdot)$ we get:

$$D_v^{+u}(\lambda^T f(w) - y_{I_0}^T g_{I_0}(w)) > 0 \tag{4.3}$$

From $(c2)$, $(c3)$ and form the feasibility of x for (P) it follows that:

$$y_{I_k}^T g_{I_k}(x)) \leqq y_{I_k}^T g_{I_k}(w)) \quad \text{for any } 1 \leqq k \leqq \nu$$

Now, from the definition of D_v^{+u} for $y_{I_k}^T g_{I_k}(\cdot)$ for any $1 \leqq k \leqq \nu$ we obtain:

$$D_v^{+u}(y_{I_k}^T g_{I_k}(w)) \leqq 0 \text{ for any } 1 \leqq k \leqq \nu \tag{4.4}$$

Using (4.4) and the sublinearity for D_v^{+u} we get that:

$$D_v^{+u}(y_{I_0^C}^T g_{I_0^C}(w)) \leqq 0 \tag{4.5}$$

By (4.5) and $(c1)$ we obtain:

$$D_v^{+u}(\lambda^T f(w) - y_{I_0}^T g_{I_0}(w)) \leqq 0$$

which contradicts (4.3). Hence, the theorem is proved.

Theorem 4.2. *(Strong Duality) Let x^0 be a weak maximum for (P) at which the constraints of (P) satisfy the conditions of Lemma 2.1. Then there exist $y \in \mathbf{R}^m$ and $\lambda \in \mathbf{R}^k$ such that (x^0, λ, y) is feasible for (D) along v. Furthermore, if for all x feasible for (P) and (w, λ, y) feasible for (D), $\lambda^T f(\cdot) - y_{I_0}^T g_{I_0} e$ is pseudoconcave, then (x^0, λ, y) is a weak minimum point for (D) along v.*

Proof: By Theorem 4.1 we have for $L = I_0$ and $M = I_0^C$ that there exist a direction $v \in F(x^0)$, and $y \in \mathbf{R}^m$, $\lambda \in \mathbf{R}^k$ such that

$$D_v^{+u}(\lambda^T f - y_{I_0}^T g_{I_0})(x^0) - D_v^{+l}(y_{I_0^C}^T g_{I_0^C})(x^0) \leqq 0 \tag{4.6}$$

$$y^T g\,(x^0) = 0 \tag{4.7}$$

$$y \geqq 0 \tag{4.8}$$

$$\lambda \geqq 0,\ \lambda^T e = 1 \tag{4.9}$$

Since $D_v^{+l} \leqq D_v^{+u}$, by (4.6) we get:

$$D_v^{+u}(\lambda^T f - y_{I_0}^T g_{I_0})(x^0) - D_v^{+u}(y_{I_0^C}^T g_{I_0^C})(x^0) \leqq 0 \tag{4.10}$$

From (4.7) we obtain:

$$y_{I_k}^T g_{I_k}(x^0)) \leqq 0 \text{ for any } 0 \leqq k \leqq \nu \tag{4.11}$$

Using (4.8)-(4.11) we have that (x^0, λ, y) is a feasible solution for (D) along v.

By (4.11) we note that the objective values of (P) at x^0 and of (D) at (x^0, λ, y) are equal.

Now we shall prove, on similar lines as in [2, p.77] that (x^0, λ, y) is a weak minimum for (D) along $v \in F(x^0)$. Suppose, *ad absurdum*, that (x^0, λ, y) is not a weak minimum for (D) along $v \in F(x^0)$. Then there exists a feasible solution (u^0, λ^0, y^0) for (D) along $v \in F(x^0)$ such that:

$$f(u^0) - y_{I_0}^{0T} g_{I_0}(u^0)e < f(x^0) - y_{I_0}^T g_{I_0}(x^0)e \tag{4.12}$$

where $x^0 = u^0 + t^0 v,\ y = y^0 + t^0 p,\ \lambda = \lambda^o + t^0 q$ with
$v = (x^0 - u^0)\left[(x^0 - u^0)^T(x^0 - u^0)\right]^{-1/2}$, $t^0 = \left[(x^0 - u^0)^T(x^0 - u^0)\right]^{-1/2}$
and $(v, p, q)^T(v, p, q) = 1$.

Since $y_{I_0}^T g_{I_0}(x^0) = 0$, $y^0 \geqq 0$ and x^0 is feasible for (P), by (4.12) we obtain:

$$f(u^0) - y_{I_0}^{0T} g_{I_0}(u^0)e < f(x^0) - y_{I_0}^{0T} g_{I_0}(x^0)e \tag{4.13}$$

Using $\lambda \geqq 0$, $\lambda^T e = 1$ and (4.13), it results that:

$$\lambda^{0T} f(u^0) - y_{I_0}^{0T} g_{I_0}(u^0) < \lambda^{0T} f(x^0) - y_{I_0}^{0T} g_{I_0}(x^0)$$

Now, by the pseudoconcavity of $\lambda^{0T} f(\cdot) - y_{I_0}^{0T} g_{I_0}(\cdot)$ we get:

$$D_v^{+u}(\lambda^{0T} f(u^0) - y_{I_0}^{0T} g_{I_0}(u^0)) > 0 \tag{4.14}$$

From (c2), (c3) for (u^0, λ^0, y^0) and from the feasibility of x^0 we obtain:

$$y_{I_k}^{0T} g_{I_k}(x^0) \leqq y_{I_k}^{0T} g_{I_k}(u^0) \quad \text{for any } 0 \leqq k \leqq \nu$$

Now, using the definition of D_v^{+u} for $y_{I_k}^{0T} g_{I_k}(\cdot)$, for $0 \leqq k \leqq \nu$, we obtain:

$$D_v^{+u}(y_{I_k}^{0T} g_{I_k}(u^0)) \leqq 0 \quad \text{for any } 0 \leqq k \leqq \nu \tag{4.15}$$

From (4.15) and (c1) for (u^0, λ^0, y^0) it results that:

$$D_v^{+u}(\lambda^{0T} f(u^0) - y_{I_0}^{0T} g_{I_0}(u^0)) \leqq 0$$

contradicting (4.14). Hence (x^0, λ, y) is a weak minimum for (D) along v.

Theorem 4.3. *(Converse Duality) Let (u^0, λ, y) be a weak minimum for (D) such that u^0 is feasible for (P). Also, we assume that $\lambda^T f(\cdot) - y_{I_0}^T g_{I_0}(\cdot)$ is pseudoconcave at u^0. Then u^0 is a weak maximum for (P).*

Proof: Since (u^0, λ, y) is feasible for (D) then (u^0, λ, y) is feasible for (D) along $v \in F(x^0)$. Let $t > 0$ be such that $u^0 + tv$ is feasible for (P) for any $0 < t \leqq \delta$. Also, we have:

$$y_{I_k}^T g_{I_k}(u^0 + tv) \leqq y_{I_k}^T g_{I_k}(u^0) \quad \text{for any } 0 \leqq k \leqq \nu$$

From here we obtain:

$$D_v^{+u}(y_{I_k}^T g_{I_k}(u^0)) \leqq 0 \quad \text{for any } 0 \leqq k \leqq \nu$$

Now, by (c1) for (u^0, λ, y) we get:

$$D_v^{+u}(\lambda^T f(u^0) - y_{I_0}^T g_{I_0}(u^0)) \leqq 0$$

Using the pseudoconcavity of $\lambda^T f(\cdot) - y_{I_0}^T g_{I_0}(\cdot)$ at u^0 it results:

$$\lambda^T f(u^0 + tv) - y_{I_0}^T g_{I_0}(u^0 + tv) \leqq \lambda^T f(u^0) - y_{I_0}^T g_{I_0}(u^0)$$

Now using the feasibility of u^0 and $u^0 + tv$ for (P) and (u^0, λ, y) for (D) along v we get:

$$\lambda^T f(u^0 + tv) \leqq \lambda^T f(u^0) \tag{4.16}$$

Now we proceed by contradicting. If u^0 is not a weak maximum for (P) then there exists a $v \in \mathbf{R}^n$ with $v^T v = 1$ such that $f(u^0 + tv) > f(u^0)$ for $0 < t \leqq \delta' \leqq \delta$, for some $\delta' > 0$. Since $\lambda \geqq 0$, $\lambda^T e = 1$ we get $\lambda^T f(u^0 + tv) > \lambda^T f(u^0)$ which contradicts (4.16). Thus the theorem is proved.

Now we consider a strict converse duality result of Mangasariam type for (P_λ) and (D).

Theorem 4.4. *(Strict Converse Duality) Let (u^0, λ, y) be a weak minimum for (D) and x^0 be a weak maximum for (P_λ) such that*

$$\lambda^T f(u^0) - y_{I_0}^T g_{I_0}(u^0)) \leqq \lambda^T f(x^0) \tag{4.17}$$

Also we assume that $\lambda^T f(\cdot) - y_{I_0}^T g_{I_0}(\cdot)$ is quasiconcave at u^0 and there exists s, $1 \leqq s \leqq v$, such that $y_{I_s}^T g_{I_s}(\cdot)$ is strictly pseudoconvex at u^0. Then $u^0 = x^0$, that is u^0 is a weak maximum for (P).

Proof: We suppose that $u^0 \neq x^0$. Since (u^0, λ, y) is feasible for (D), there exist $v \in \mathbf{R}^n$, $v^T v = 1$, $\delta > 0$ such that $u^0 + tv \in X$ for any $0 < t \leqq \delta$. By the feasibility of $u^0 + tv$ for (P) and (u^0, λ, y) for (D) we get:

$$y_{I_k}^T g_{I_k}(u^0 + tv) \leqq y_{I_k}^T g_{I_k}(u^0) \quad \text{for any } 0 \leqq k \leqq \nu \tag{4.18}$$

By (4.18) for $k = s$ and strict pseudoconvexity of $y_{I_s}^T g_{I_s}(\cdot)$ at u^0 we get:

$$D_v^{+u}(y_{I_s}^T g_{I_s}(u^0)) < 0 \tag{4.19}$$

From (4.18) for $1 \leqq k \leqq \nu$, $k \neq s$ and the definition of D_v^{+u} we obtain:

$$D_v^{+u}(y_{I_k}^T g_{I_k}(u^0)) \leqq 0 \quad \text{for any } 0 \leqq k \leqq \nu,\ k \neq s \tag{4.20}$$

Using (4.19), (4.20) and the sublinearity of D_v^{+u} we obtain:

$$D_v^{+u}(y_{I_0^C}^T g_{I_0^C}(u^0)) < 0$$

Now, by (c1) for (u^0, λ, y) we get

$$D_v^{+u}(\lambda^T f - y_{I_0}^T g_{I_0})(u^0) < 0 \tag{4.21}$$

Using (4.21) and the assumption of quasiconcavity of $\lambda^T f(\cdot) - y_{I_0}^T g_{I_0}(\cdot)$ at u^0 we get:

$$\lambda^T f(u^0 + tv) - y_{I_0}^T g_{I_0}(u^0 + tv) \leqq \lambda^T f(u^0) - y_{I_0}^T g_{I_0}(u^0)$$

for some $0 < t \leqq \delta''$. Now we obtain:

$$\lambda^T f(u^0 + tv) < \lambda^T f(u^0) - y_{I_0}^T g_{I_0}(u^0)$$

Now by (4.17) we get:

$$\lambda^T f(u^0 + tv) < \lambda^T f(u^0)$$

which contradicts the optimality of x^0 for (P_λ).

<u>**Remark 4.1:**</u> The dual problem (D) contains as special cases a number of interesting dual problems for problem (P). These duals can be identified by

appropriately specializing the partitioning sets $I_0, I_1, \ldots, I_r$ and by altering the assumptions in the above theorems accordingly.

Remark 4.2: The dual problem (D) is a Mond-Weir type dual [8].

References

[1] Aggarwal,S., D. Bhatia: Pseudolinearity and efficiency via Dini derivatives. Indian J. Pure Appl. Math. 20 (1989) 12, 1173-1183

[2] Bector, C.R., D. Bhatia, Pushp Jain: Generalized concavity and duality in multiobjective nonsmooth programming. Utilitas Math. 43 (1993) 71-78

[3] Bhatia, D., S. Aggarwal : Optimality and duality for multiobjective nonsmooth programming. European J. Oper. Res. 57 (1992) 360-367

[4] Diewert, W.E.: Alternative characterizations of six kinds of quasiconcavity in the non-differentiable case with applications to nonsmooth programming. In: S.Schaible and W.T.Ziemba (eds.): *Generalized Concavity in Optimization and Economics.* Academic Press, New York 1981, pp. 51-93

[5] Egudo, R.R.: Proper efficiency and multiobjective duality in nonlinear program. J. Inform. Optim. Sci. 8 (1987) 2, 155-166

[6] Geoffrion, A.M.: Proper efficiency and the theory of vector maximization. J. Math. Anal. Appl. 22 (1968) 3, 618-630

[7] Mangasarian, O.L.: *Nonlinear Programming.* McGraw- Hill Book Co., New York etc. 1969

[8] Mond, B., T. Weir: Generalized concavity and duality. In : S.Schaible and W.T.Ziemba (eds.): *Generalized Concavity in Optimization and Economics.* Academic Press, New York 1981, pp. 263-279

[9] Preda, V., I.M. Stancu-Minasian: Generalized Mond-Weir duality for multiobjective nonsmooth programming. In: G.Fandel, T.Gal (eds.) In collaboration with T. Hanne: *Multiple Criteria Decision Making.* Proc. of the Twelfth International Conference on MCDM, Hagen, Germany June 19-23, 1995, Lecture Notes in Econ. and Math. Syst. No 448, Springer, 1997, pp. 55-63

[10] Weir, T., B. Mond: Generalized convexity and duality in multiple programming. Bull. Austral. Math. Soc. 39 (1989) 287-299

Sensitivity in Multiobjective Programming by Differential Equations Methods. The case of Homogeneous Functions

A. Balbás[1] and P. Jiménez Guerra[2]

[1]Departamento de Economía de la Empresa. Universidad Carlos III de Madrid. c) Madrid 126. 28903 Getafe (Madrid, Spain).

[2]Departamento de Matemáticas Fundamentales. Facultad de Ciencias. UNED. c) Senda del Rey s\n. 28040 Madrid (Spain).

Abstract. The purpose of this paper is to characterize for convex multiobjective programming, the situations in which the sensitivity with respect to the right side vector of the constraints can be obtained as a solution of a dual program.

Key words and phrases. Convex program, dual program, homogeneous function, sensitivity.

1 Introduction

It is well known that for convex scalar minimization programs and under very general conditions, the differential of the optimum value of the objective with respect to the right side vector of the constraints is a solution of a dual program. The extensions of this result to the multiobjective programming have been slight in general, although we can refer for instance to [9] where the problem is studied for finite dimensional linear programs assuming that the decisor chooses always optimum solutions which are also basic feasible solutions.

Many papers have stated important contributions to the development of a duality theory for the multiobjective convex programming (see for instance [10], [13], [15] and [17]) but a result about sensitivity similar to that of the scalar case has never been obtained although several sets have been studied which can be determined by duality properties and which contain the mentioned differential (see for instance [14]). In [3] it is proved that in general for a multiobjective linear program its sensitivity depends not only on the dual solution but also on the differential of this dual solution. It is also proved that this result is a genuine extension of the standard one in scalar programming, and several examples are presented where the differential of the dual solution does not vanish. Last results have been extended for convex programs in [4] and for differential programs in [2], where the sensitivity is measured with respect to any parameter appearing in the program (and not only with respect to the right side vector of the constraints).

We use here the dual introduced in [4] since it extends the classic ones of the linear multiobjective programming and the scalar convex programming. In Section 3 the situations in which the primal sensitivity coincides with the dual solution are characterized. The main result is Theorem 3.4 which states a condition of homogeneity which solves completely the problem using the Euler's Formula (which characterizes the homogeneous functions). Afterwards it is studied the particular case of programs defined by homogeneous functions obtaining in this case that the dual solution measures the sensitivity if and only if the optimum value of the objective is an homogeneous function of the right side vector of the constraints (see Theorem 3.7). An important particular case is the linear programming which is studied in Section 4 where Theorem 4.2 characterizes the property when the optimum solution is chosen to be a convex linear combination of basic feasible optimum solutions, stating that in this case the primal sensitivity is a dual solution if and only if the weights applied over every basic feasible solution are homogeneous functions of degree zero.

2 Preliminaries and notations

Let Ω be a convex subset of a Banach space X, Y a Banach space ordered by a closed convex pointed cone Y_+, Z a Banach lattice with positive cone Z_+ and W an order complete Banach lattice with positive cone W_+. Let also consider two convex functions $f:\Omega \to Y$ and $g:\Omega \to Z$, and a surjective linear and continuous mapping $T:Y \to W$ such that $T(Y_+ - \{0\}) \subseteq W_+ - \{0\}$ and KerT has a topological supplement Y_T (in Y) (last condition holds in particular, if Y is a Hilbert space or W is finite dimensional). Let us denote by T^* the restriction of T to Y_T (it follows from the open mapping theorem that T^{*-1} is continuous), by π the natural projection from Y onto KerT and by $L(Z, W)$ the space of all linear and continuous functions from Z into W, endowed with the usual norm and order. Similar notations will be used in other similar cases. Finally let B be an open subset of Z and consider for every $b \in B$ the following multiobjective minimization program:

$$\begin{array}{c} \text{Min } f(x) \\ x \in \Omega,\ g(x) \le b \end{array} \qquad (2,1b)$$

For every $L \in L(Z, W)$, we write $L \in \Gamma_T$ if $L \ge 0$ and $\{Tf(x) + Lg(x) : x \in \Omega\}$ is an order bounded from below subset of W, and in this case its infimum will be denote by $\varphi(T, L)$. Following [4], for every $b \in B$ and $G \in L(Z, Y)$ such that $TG \in \Gamma_T$ we consider the dual function $\psi(T, G) = T^{*-1}\varphi(T, TG) - G(b)$ and the dual program

$$\begin{array}{c} \text{Max } \psi(T, G) \\ G \in L(Z, Y),\ TG \in \Gamma_T \end{array} \qquad (2,2b)$$

We say that $x_b \in \Omega$ is a T-optimal solution of (2,1b) if x_b is a feasible solution of this program and $Tf(x) \ge Tf(x_b)$ for every feasible solution $x \in \Omega$. Analogously a feasible solution $G_b \in L(Z, Y)$ of (2,2b) is said to be a T-optimal solution of this program if $T\psi(T, G) \le T\psi(T, G_b)$ for every feasible solution $G \in L(Z, Y)$. If x_b is a primal T-optimal solution, G_b is a dual T-optimal solution and $f(x_b) = \psi(T, G_b)$ then x_b and G_b are said to be associated solutions. Several conditions for the existence of dual T-optimal solutions associated with a primal T-optimal solution can be fond in [4].

3 Homogeneity conditions

Throughout this paper we will suppose that for every $b \in B$ there exists a

T-optimal solution x_b of (2,1b) and $F: B \to Y$ will be the function defined by $F(b) = f(x_b)$ for every $b \in B$. Theorem 12 of [4] states that, " if there exists a Fréchet differentiable function $\mathbf{G}: B \to L(Z, Y)$ such that $\mathbf{G}(b)$ is a T-optimal solution of (2,2b) associated to x_b for every $b \in B$, then F is also Fréchet differentiable (on B) and $F'(b, z) = -\mathbf{G}_b(z) - \pi(\mathbf{G}'(b, z)(b))$ for every $b \in B$ and $z \in Z$ ", where $F'(b, z)$ and $\mathbf{G}'(b, z)$ denote the images of z by the Fréchet differentials at b of F and **G**, respectively, and $\mathbf{G}'(b, z)(b)$ is the image of b by $\mathbf{G}'(b, z)$. Throughout the paper $F'(b, -)$ will denote the Fréchet differential of F at $b \in B$.

In this Section we characterize the situations where the differential $F'(b, -)$ is obtained as a dual T-optimal solution associated to x_b, for every $b \in B$.

Lemma 3.1. Let us assume that F is Fréchet differentiable on B and at least one of the following conditions holds:

3.1.1. B is convex and $g(\Omega) \subseteq B$.

3.1.2. There exists a continuous function $\mathbf{G}: B \to L(Z, Y)$ such that $\mathbf{G}(b)$ is a T-optimal solution of (2,2b) associated to x_b, for every $b \in B$.

Then, for every $b \in B$, $-F'(b, -)$ is a dual T-optimal solution and

$$\varphi(T, -TF'(b, -)) + TF'(b, b) = TF(b).$$

Proof. Suppose that 3.1.1 holds and let $b \in B$ and $z \in Z_+$ be. Then, there exists $\varepsilon > 0$ such that $b + tz \in B$ for every $0 \le t < \varepsilon$, and since x_b is a feasible solution of (2,1(b + tz)), we have that $TF(b + tz) \le TF(b)$ and

$$TF'(b, z) = \lim_{t \to 0} \frac{TF(b + tz) - TF(b)}{t} \le 0. \qquad (3.1)$$

Let us see now that TF is a convex function. In fact, if $b, b' \in B$, $\alpha, \alpha' \ge 0$ and $\alpha + \alpha' = 1$, then since g is convex, we have that

$$g(\alpha x_b + \alpha' x_{b'}) \le \alpha g(x_b) + \alpha' g(x_{b'}) \le \alpha b + \alpha' b'$$

and therefore $TF(\alpha b + \alpha' b') \le Tf(\alpha x_b + \alpha' x_{b'}) \le \alpha TF(b) + \alpha' TF(b')$.

Consider now $x \in \Omega$ and $h = g(x) - b$, then $b + h = g(x) \in B$ and for every $0 \le t \le 1$ we have that

$$TF(b + th) = TF((1 - t)b + t(h + b)) \le (1 - t)TF(b) + tTF(h + b),$$

from where we obtain that

$$TF(b + th) - TF(b) - TF'(b, th) \le t[TF(b + h) - TF(b) - TF'(b, h)].$$

Therefore

$$\frac{TF(b + th) - TF(b) - TF'(b, th)}{t} \le TF(b + h) - TF(b) - TF'(b, h)$$

and taking now limits when $t \to 0$, it follows that

$$TF(g(x)) - TF'(b, g(x)) \ge TF(b) - TF'(b, b).$$

Now, since x is a feasible solution of (2,1g(x)), we have that $TF(g(x)) \le Tf(x)$ and therefore

$$Tf(x) - TF'(b, g(x)) \ge TF(b) - TF'(b, b). \qquad (3.2)$$

From (3.1) and (3.2) it is immediately deduced that $-F'(b, -) \in \Gamma_T$ and

$$\varphi(T, -TF'(b,-)) \geq TF(b) - TF'(b,b). \qquad (3.3)$$

To prove the opposite of inequality (3.3) it is enough to see that the inequality of (3.2) changes its sense if $x = x_b$, and this is a consequence of the inequality $-TF'(b, g(x_b)) \leq -TF'(b,b)$, which follows immediately from $g(x_b) \leq b$ and $-TF'(b,-) \geq 0$.

Suppose now that 3.1.2 holds and let $b \in B$ be. Then it follows from Theorem 10 of [4] that **TG**(b) is a Lagrangian T-multiplier of (2,1b) (in the sense of Definition 1 of [4]) and therefore, since x_b is a T-optimal solution of (2,1b), we have that

$$\begin{aligned} TF(b) &= Tf(x_b) = \operatorname{Inf}\{Tf(x);\ x \in \Omega, g(x) \leq b\} \\ &= \operatorname{Inf}\{Tf(x) + \mathbf{TG}(b)(g(x) - b);\ x \in \Omega\} \\ &= \operatorname{Inf}\{Tf(x) + \mathbf{TG}(b)(g(x));\ x \in \Omega\} - \mathbf{TG}(b)(b). \qquad (3.4) \end{aligned}$$

Moreover, it follows from Theorem 4 of [4] that $TF'(b,-) = -\mathbf{TG}(b)$, and thus it is obtained from (3.4) that $TF(b) = \varphi(T, -TF'(b,-)) + TF'(b,b)$.

Definition 3.2. Let U and V be two Banach spaces, $p \in \mathbf{R}$ and A an open subset of U. We say that a function $h: A \to V$ is homogeneous of degree p if for every $a \in A$ there exists a neighborhood $U^a \subset (0, +\infty)$ of 1 (in the real line **R**) such that $ta \in A$ and $h(ta) = t^p h(a)$ for every $t \in U^a$.

Theorem 3.3. (Euler's Formula). Let U and V be two Banach spaces, $p \in \mathbf{R}$, A an open subset of U and $h: A \to V$ a Fréchet differentiable function. Then h is homogeneous of degree p if and only if $h'(a,a) = ph(a)$ for every $a \in A$.

Theorem 3.4. With the already established notations, let us assume that F is Fréchet differentiable on B and that at least one of the conditions 3.1.1 and 3.1.2, holds. Then, $-F'(b,-)$ is a dual T-optimal solution associated to x_b for every $b \in B$ if and only if πF is homogeneous of degree 1.

Proof. Suppose that $-F'(b,-)$ is associated to x_b for every $b \in B$. Then we have that $F(b) = f(x_b) = \psi(T, -F'(b,-)) = T^{*-1}\varphi(T, -TF'(b,-)) + F'(b,b)$ for every $b \in B$, and since clearly $\pi T^{*-1} = 0$, applying π to the terms of last equalities we obtain that $\pi F(b) = \pi F'(b,b)$ and now, it results immediately from the Euler's Formula that πF is homogeneous of degree 1.

Reciprocally, if πF is supposed to be homogeneous of degree 1 then it follows from the Euler's Formula that $\pi F(b) = \pi F'(b,b)$. Moreover, it is deduced from Lemma 3.1 that $-F'(b,-)$ is a feasible dual solution (and a dual T-optimal solution) and $\varphi(T, -TF'(b,-)) + TF'(b,b) = TF(b)$.

Therefore, since $y = T^{*-1}T(y) + \pi(y)$ for every $y \in Y$, it results that

$$\begin{aligned} \psi(T, -TF'(b,-)) &= T^{*-1}\varphi(T, -F'(b,-)) + F'(b,b) \\ &= T^{*-1}(TF(b) - TF'(b,b)) + F'(b,b) \\ &= F(b) - F'(b,b) - \pi F(b) + \pi F'(b,b) + F'(b,b) = F(b) = f(x_b) \end{aligned}$$

and thus $-F'(b,-)$ is associated to x_b.

Remark. Concrete examples at which $-F'(b,-)$ is not a dual solution associated to x_b, can be fond in [4]. This fact and Theorem 3.4 point out that the homogeneity of πF does not hold in general. In the particular case of the scalar programming clearly $\pi = 0$ and so the homogeneity of πF is obvious and this shows that Theorem 3.4 is a genuine extension of the classic results known for the scalar programming.

Definition 3.5. Let $b \in B$ be and $p, q \in \mathbf{R}$. We say that (2,1b) is a (p,q)-homogeneous program if Ω is a convex cone of X and the equalities $f(tx) = t^p f(x)$ and $g(tx) = t^q g(x)$ hold for every $x \in \Omega$ and every $0 < t (\in \mathbf{R})$. If moreover $p = q$ we merely say that (2,1b) is p-homogeneous.

Proposition 3.6. If the program (2,1b) is (p,q)-homogeneous and $q \neq 0$ then the function TF is homogeneous of degree p/q.

Proof. It follows easily proceeding in standard way.

Theorem 3.7. Suppose that F is Fréchet differentiable on B, $p \in \mathbf{R} - \{0\}$, the program (2,1b) is p-homogeneous for every $b \in B$ and at least one of the conditions 3.4.1 and 3.4.2 holds. Then $-F'(b,-)$ is a T-optimal solution associated to x_b for every $b \in B$ if and only if F is homogeneous of degree 1.

Proof. It is an immediate consequence of Theorem 3.4 and Proposition 3.6 since $F = T^{*-1}TF + \pi F$.

Remark. In [3] it is given an example of a linear program for which $-F'(b,-)$ is not a dual associated solution and this proves that in general F is not homogeneous . In the particular case of the scalar programming, Theorem 3.7 states that if the function F is Fréchet differentiable then it is necessarily homogeneous of degree 1 (even for infinite dimensional problems).

4 Linear programs

We will say that (2,1b) is linear if Ω is a convex cone of X and the functions f and g can be extended to all X in such a way that their extensions, which will be also denoted by f and g , belong to $L(X, Y)$ and $L(X, Z)$, respectively. In particular, linear programs are 1-homogeneous and as usual we will write $x \geq 0$ instead of the constraint $x \in \Omega$.

Throughout this section (2,1b) will be supposed to be a linear program for every $b \in B$, and since in linear programing it is more convenient to work with equality constraints, we will change (2,1b) by the following program:

$$\begin{array}{c} \text{Min } f(x) \\ g(x)+z=b,\ x \in \Omega,\ z \in Z_+ \end{array} \qquad (4,1b)$$

Clearly there is a natural bijection between the feasible sets of (2,1b) and (4,1b), preserving the optimums, the T-optimal solutions and the extremal points of both feasible sets.

Theorem 14 of [4] proves that if the program (2,1b) is linear, then its dual has the following more simple formulation:

$$\text{Max } -G(b)$$
$$G \in L(Z, Y), \ TG \geq 0, \ TGg \geq -Tf \qquad (4,2b)$$

Last dual extends the duals given in [1] for scalar linear programs, in [9] for linear multiobjective programs in finite dimensional spaces and in [3] for linear programs between spaces of arbitrary dimensions. Moreover (4.2b) is quite similar to the dual introduced in [8] for goal programming problems and also to the dual stated in [10] for general linear problems. Throughout this section we will suppose that for every $b \in B$, $\{x^i{}_b\}_{i=1,2,\ldots,k}$ is a family of basic feasible solutions (in the sense of [1]) which are T-optimal solutions of (4,1b). Therefore $Tf(x^i{}_b) = Tf(x^j{}_b)$ for every $b \in B$ and $i, j \in \{1, \ldots, k\}$, and every convex linear combination of $\{x^1{}_b, \ldots, x^k{}_b\}$ is also a T-optimal solution of (4,2b) for every $b \in B$.

If $z^i{}_b = b - g(x^i{}_b)$ for every $1 \leq i \leq k$ and $b \in B$, we have that $\beta_0(x^i{}_b) = \{(x, z) \in XxZ : (x^i{}_b, z^i{}_b) + \lambda(x, z), (x^i{}_b, z^i{}_b) - \lambda(x, z) \in \Omega x Z_+$ for some scalar $\lambda > 0\}$ is a vector subspace of XxZ such that $(x^i{}_b, z^i{}_b) \in \beta_0(x^i{}_b)$ and $\beta_0(x^i{}_b) \cap \text{Ker } g^* = \{0\}$, where $g^* : XxZ \to Z$ is defined by $g^*(x, z) = g(x) + z$ for every $(x, z) \in XxZ$. We will assume that $\beta_0(x^i{}_b)$ is contained in a closed subspace β_i which is a topological supplement of Ker g^* and it does not depend on $b \in B$.

Clearly, if we assume in this section that g is surjective then the restriction $g_i{}^*$ of g^* to β_i is a topological isomorphism such that $g_i^*(x^i{}_b, z^i{}_b) = b$ for every $i = 1,\ldots,k$ and $b \in B$.

Consider also k functions $h_i : B \to [0,1]\ (i = 1,\ldots,k)$ such that $\sum_{i=1}^{k} h_i(b) = 1$ for every $b \in B$, and let us fix the following notations: $x_b = \sum_{i=1}^{k} h_i(b)\, x^i{}_b$, $F(b) = f(x_b)$ and $z_b = b - g(x_b) = \sum_{i=1}^{k} h_i(b)\,(b - g(x^i{}_b)) = \sum_{i=1}^{k} h_i(b)\, z^i{}_b$ for every $b \in B$. Finally assume throughout this section that at least one of the conditions 3.1.1 and 3.1.2, holds.

Proposition 4.1. If F is Fréchet differentiable at B and h_i ($i = 1,\ldots,k$) is homogeneous of degree 0, then $-F'(b, -)$ is a dual T-optimal solution associated to x_b for every $b \in B$.

Proof. It follows from Theorem 3.7 that it is enough to prove that the function F is homogeneous of degree 1, and this can be easily proved in standard way.

Theorem 4.2. Suppose that the functions F and h_i ($i = 1,\ldots,k$) are Fréchet differentiable on B and $\{f(x^i{}_b) \ ; i = 1,\ldots,k\}$ is a linearly independent system for every $b \in B$. Then $-F'(b, -)$ is a dual T-optimal solution associated to x_b for every $b \in B$ if and only if h_i is homogeneous of degree 0 for every $i = 1,\ldots,k$.

Proof. Like in the proof of Proposition 4.1, let us denote by f the natural extension of f to XxZ (such that $(x, z) \to f(x)$). If $-F'(b,-)$ is a dual T-optimal solution associated to x_b for every $b \in B$, then it results from Theorem 3.7 that F is homogeneous of degree 1 and then it follows from the Euler´s Formula that $F(b) = F'(b, b)$ for every $b \in B$. Furthermore

$$F(b) = f(x_b, z_b) = f\left(\sum_{i=1}^{k} h_i(b)\,(x^i_b, z^i_b)\right) = f\left(\sum_{i=1}^{k} h_i(b)\, g_i^{*-1}(b)\right) \quad (4.1)$$

and

$$F'(b,b) = f\left(\sum_{i=1}^{k} h_i'(b,b)\, g_i^{*-1}(b) + \sum_{i=1}^{k} h_i(b)\, g_i^{*-1}(b)\right)$$

$$= f\left(\sum_{i=1}^{k} h_i'(b,b)\, g_i^{*-1}(b)\right) + f\left(\sum_{i=1}^{k} h_i(b)\, g_i^{*-1}(b)\right) \quad (4.2)$$

for every $b \in B$.Therefore since $F(b) = F'(b, b)$ for every $b \in B$, it results from (4.1) and (4.2) that

$$0 = \sum_{i=1}^{k} h_i'(b,b)\, f(g_i^{*-1}(b)) = \sum_{i=1}^{k} h_i'(b,b)\, f(x^i_b, z^i_b)$$

for every $b \in B$, from where it is immediately deduced that $h_i'(b, b) = 0$ for every $b \in B$ and $i = 1,...,k$, since the system $\{f(x^i_b, z^i_b)\ ;\ i = 1,...,k\}$ is linearly independent. Now it results from the Euler´s Formula that h_i is homogeneous of degree 0 for every $i = 1,...,k$. The converse follows trivially from Proposition 4.1.

Remark. The assumption about the linearly independence of the system $\{f(x^i_b, z^i_b)\ ;\ i = 1,...,k\}$ $(b \in B)$ cannot be avoid in Theorem 4.2 since it is known that in the scalar programming (i.e., $Y = \mathbf{R}$) the differential $-F'(b,-)$ is the dual solution independently of the choice of the primal optimum. Actually what happens in this case is that F does not depend on the "weights" h_i applied to every basic feasible solution.

Many papers (like [3], [6] and [9]) prove that if the decisor chooses an optimum solution which is a basic feasible solution (or a convex linear combination of basic feasible solutions whose scalar coefficients are fixed and independent of $b \in B$) then $-F'(b,-)$ can be fond as a dual associated solution. Theorem 4.2 states that $-F'(b,-)$ continues being a dual solution associated to x_b even though the weights are not constant and they are homogeneous functions (of b) of degree 0.

5 The basic differential equation

Let us remark that if $-F'(b,-)$ is a T-optimal solution associated to x_b, for every $b \in B$, and we consider the function $\mathbf{G}: B \to L(Z, Y)$ such that $\mathbf{G}(b) = -F'(b,-)$ for every $b \in B$, then it follows from Theorem 12 of [4] that if $\mathbf{G}$ is Fréchet differentiable (or equivalently, if F is twice Fréchet differentiable) then

$$(\pi \mathbf{G})'\,(b, z)\,(b) = 0 \qquad (5.1)$$

for every $b \in B$ and $z \in Z$. Thus in many situations (stated in Sections 3 and 4) the differential equation (5.1) (jointly with some initial and optimal conditions) let to obtain F'. In particular, if $Y = \mathbf{R}^m$ and $Z = \mathbf{R}^n$, then $\pi \mathbf{G}(b)$ will be a functional matrix $(g_{ij}(b))_{i=1,\dots,m,\, j=1,\dots,n}$, and (5.1) leads to the following system of partial differential equations:

$$\sum_{j=1}^{n} b_j \frac{\partial g_{ij}}{\partial b_k} = 0 \quad i = 1,\dots,m,\; k = 1,\dots,n. \qquad (5.2)$$

6 Conclusions

If b denotes the right side vector of the constraints of a convex multiobjective program and F(b) is the optimum value of that program, then the differential of F with respect to b cannot be obtained in general as a solution of a dual program (see [3] or [4]) and it appears the natural question of characterizing the situations where this differential (denoted by $F'(b, -)$ in this paper) solves the dual program. Theorem 3.4 gives an answer to this question by means of a condition of homogeneity, and Theorem 3.7 states for a particular type of programs (containing the linear ones) that a necessary and sufficient condition to be $-F'(b, -)$ an associated dual solution is that F will be (as a function of b) homogeneous of degree 1. In particular, for the scalar linear programming it is verified that if F is differentiable then it is necessarily homogeneous of degree 1, result which is false in general in the multiobjective case.

In the linear case, when the optimum solution is chosen to be a convex linear combination of basic feasible solutions (what is the usual situation in programs described between finite dimensional spaces, and also in many examples of infinite dimensional programs like the Mass-Transfer type problems) the necessary and sufficient condition for $-F'(b, -)$ to be an associated dual solution is that the weights applied on the basic feasible solutions will be homogeneous functions (of b) of degree 0 (as Theorem 4.2 states).

Last results can be proved, for programs described between finite dimensional spaces, solving the Euler´s partial differential equations which characterize the homogeneous functions. Furthermore, if F is homogeneus of degree 1, then the system of partial differential equations (5.2) may lead in practical situations to a general expresion of F. When F is not a homogeneous function (of b) then the sensitivity of the program can be obtained solving a more general and complex system of partial differential equations which can be deduced from Theorem 12 of [4].

References

[1] Anderson, E.J. and P. Nash: Linear programming in infinite dimensional spaces. John Wiley & Sons, New York, 1987.

[2] Balbás, A., F. Fernández, and P. Jiménez Guerra: On the envolvent theorem in multiobjective programming. Indian Journal of Pure and Applied Mathematics, 26 (1995), 1035-1047.

[3]Balbás, A. and A. Heras,: Duality theory for infinite dimensional multiobjective programming. European Journal of Operational Research, 68 (1993), 379-388.

[4] Balbás, A. and P. Jiménez Guerra: Sensitivity analysis for convex multiobjective programming in abstract spaces. Journal of Mathematical Analysis and Applications, 202 (1996), 645-658.
[5] Balbás, A., P. Jiménez Guerra and C. Núñez: Strongly proper optimums and maximal optimization in multiobjective programming. Revista Real Academia de Ciencias de Madrid, 86 (1992), 289-295.
[6] Henggeler, C. and J. Cimaco: Sensitivity analysis in MCDM using the weight space. Operations Research Letters, 12 (1992), 187-196.
[7] Ignizio, J.P.: Introduction to linear goal programming. Sage University Papers, London, 1976.
[8] Ignizio, J.P. : An algorithm for solving the linear goal programming problem by solving its dual. Journal of the Operations Research Society, 36 (1985), 507-515.
[9] Isermann, H.: On some relations between a dual pair of multiobjective linear programs. Zeitschrift für Operations Research, 22 (1978), 33-41.
[10] Jahn, J.: Mathematical vector optimization in partially ordered linear spaces. Verlag Peter Lang, Frankfurt , 1986.
[11] Jiménez Guerra, P. and B. Rodríguez-Salinas: A general solution of the Monge-Kantorovich Mass-Transfer problem. Journal of Mathematical Analysis and Applications, 202 (1996), 402-510.
[12] Khanh, P.O.: Sufficient optimality conditions and duality in vector optimization with invex-convexlike functions. Journal of Optimization. Theory and Applications, 87 (1995), 359-368.
[13] Khanh, P.O.: Proper solutions of vector optimization problems. Journal of Optimization.Theory and Applications, 74 (1993), 105-130.
[14] Klose, J.: Sensitivity analysis using the tangent derivative. Numerical Functional Analysis and Optimization, 13 (1992), 143-153.
[15] Luc, D.T.: About duality and alternative in multiobjective optimization. Journal of Otimization. Theory and Applications, 53 (1987), 303-307.
[16] Schaefer, H.H.: Banach lattices and positive operators. Springer-Verlag, Berlin, 1974.
[17] Tanino, J., Y. Sawaragy and Y. Nakayama: Theory of multiobjective optimization. Academic Press, 1985.
[18] Zeleny, Y.: Multiobjective linear programming. Springer-Verlag, Berlin, 1974.
[19] Zowe, S.: A duality theorem for a convex programming problem in order complete vector lattices. Journal of Mathematical Analysis and Applications, 50 (1975), 273-287.

On Pseudo-Boolean Multicriteria Optimization Problems with Incomplete Information

Vladimir Donskoy and Anna Perekhod

Department of Mathematics, Simferopol State University
4, Yaltinskaya Street, Simferopol 333036 Crimea, Ukraine

Abstract

This paper presents a new approach for Intelligent Decision Support Systems (IDSS) development based on the pseudo-Boolean multicriteria optimization with incomplete information for decision making in the complicated, weakly formalized sphere. This approach uses an interactive procedure to form the learning set $W = \{W_0 \bigcup W_1\}$, where W_0 contains only Pareto, and W_1 contains only non-Pareto points. This set W is used for inductive learning to approximate the Pareto set.

Keywords. Multiple criteria optimization, pseudo-Boolean functions, machine learning, Pareto set approximation, decision support systems.

1 Introduction

We consider decision making as a choice of some element $\tilde{x} = (x_1, \ldots, x_n)$ from the admissible solutions set Ω. We also suppose $x_i \in \{0,1\}, i = \overline{1,n}$, and consider the pseudo-Boolean models, but the problems of the variable binarization are excluded from consideration in this paper. In other words, we consider given or chosen feature predicates set $\{x_i: Y \to \{0,1\}\}$, where Y is the set of states discribing the problem region.

Let us denote $B^n = \{0,1\}^n$. We suppose that decision making is realized on the basis of criteria

$$\{f_j: B^n \to R, j = \overline{1,r}\}.$$

These criteria are pseudo-Boolean functions according to the model:

$$\begin{cases} extr\ f_1(\tilde{x}), \ldots, extr\ f_r(\tilde{x}); \\ \tilde{x} \in \Omega \subset B^n \end{cases} \tag{1}$$

where *extr* stands for "extremum". As a rule, in any difficult practical problem, there is no complete information about the criteria $f_1,\ldots,f_r$ and the admissible solutions set Ω.

Nevertheless, we assume the model components $f_1,\ldots,f_r$ and Ω exist, the choice $\tilde{x}^* \in B^n$ being realized in spite of initial information incompleteness and afterwards can be estimated.

In these assumptions we only have information about choices already made and their estimations. We suppose that experts a posteriori estimate solutions (choices) $\left\{\tilde{x}^*\right\}$. The expert estimation results are used to synthesize a decision making model approximating the model (1).

2 Initial information acquisition

Decision making is accompanied with accumulation of the choices, which are made. We denote $\mathfrak{R}$ a set of these choices and use it for the inductive learning. Let any $\tilde{x} \in \mathfrak{R}$ be called a solution.

For any $\tilde{x} \in \mathfrak{R}$ experts can perform:

- definition of admissibility of this solution, i.e. the predicate "$\tilde{x} \in \Omega$" is true;
- a posteriori estimation of the criterion values $\hat{f}_j(\tilde{x}),\ j=\overline{1,r}$;
- mutual comparison of the solutions from $\mathfrak{R}$ by a single criterion or by all criteria;
- definition of the Pareto unimprovability of the solution $\tilde{x}$.

Let P be the Pareto set of problem (1). Additional information about criteria $f_1,\ldots,f_r$ and admissible solutions set Ω is significant. For example, let the additional information be:

$$\left(x_{i_1}^{\sigma_1} \& \cdots \& x_{i_s}^{\sigma_s} = 1\right) \Rightarrow \left(\tilde{x} \notin \Omega\right),$$

where

$$z^{\sigma} = \begin{cases} z, \sigma = 1, \\ \bar{z}, \sigma = 0, \end{cases} \qquad z \in \{0,1\}; \sigma \in \{0,1\}.$$

This additional information may be used to eliminate some solutions, which are non-admissible. At the same time

$$\left(\tilde{x} \in \Omega\right) \Rightarrow \overline{x_{i_1}^{\sigma_1} \& \cdots \& x_{i_s}^{\sigma_s}} = x_{i_1}^{\bar{\sigma}_1} \vee \cdots \vee x_{i_s}^{\bar{\sigma}_s}, \tag{2}$$

and we have the necessary condition for any admissible solution. The clause in the right part in (2) may be used for the productions acquisition and knowledge base completion:

$$x_{i_1}^{\bar{\sigma}_{i_1}} \vee \cdots \vee x_{i_{s-1}}^{\bar{\sigma}_{s-1}} \vee x_{i_s}^{\bar{\sigma}_s} = x_{i_1}^{\sigma_1} \& \cdots \& x_{i_{s-1}}^{\sigma_{s-1}} \rightarrow x_{i_s}^{\bar{\sigma}_s}.$$

Using (2), we can obtain no more than s such productions [1].

The set of solutions $\mathfrak{R}$ contains subsets $\mathfrak{R}_0$ and $\mathfrak{R}_1$ such that any $\tilde{x} \in \mathfrak{R}_0$ is admissible and any $\tilde{x} \in \mathfrak{R}_1$ is a non-admissible solution; $\mathfrak{R}_0 \bigcup \mathfrak{R}_1 = \mathfrak{R}; \mathfrak{R}_0 \bigcap \mathfrak{R}_1 = \varnothing$. Using the set W, we can synthesize the approximate rule to compute predicate "$\tilde{x} \in \Omega$". Learning algorithm based on Binary Decision Tree (BDT) construction may be used for this purpose. Thereupon, using this rule, we can test any $\tilde{x} \in B^n$ to determine the admissibility of $\tilde{x}$. If the set Ω is exactly defined (for example, the inequality system is given), a test of admissibility is realized by substitution $\tilde{x}$ in this system and testing of the compatibility. Note that we are appealing to some "oracle" in both cases while testing.

3 The approach to the problem solving

Evidently, using model (1), it is necessary to choose solutions only from the Pareto set $P \subset \Omega$. Therefore, approximation of P based on the initial information is important. We suggest the following synthetic procedure.

1. Form the set of solutions $\mathfrak{R} = \mathfrak{R}_0 \bigcup \mathfrak{R}_1$.
2. Build the rule for recognition of the property "$\tilde{x} \in \Omega$".
3. Separate the subset $\mathfrak{R}_0 \subset \Omega$ by two subsets W_0 and W_1 such that W_0 contains only Pareto (unimprovable by expert opinion) points, and W_1 contains only non-Pareto points. On this step, expert knowledge is used to extract solutions from the set W_1 and obtain a new subset W_1^* of Pareto points. The process of addition new Pareto points is realized by experts, who consider the initial information and oracle estimations. As the result of this step, the sets W_1 (non-Pareto) and $W_0^* = W_0 \bigcup W_1^*$ (Pareto) are obtained.
4. Build a decision rule using the sets W_0^* and W_1 as learning information through inductive learning to check the solution $\tilde{x}$ membership to the Pareto set: $\hat{\varphi}(\tilde{x}) = "\tilde{x} \in P"$. The mark " ^ " denotes the *approximate* rule $\hat{\varphi}(\tilde{x})$. This rule may be represented as Disjunctive Normal Form (DNF) [2,3]. If $\hat{\psi}(\tilde{x}) = "\tilde{x} \in \Omega"$ then the final rule of the Pareto solution choice is presented in the form $F(\tilde{x}) = \hat{\varphi}(\tilde{x}) \& \hat{\psi}(\tilde{x})$.

The DNF representation

$$F(\widetilde{x}) = \bigvee_{\mu=1}^{m} x_{\mu 1}^{\sigma_{\mu 1}} \& \cdots \& x_{\mu r_\mu}^{\sigma_{\mu r_\mu}} \tag{3}$$

makes possible to realize the final choice of the solution $\widetilde{x}^*: F(\widetilde{x}^*) = 1$ considering minimization of the expenditures, which are spent to choose the solution.

Let $c_{\mu i} > 0$ be a value of the expenditures spent to get the value $\sigma_{\mu i}$ for variable $x_{\mu i}$. According to (3),

$$\tau = \arg\min_{1 \le \mu \le m} \sum_{i=1}^{r_\mu} c_{\mu i}$$

Then the final solution is a Pareto point $\widetilde{x}^*$ with components

$$x_{\tau i}^* = \sigma_{\tau i}, i = \overline{1, r_\tau}$$

Every component x_α^*, where $\alpha \in \{1,2,\ldots,n\} \setminus \{\tau 1, \tau 2, \ldots, \tau r_\tau\}$, have arbitrary values from $\{0,1\}$.

4 Conclusions

The adequacy of the suggested approach to the modern IDSS building is explained by initial information incompleteness. Only a set of precedents and expert estimation are required to synthesize decision rules in the model. Similar approaches demonstrated their convenience for applications [4].

Acknowledgements

We thank Ralph Steuer for his helpful suggestions on this paper.

References

[1] Zakrevski, A.: *Logic of the Recognition.* Nauka, Minsk. 1988

[2] Donskoy V.I.: Decision Trees Learning Algorithms. Journal of Computational Mathematics and Mathematical Physics. Moscow, 22, 4 (1982) 963-974

[3] Donskoy V.I.: Weakly Defined Problems of Boolean Linear Programming. Journal of Computational Mathematics and Mathematical Physics Moscow, 28, 9 (1988) 1379-1385

[4] Jog V., Michalowski W.: An Interactive Procedure for Learning about Preferences: Case Study of a Portfolio Manager. Journal of Multi-Criteria Decision Analysis, 3 (1994) 27-40

Properties of Efficient Points Sets and Related Topics

Vasile Postolică

Romanian Academy of Scientists,
Bacău State University
Department of Mathematical Sciences
B-dul Traian nr. 11, bl. A1, sc. A, apt. 13
5600-Piatra Neamt, România

Abstract. This research work is devoted to the study of the properties for the efficient points sets in separated locally convex spaces, that is, in the framework vector spaces with unknown geometries. Firstly, we present general new properties which illustrate immediate links between Strong Optimization and Vector Optimization and a dual characterization. Afterwards we mention our main results obtained in this field using supernormal cones and the completeness instead of compactness, specifying in the beginning a coincidence between the corresponding efficient points sets for compact subsets with respect to convex , closed and pointed cones and a special kind of Choquet boundaries. This last result gives emphasis to an important connection between Vector Optimization and Potential Theory, being useful also for future developments of these fields in Mathematics. Finally, we give some considerations concerning best approximation simultaneous and vectorial in H-locally convex spaces.

Keywords. Efficient (Pareto minimal) point, supernormal cone, Choquet boundary, H-locally convex space, best approximation.

Mathematics Subject Classification 1991: 90C29.

1. INTRODUCTION

One of the main directions of research of research in vector optimization is the study of the properties for the sets of the efficient (Pareto) points. Since the finite dimensionality of the framework vector space is, in general an essential condition with strong implications on the structure of sets such as these (see, for instance [14] and other recent connected papers), following a continuation of our research works [4], [17], [18], [19] and [20] and taking into account important concepts and results indicated by the references, we shall give for separated vector (locally convex) spaces appropriate conditions on the one hand improved upon the ordering cone and on the other hand on the sets for which one examines the existence of the efficient points in order to obtain interesting and useful properties for sets of efficient points accompanied by pertinent comments. We present also here the best approximation simultaneous and vectorial by spline functions introduced by us in [16] as the unique solutions of several special optimization problems in H-locally convex spaces. All the elements of ordered topological vector spaces used in this work are in accordance with [15].

2. EFFICIENT POINTS SETS AND RELATED TOPICS

Let **X** be a vector space ordered by a convex cone **K**. If **A** is a non-empty subset of **X** and $a_0 \in A$, then

Definition 2.1 *We say that* a_0 *is an efficient (Pareto minimal) point for* **A** *with respect to* **K**, *in notation,* $a_0 \in \mathbf{MIN_K(A)}$ *if it satisfies one of the following equivalent conditions:*

(i) $\mathbf{A} \cap (a_0\mathbf{-K}) \subseteq a_0\mathbf{+K}$; (ii) $\mathbf{K} \cap (a_0\mathbf{-A}) \subseteq \mathbf{-K}$;

(iii) $(\mathbf{A+K}) \cap (a_0\mathbf{-K}) \subseteq a_0\mathbf{+K}$; (iv) $\mathbf{K} \cap (a_0\mathbf{-A-K}) \subseteq \mathbf{-K}$;

When K is pointed, that is, $\mathbf{K} \cap (\mathbf{-K}) = \{0\}$, then $a_0 \in \mathbf{MIN_K(A)}$ means $\mathbf{A} \cap (a_0\mathbf{-K}) = \{a_0\}$, or equivalently, $\mathbf{K} \cap (a_0\mathbf{-A}) = \{0\}$.

In a similar manner one defines the (Pareto) maximal elements of **A**. In fact, a_0 is a maximal point of A with respect to **K** iff it is a minimal point for **A** with respect to -**K**, that is $a_0 \in \mathbf{MIN_{-K}(A)}$.

Remark 2.1 It is obvious that $a_0 \in \mathbf{MIN_K(A)}$ if and only if a_0 is a fixed point for one of the following multifunctions:

$F_1:\mathbf{A} \to \mathbf{A}$, $F_1(t)=\{\alpha \in \mathbf{A}: \mathbf{A} \cap (\alpha\mathbf{-K}) \subseteq t+K\}$;

$F_2:\mathbf{A} \to \mathbf{A}$, $F_2(t)=\{\alpha \in \mathbf{A}: \mathbf{A} \cap (t\mathbf{-K}) \subseteq \alpha+K\}$;

$F_3:\mathbf{A} \to \mathbf{A}$, $F_3(t)=\{\alpha \in \mathbf{A}: (\mathbf{A+K}) \cap (\alpha\mathbf{-K}) \subseteq t+K\}$;

$F_4:\mathbf{A} \to \mathbf{A}$, $F_4(t)=\{\alpha \in \mathbf{A}: (\mathbf{A+K}) \cap (t\mathbf{-K}) \subseteq \alpha+K\}$;

that is, $a_0 \in F_i(a_0)$ for some $i=\overline{1,4}$.

Theorem 2.1 *If we denote* $\mathbf{S(A)} = \{a_1: a - a_1 \in \mathbf{K}, \forall a \in \mathbf{A}\}$ *and* $\mathbf{S(A)} \neq \varnothing$, *then* $\mathbf{MIN_K(A)} = \mathbf{S(A)}$.

Proof. Clearly $\mathbf{S(A)} \subseteq \mathbf{MIN_K(A)}$. Indeed, if $a_0 \in \mathbf{S(A)}$ and $a \in \mathbf{A} \cap (a_0\mathbf{-K})$, then $a \in (a_0\mathbf{+K})$. Assume now that $\mathbf{S(A)} \neq \varnothing$ and there exists $a_0 \in \mathbf{MIN_K(A)} \setminus \mathbf{S(A)}$. Then, $a^* - a_0 \notin \mathbf{K}$ for some $a^* \in \mathbf{A}$. Let $\bar{a} \in \mathbf{S(A)}$. Then $a_0 - \bar{a} \in \mathbf{K} \setminus \{0\}$ because if $a_0 = \bar{a}$ one obtains $a_0 \in \mathbf{S(A)}$ in contradiction with $a^* - a_0 \notin \mathbf{K}$. Therefore $a_0 - \bar{a} \in \mathbf{K} \setminus \{0\}$ and $a_0 \in \mathbf{MIN_K(A)}$ imply $\bar{a} \in a_0 + K$ which together with $\bar{a} \in \mathbf{S(A)}$ leads to $a_0 \in \mathbf{S(A)}$, a contradiction. This completes the proof.

Remark 2.2 The above theorem shows that for a non-empty subset of an arbitrary vector space the set of simultaneous minimal elements with respect to any convex cone coincides with its corresponding set of efficient points whenever there exists at least a simultaneous minimal element. Using this result and through the agency of our abstract construction given in [16] for **H**-locally convex spaces we shall see in the last part of this paper that the only best simultaneous and vectorial approximation for every element in the direct sum of a (closed) linear subspace and its orthogonal with respect to a linear (continuous) operator between two **H**-locally convex spaces is its spline function. We also note that is possible to have $\mathbf{S(A)} = \varnothing$ and $\mathbf{MIN_K(A)} = \mathbf{A}$. Thus, for example, if one considers $\mathbf{X} = \mathbf{R}^2$ endowed with the topology generated by the seminorms $p_1, p_2: \mathbf{X} \to \mathbf{R}_+$, $p_1(x,y) = |x|$, $p_2(x,y) = |y|$, $\mathbf{K} = \mathbf{R}^2_+ = \{(x,y) \in \mathbf{R}^2: x, y \geq 0\}$ and $\mathbf{A} = \{(\lambda, 1-\lambda): 0 \leq \lambda \leq 1\}$, then it is clear that $\mathbf{S(A)} = \varnothing$ and $\mathbf{MIN_K(A)} = \mathbf{A}$.

In all our further considerations we suppose that **X** is a Hausdorff locally convex space with the topology induced by a family $\mathcal{P} = \{p_\alpha: \alpha \in I\}$ of seminorms, ordered by a convex cone **K** and its topological dual space is $\mathbf{X}^*$.

The following theorem offers a dual characterization for the efficient points useful in the scalarization of vectorial optimization problem.

Theorem 2.2 *If* **K** *is a closed convex cone in* **X** *and* **A** *is a non-empty subset, then* $a_0 \in \mathbf{MIN_K(A)}$ *if and only if for every* $a \in \mathbf{A} \setminus (a_0 + \mathbf{K})$, *there exists* $x^* \in \mathbf{X}^*$ *such that* $x^*(a) < x^*(a_0)$ *and* $x^* \in \mathbf{K}^0$ *where* $\mathbf{K}^0$ *denotes the usual polar cone of* **K**.

Proof. Let $a_0 \in \mathbf{MIN_K(A)}$ and $a \in \mathbf{A} \setminus (a_0 + \mathbf{K})$. Then $a \notin a_0 - \mathbf{K}$ and by virtue of an usual separation theorem there exists $x^* \in \mathbf{X}^*$ such that $x^*(a) < x^*(a_0) - x^*(y)$ for every $y \in \mathbf{K}$. Hence $x^*(a) < x^*(a_0)$. On the other hand, if one assumes that there exists $y \in \mathbf{K}$. with $x^*(y) > 0$, then $x^*(a) < x^*(a_0) - \lambda x^*(y)$ whenever $\lambda > 0$ which for λ sufficiently large leads to a contradiction. Therefore $x^*(y) \leq 0$ for all $y \in \mathbf{K}$ which shows that $x^* \in \mathbf{K}^0$.

Conversely, suppose now that for every $a \in \mathbf{A} \setminus (a_0 + \mathbf{K})$ there exists $x^* \in \mathbf{K}^0$ with $x^*(a) < x^*(a_0)$ and $\mathbf{A} \cap (a_0 - \mathbf{K}) \not\subset a_0 + \mathbf{K}$, that is, there exists $\bar{a} \in \mathbf{A} \cap (a_0 - \mathbf{K})$ and $\bar{a} \notin a_0 + \mathbf{K}$. Therefore $\bar{a} \in \mathbf{A} \setminus (a_0 + \mathbf{K})$ and by hypothesis there exists $y^* \in \mathbf{K}^0$ such that $y^*(\bar{a}) < y^*(a_0)$ which contradicts $y^*(a_0) \leq y^*(\bar{a})$ obtained from $a_0 - \bar{a} \in \mathbf{K}$. The result follows.

An important connection between Vector Optimization and Potential Theory which can not be obtained as a consequence of the Axiomatic Potential

Theory is the next coincidence of the efficient points sets and Choquet boundaries given and discussed in [4] and [20] respectively.

Theorem 2.3 [4] *If* **K** *is closed, pointed and* **A** *is a non-empty, compact subset of* **X**, *then* $\mathbf{MIN_K(A)}$ *coincides with the Choquet boundary of* **A** *with respect to the convex cone of all real, continuous functions on* **A** *which are increasing in comparison with the order relation induced by* **K**. *Consequently, the set* $\mathbf{MIN_K(A)}$ *endowed with the trace topology* τ_A *generated on* **A** *by the locally convex topology* τ *of* **X** *is a Baire space. Moreover, if* (A, τ_A) *is metrizable, then* $\mathbf{MIN_K(A)}$ *is a* G_δ *subset of* (A, τ_A).

Remark 2.3 As we have already seen in [6] and [18], there exist more general conditions imposed upon a non-empty set A ensuring that $\mathbf{MIN_K(A)}$ is non-void. Perhaps our coincidence result suggests an extension of the concept of Choquet boundary at least in these cases. Anyhow, the above theorem was appreciated at the 4th Workshop of the German Group on Decision Theory in Holzhau, Germany, March 14 - 18, 1994 as a significant connection between Vector Optimization and Potential Theory.

Definition 2.2 [8] *The convex cone is supernormal if for every* $p_\alpha \in \mathcal{P}$ *there exists* $f_\alpha \in \mathbf{X}^*$ *such that* $p_\alpha(x) \leq f_\alpha(x)$ *for all* $x \in \mathbf{K}$.

Remark 2.4 The concept of supernormal cone in normed spaces was introduced by Krasnoselski in fifties (see, for example, [13] and other connected papers). G. Isac has extended this notion for separated locally convex spaces in 1981 [8] and published it in 1983 [9]. Such a convex cone was initially called "nuclear cone" because in every nuclear space any normal cone is a nuclear cone in the Isac sense (Proposition 6 of [9]). Afterwards, since the nuclear cone introduced by Isac appears as a reinforcement of the normal cone, it was called supernormal. The class of supernormal cones in Hausdorff locally convex spaces was initially imposed by the theory and the applications of the efficient points (especially existence conditions based on completeness instead of compactness were decisive together with the main properties of the efficient points sets), the study of critical points for dynamical systems and conical support points and their importance was very well illustrated in [1], [2], [8], [9],[10]. [11], [12], [17], [18], [19] and so on.

Definition 2.3 *A non-empty set* $\mathbf{B} \subset \mathbf{X}$ *is* **K**-*bounded* [19] *if there exists a bounded set* $\mathbf{B_0} \subset \mathbf{X}$ *such that*
$\mathbf{B} \subseteq \mathbf{B_0} + \mathbf{K}$ *and* ***B*** *is said to be* **K**-*closed* [14] *if the conical extension* $\mathbf{B} + \mathbf{K}$ *is closed.*

The next theorem summarizes some of our existence results concerning the efficient points based on supernormality and properties for sets of efficient points.

Theorem 2.4 [18]

(i) *Let* $\mathbf{A} \subseteq \mathbf{B} \subseteq \mathbf{A+K}$. *If* **K** *is supernormal and* $\mathbf{B} \cap (\mathbf{A_0\text{-}K})$ *is bounded and complete for some non-empty set* $\mathbf{A_0} \subseteq \mathbf{A}$, *then* $\mathbf{MIN_K(A)} \neq \emptyset$;

(ii) *If* **X** *is quasi-complete, that is, every non-empty, bounded and closed subset of* **X** *is complete and* **K** *is closed and supernormal, then for every non-void* **K**-*bounded and* **K**-*closed subset* **A** *we have*:
$\mathbf{MIN_K(A)} \neq \emptyset$, $\mathbf{A} \subseteq \mathbf{MIN_K(A)+K}$, $\mathbf{MIN_K(A)+K} = \mathbf{A+K}$ *and* $\mathbf{MIN_K(A)}$ *is* **K**-*bounded and* **K**-*closed*;

(iii) *Under the same hypotheses of* (ii), $\mathbf{MIN_K(A)} \neq \emptyset$ *whenever the set* $\mathbf{B} \cap (\mathbf{A_0\text{-}K})$ *is* **K**-*bounded and* **K**-*closed for some non-empty sets* **B** *and* $\mathbf{A_0} \subseteq \mathbf{A} \subset \mathbf{X}$ *with* $\mathbf{A} \subseteq \mathbf{B} \subseteq \mathbf{A+K}$.

By (i) in the above theorem we obtain

Corollary 2.4.1 *Let* $\mathbf{A} \subseteq \mathbf{B} \subseteq \mathbf{A+K}$. *If* **K** *is weakly supernormal and* $\mathbf{B} \cap (\mathbf{A_0\text{-}K})$ *is bounded and weakly complete for some non-empty set* $\mathbf{A_0} \subseteq \mathbf{A}$, *then* $\mathbf{MIN_K(A)} \neq \emptyset$. *In particular, if* **K** *is a weakly supernormal cone in* **X** *and* $\mathbf{A} \cap (\mathrm{a}\text{-}\mathbf{K})$ *or* $(\mathbf{A+K}) \cap (\mathrm{a}\text{-}\mathbf{K})$ *is bounded and weakly complete for some* $\mathrm{a} \in \mathbf{A}$, then $\mathbf{MIN_K(A)} \neq \emptyset$. *When the boundedness and weak completeness properties hold for every* $\mathrm{a} \in \mathbf{A}$, *then we have the following domination relation* $\mathbf{A} \subseteq \mathbf{MIN_K(A)+K}$.

Since in every separated locally convex space any normal cone is weakly supernormal (Proposition 2 of [10]), the conclusion of the above corollary remains valid whenever K is normal.

Using this remark one obtains

Corollary 2.4.2 $\mathbf{MIN_K(A)} \neq \emptyset$ *if one satisfies one of the following conditions*:

(i) **K** *is closed, normal, weakly complete and* **A** *is weakly closed such that* $\mathbf{A} \cap (\mathrm{a}\text{-}\mathbf{K})$ *is bounded for some* $\mathrm{a} \in \mathbf{A}$. *We also have* $\mathbf{A} \subseteq \mathbf{MIN_K(A)+K}$ *if, under the above hypotheses* $\mathbf{A} \cap (\mathrm{a}\text{-}\mathbf{K})$ *is bounded for every* $\mathrm{a} \in \mathbf{A}$;

(ii) **K** *is closed, normal and* **A** is *bounded and weakly complete. The domination property holds again*;

(iii) **K** *is closed, normal, weakly complete and* **A+K** *is weakly closed such that* $(\mathbf{A+K}) \cap (\mathrm{a}\text{-}\mathbf{K})$ *is bounded for some* $\mathrm{a} \in \mathbf{A}$. *The domination property holds if, in addition,* $(\mathbf{A+K}) \cap (\mathrm{a}\text{-}\mathbf{K})$ *is bounded for any* $\mathrm{a} \in \mathbf{A}$.

Remark 2.5 For all we know, the weakly complete cones are very important in Choquet's Theory (see [5] and so on) and in the Axiomatic Theory of Potential. By virtue of the Corollary of Proposition 2 in [10] it is clear that all the above results remain valid if one replaces the hypothesis of normality on **K** with the weak normality.

Corollary 2.4.3 *If* **A** *is a non-empty bounded and closed subset of* **X** *and* **K** *is well based (that is, it is generated by a convex, bounded set which does not contain the origin in its closure) by a complete set, then* $\mathbf{MIN_K(A)} \neq \varnothing$ *and* $A \subseteq \mathbf{MIN_K(A)+K}$.

Corollary 2.4.4 *If* **A** *is a non-empty, bounded and closed subset of a Banach space ordered by a convex cone* **K** *well based by a closed set, then* $\mathbf{MIN_K(A)} \neq \varnothing$ *and* $\mathbf{A} \subseteq \mathbf{MIN_K(A)+K}$.

Remark 2.6 The Corollary 2.4.4 is in fact the Theorem 2.7 in [3] obtained by us in a different manner. Since in a normed space a convex cone is supernormal if and only if it is well based, the last corollary offers the same conclusion in a Banach space whenever **K** is supernormal.

Remark 2.7 The largest class of convex cones ensuring the existence of the efficient points in compact sets was defined in [23]. A new direction not only for the existence, but also for important investigations about Pareto type optimums in ordered locally convex spaces through the agency of completeness instead of compactness was offered in 1981 by the supernormal (nuclear) cones. Under appropriate conditions, in [6] and [7], respectively, it is shown that in every separated topological vector space, the largest class of convex cones ensuring the existence of the efficient points in any bounded and complete subset coincides with the class of cones **K** satisfying one of the following equivalent conditions:

(i) any bounded increasing net which is contained in **K** and in a complete subset of **X** has a limit;

(ii) any bounded monotone net which is contained in a complete subset of **X** has a limit.

Every supernormal cone has the properties (i), (ii), but there exists convex cones having these properties which are not supernormal. Thus, in the classical Banach spaces $\mathbf{L}^p([a,b])$ $(p>1)$ the usual positive cone is closed, convex, it has the properties (i), (ii), but it is not supernormal. The same conclusion is valid for the cone of nonnegative functions in an Orlicz space.

Theorem 2.5 $a_0 \in \mathbf{MIN_K(A)}$ *if for every* $p_\alpha \in \mathcal{P}$ *and* $\varepsilon \in (0,1)$, *there exists* x^* *in the polar cone* $\mathbf{K}^0$ *of* **K** *such that* $p_\alpha(a_0-a) \leq x^*(a_0-a)+\varepsilon,\ \forall\, a \in \mathbf{A}$.

Proof. Let us suppose that $\mathbf{K} \cap (a_0-\mathbf{A}) \subseteq -\mathbf{K}$. Then, there exists $a \in \mathbf{A}$ such that $a_0-a \in \mathbf{K}$ and $a_0-a \notin -\mathbf{K}$. Since **X** is separated, $p_\alpha(a_0-a)>0$ for some

$p_\alpha \in \mathcal{P}$. On the other hand, there exists $n \in \mathbf{N}$ sufficiently large such that $p_\alpha\ (a_0-a)/n \in (0,1)$ and the relation given by the theorem leads to $p_\alpha\ (a_0-a) \leq x^*(a_0-a) + p_\alpha\ (a_0-a)/n$ with $x^* \in \mathbf{K}^0$, which implies that $p_\alpha\ (a_0-a) \leq 0$, a contradiction and the theorem is proved.

Remark 2.8 The above theorem is a natural generalization for Hausdorff locally convex spaces of Precupanu's result contained in Proposition 1.2 of [22].The converse of this theorem is not valid, in general, as we can see from the example considered in the framework of Remark 2.2.

Let $(\mathbf{X}, \mathcal{P} = \{p_\alpha : \alpha \in \mathbf{I}\})$ be a H-locally convex space [21] with each seminorm p_α being induced by a scalar semiproduct $(.,.)_\alpha$ $(\alpha \in \mathbf{I})$ and $\mathbf{M}$ a closed linear subspace of $\mathbf{X}$ for which there exist a H-locally convex space $(Y, \mathcal{Q} = \{ q_\alpha : \alpha \in \mathbf{I}\})$ with every seminorm $q_\alpha \in \mathcal{Q}$ generated by a scalar semiproduct $<.,.>_\alpha$ $(\alpha \in \mathbf{I})$ and a linear (continuous) operator $U: X \rightarrow Y$ such that

$$\mathbf{M} = \{x \in X: (x,y)_\alpha = <Ux,Uy>_\alpha,\ \forall\ \alpha \in \mathbf{I}\}.$$

The space of spline functions with respect to U was defined in [16] as the U-orthogonal of **M**, that is

$$M^\perp = \{x \in X: <Ux,U\zeta>_\alpha = 0,\ \forall\ \zeta \in \mathbf{M}, \alpha \in \mathbf{I}\}.$$

Clearly, $M^\perp$ is the orthogonal of **M** in the H-locally convex sense.

Let $x_0 \in \mathbf{X}$ and **G** a non-empty subset of **X**.

Definition 2.3. [11] *$g_0 \in \mathbf{G}$ is said to be a best simultaneous approximation for x_0 by the elements of* **G** *with respect the family $\mathcal{P}$ (abbreviated g_0 is a $\mathcal{P}$-b.s.a. of x_0) when*

$p_\alpha\ (x_0 - g_0) \leq p_\alpha(x_0 - g)$ *for all* $g \in \mathbf{G}$ *and* $p_\alpha \in \mathcal{P}$.

If, in addition, each element $x \in \mathbf{X}$ *possesses at least one $\mathcal{P}$-b.s.a. in* **G**, *then the set* **G** *is called $\mathcal{P}$-simultaneous proximinal.*

Definition 2.4. [11] *$g_0 \in \mathbf{G}$ is said to be a best vectorial approximation of x_0 by* **G** *with respect to $\mathcal{P}$ (abbreviated g_0 is a $\mathcal{P}$-b.v.a. of x_0) if*

$(p_\alpha\ (x_0 - g_0)) \in \mathbf{MIN}_K\ (\{(p_\alpha(x_0 - g)) : g \in \mathbf{G}\})$ *where* $\mathbf{K} = R^I_+$.

Following the Theorem 2.1 of the present research work, the results given in Chapter 3 of [11] and the Theorem 3 in [16], we have

Theorem 2.6 *If* $\mathbf{X}'=\mathbf{M}\oplus M^{\perp}$, *and for every* $\mathbf{x}\in\mathbf{X}'$ *we denote its projection onto* $M^{\perp}$ *by* s_x, *then*:

(i) *for every* $\mathbf{x}\in\mathbf{X}'$ *the only elements of best simultaneous and vectorial approximation with respect to any family of seminorms which generates the H-locally convex topology on* **X** *by the linear subspace of splines are the spline functions* s_x. *Moreover, if* **M** *and* $M^{\perp}$ *supply an orthogonal decomposition for* **X**, *that is* $\mathbf{X}=\mathbf{M}\oplus M^{\perp}$, *then* $M^{\perp}$ *is simultaneous and vectorial proximinal;*

(ii) if $\mathbf{K}=R^1_+$, *then for each* $s\in M^{\perp}$, *every* $\sigma\in M^{\perp}$ *is the only solution of following optimization problem*

$\mathbf{MIN_K}(\{(q_\alpha(U(\eta\text{-}s))): \eta\in\mathbf{X}' \textit{ and } \eta-\sigma\in\mathbf{M}\})$;

(iii) *for every* $\mathbf{x}\in\mathbf{X}'$ *its spline function* s_x *is the only solution for the next vectorial optimization problems:*

$\mathbf{MIN_K}(\{(q_\alpha(U(\eta\text{-}x))): \quad \eta\in M^{\perp}\})$, $\quad\mathbf{MIN_K}(\{(p_\alpha(x\text{-}y)): \quad y\in M^{\perp}\})$, $\mathbf{MIN_K}(\{(q_\alpha(Uy)): \ y\text{-}x\in\mathbf{M}\})$.

REFERENCES

1. Bahya A.O. - Ensembles coniquement bornés et cônes nucléaires dans les éspaces localement convexes separés. Théses de Doctorat de 3-ème Cycle en Mathématiques. L'école Normale Superieure - Takaddoum Rabat (Maroc), 1989.
2. Bahya A.O. - Etude des cônes nucléaires. Ann. Sci. Math. Québes 15, nr. 2, 1992, 123 - 134.
3. Borwein J.M., Zhuang D. - Super efficiency in vector optimization. Transactions of the American Mathemathical Society, vol. 338, nr. 1, 1993, 105 - 122.
4. Bucur I., Postolicã V. - A coincidence result between sets of efficient points and Choquet boundaries in separated locally convex spaces. Journ. Optimization, 36, 1996, 231 - 234.
5. Choquet G. - Lecture on Analysis 1 - 3 in "Mathematics Lecture Notes Series", Benjamin, New York/Amsterdam, 1969.
6. Ha T.X.D. - On the existence of efficient points in locally convex spaces. Research Paper, 1993 (to appear in Journal of Global Optimization, Kluwer Academic Publishers, The Netherlands).

7. Ha T.X.D. - A note on a class of cones ensuring the existence of efficient points in bounded complete sets. Research Paper, 1993 (to appear in Journal Optimization, Gordon and Breach Science Publishers).
8. Isac G. - Points critiques des systèmes dynamiques. Cones nucléaires et optimum de Pareto. Research Report, Royal Military College of St. Jean, Québec, Canada, 1981.
9. Isac G. - Sur l'existence de l'optimum de Pareto. Riv. Mat. Univ. Parma, 4, 9, 1983, 303 - 325.
10. Isac G. - Supernormal cones and absolute summability. Libertas Mathematica, vol. 5, 1985, 17 - 32.
11. Isac G., Postolică V. - The Best Approximation and Optimization in Locally Convex Spaces. Verlag Peter Lang, Frankfurt am Main, Germany, 1993.
12. Isac G. - Pareto optimization in infinite dimensional spaces: the importance of nuclear cones. Journ. of Math. Anal. And Appl., 182, no. 2, 1994, 393 - 404.
13. Krasnoselski, M. A. - Positive solutions of operator equations. Groningen, Noordhoff, 1964.
14. Luc D.T. - Theory of Vector Optimization. Springer Verlag, 1989.
15. Peressini, A. L. - Ordered Topological Vector Spaces. Harper & Row, New York, 1967.
16. Postolică V. - Spline functions in H-locally convex spaces. An. St. Univ. "Al. I. Cuza" Iasi, Romania, 27, 1981, 333 - 338.
17. Postolica V. - Vectorial optimization programs with multifunctions and duality. Ann. Sci. Math. Québec 10, nr. 1, 1986, 85 - 102.
18. Postolică V. - New existence results for efficient points in locally convex spaces ordered by supernormal cones. Journ. of Global Optimization 3, 1993, 233 - 242, Kluwer Academic Publishers, The Netherlands.
19. Postolică V. - Vectorial optimization in locally convex spaces ordered by supernormal cones and extensions. Research Report at the 4th Workshop of the German Working Group on Decision Theory, Hotel Talblick, Holzhau, Germany, March 14-18, 1994, 17 pp.
20. Postolică V. - Properties of Pareto sets in locally convex spaces. Journ. Optimization 34, 1995, 223 - 229.
21. Precupanu Th. - Espaces linéaires à semi-normes hilbertiennes. An. St. Univ. "Al. I. Cuza" Iasi, Romania, 15, 1969, 83 - 93.
22. Precupanu Th. - Scalar minimax properties in vectorial optimization. International Series of Numerical Mathematics, Birkhäuser Verlag Basel, vol. 107, 1992, 299 - 306.
23. Sterna - Karwat, A. - On the existence of cone maximal points in real topological linear spaces. Israel Journal of Mathematics, 54, 1, 1986, 33 - 41.

Existence Theorems for Cone Saddle Points and Vector-Valued Minimax Theorems[1]

Tamaki Tanaka

Department of Information Science, Faculty of Science, Hirosaki University, Hirosaki 036, Japan.

Abstract. This paper is concerned with vector-valued minimax theorems. We introduce a vector-valued version of lower semicontinuity, which is a generalization of the ordinary lower semicontinuity on real-valued functions in order to characterize efficient solutions. We give some conditions assuring the existence of cone saddle points of a vector-valued function. We present minimax theorems for vector-valued functions.

Keywords. Minimax theorems, multicriteria games, vector optimization.

1. Introduction

To begin with, we recall that saddle point theorem in usual game theory says: a real-valued payoff function possesses a saddle point if and only if the minimax value and the maximin value of the function are coincident; and that minimax theorems say: the minimax and maximin values are coincident under certain conditions. In the recent decade, many researchers have studied vector-valued minimax theorems. The common topic is whether or not games with multiple noncomparable criteria have an acceptable theory similar to standard results for scalar games, in particular, what type of minimax equation or inequality holds. In 1983, Nieuwenhuis gave his pioneer idea [10] to this area, and then Corley and Ferro presented important results; [3] and [4, 5, 6]. The author have separately researched such minimax problems in general setting and proved minimax theorems, existence theorems for saddle points, and saddle point theorems in [11, 12, 13, 14, 15, 16, 17]. These results have been approached by vector optimization method.

In Russia, on the other hand, vector-valued minimax problems have been researched by means of differential game's approach. In particular, Zhukovskiy and Salukvadze have dedicated to one-step antagonistic games with vector-valued payoff functions and positional differential games with vector-valued payoff functions in [19].

The aim of the paper is to explain recent author's minimax theorems for vector-valued functions. To this end, we recall efficient points and their properties, and then define and characterize vector-valued minimax and maximin, and vector-valued saddle points. We encounter a few difficulties of interpretation for minimax and maximin strategies and their values as well as those

[1] This paper was written when the author was a visitor at the School of Mathematical Studies, University of Portsmouth in England, and it was based on research 08740129 supported by Grant-in-Aid for Scientific Research from the Ministry of Education, Science and Culture of Japan.

of guaranteeing and checking existences of their values. Under suitable conditions, however, we observe that vector-valued minimax theorems and saddle point problems are closely connected with each other, whose result is similar to standard one for scalar games. Some elementary results and tools in this paper are based on [15], [16], and [17].

The organization of the paper is as follows. In Section 2, we give the preliminary terminology used throughout the paper, and then give an formulation for a multicriteria two-person zero-sum game, and define an optimal response strategy. In Section 3, we give some generalization of the saddle point concept, which are called "cone saddle points," and three types of existence theorems for such saddle point. In Section 4, we state a saddle point theorem and three types of minimax theorems for vector-valued function.

2. Vector-Valued Miminax and Maximin

Throughout this paper, let Z be an ordered t.v.s. with an ordering defined by a convex cone C. The convex cone C is assumed to be pointed, i.e., $C \cap (-C) = \{0\}$, and hence the ordering is antisymmetric and $C \ni 0$. Let Z^* denote the topological dual space of Z. We define the nonnegative polar cone C^* by $C^* := \{z^* \in Z^* \mid \langle z^*, z\rangle \geq 0, \forall z \in C\}$.

Moreover, for the convenience, we assume that int C is nonempty, and hence $C^0 := (\text{int } C) \cup \{0\}$ is a pointed convex cone and induces another (antisymmetric) vector ordering $\leq_{C^0}$ weaker than $\leq_C$ in Z. With respect to each of the orderings $\leq_C$ and $\leq_{C^0}$, we define minimal elements and maximal elements of a subset A of Z, i.e., lower efficient points and upper efficient points for the case of a pointed convex cone C. As the concept of efficient point, we will adopt "cone extreme point," the concept of which was proposed by Yu in [18]. An element z_0 of a subset A of Z is said to be a C-minimal point of A if $\{z \in A \,|\, z \leq_C z_0\,,\ z \neq z_0\} = \emptyset$, and a C-maximal point of A if $\{z \in A \,|\, z_0 \leq_C z, z \neq z_0\} = \emptyset$; which are equivalent to $A \cap (z_0 - C) = \{z_0\}$ and $A \cap (z_0 + C) = \{z_0\}$, respectively. We denote the set of such all C-minimal [resp. C-maximal] points of A by Min$A$ [resp. Max$A$]. Also, C^0-minimal and C^0-maximal points of A are defined similarly, and denoted by $\text{Min}_{\text{w}}A$ and $\text{Max}_{\text{w}}A$, respectively. These C^0-minimality and C^0-maximality are weaker concepts than C-minimality and C-maximality, respectively: it should be remarked that $\text{Min}A \subset \text{Min}_{\text{w}}A \subset A$ and $\text{Max}A \subset \text{Max}_{\text{w}}A \subset A$. Moreover, $\text{Min}\alpha A = \alpha\text{Min}A$, $\text{Max}\alpha A = \alpha\text{Max}A$, $\text{Min}_{\text{w}}\alpha A = \alpha\text{Min}_{\text{w}}A$, $\text{Max}_{\text{w}}\alpha A = \alpha\text{Max}_{\text{w}}A$ for any $\alpha > 0$; and $\text{Min}(A + a) = \text{Min}A + a$, $\text{Max}(A + a) = \text{Max}A + a$, $\text{Min}_{\text{w}}(A + a) = \text{Min}_{\text{w}}A + a$, $\text{Max}_{\text{w}}(A + a) = \text{Max}_{\text{w}}A + a$ for any $a \in Z$.

We say that the set A has the "domination property" (e.g., see page 697 in [8] and page 53 in [9]) if $\text{Min}A \neq \emptyset$ [resp. $\text{Max}A \neq \emptyset$], and $A \subset \text{Min}A + C$ [resp. $A \subset \text{Max}A - C$]. We need the following lemmas; see Lemmas 5.2 and 5.4 in [16] or Lemmas 1.2.3 and 1.2.5 in [15]:

Lemma 2.1 Let A be a subset of Z. If the convex cone C of Z satisfies the condition

$$\text{cl}\, C + (C \setminus \{0\}) \subset C \tag{2.1}$$

and if A is nonempty and compact, then $\text{Min}A \neq \emptyset$, $A \subset \text{Min}A + C$ and $\text{Max}A \neq \emptyset$, $A \subset \text{Max}A - C$.

Lemma 2.2 Let A be a subset of Z. If A is nonempty and compact, then $\mathrm{Min}_{\mathrm{w}}A \neq \emptyset$, $A \subset \mathrm{Min}_{\mathrm{w}}A + C^0$ and $\mathrm{Max}_{\mathrm{w}}A \neq \emptyset$, $A \subset \mathrm{Max}_{\mathrm{w}}A - C^0$. Moreover, $\mathrm{Min}_{\mathrm{w}}A$ and $\mathrm{Max}_{\mathrm{w}}A$ are compact sets.

Next, we introduce certain concepts of vector-valued minimax and maximin and vector-valued saddle points. A multicriteria two-person zero-sum game, or a two-person vector-valued zero-sum game, is a 4-tuple $\Gamma := (X, Y, -f, f)$, where X and Y are nonempty sets and f is a mapping $f : X \times Y \to Z$. The set X (resp. Y) is the set of strategies of player 1 (resp. player 2) and the mapping $-f$ (resp. f) is the payoff function of this player. (Possibly such strategies are mixed strategies and such payoff functions are the expected payoffs.) When players 1 and 2 choose strategies $x \in X$ and $y \in Y$, respectively, the payoffs with respect to players 1 and 2 are given by vectors $-f(x, y)$ and $f(x, y)$, respectively. Each player chooses a strategy in order to increase his payoff and wants to find a strategy maximizing his payoff. Thus, player 1 is regarded as a minimizer and player 2 as a maximizer.

Under this game, there are two approaches for finding equilibrium strategies. One is to find optimal response strategy pairs, and the other is to find optimal security strategy pairs. These concepts are coincident when $Z = \boldsymbol{R}$, i.e., the payoff function f is scalar-valued. The concepts of optimal response strategy and optimal security strategy are given in [7].

In this paper, we will adopt the approach of finding optimal response strategy pairs. To this end, we generalize the concept of optimal response strategy to more general type.

Definition 2.1 A strategy $x_0 \in X$ is said to be an optimal [resp. weak optimal] response strategy for player 1 against a strategy $y \in Y$ of player 2 if $f(x_0, y) \in \mathrm{Min} f(X, y)$ [resp. $f(x_0, y) \in \mathrm{Min}_{\mathrm{w}} f(X, y)$]. Similarly, a strategy $y_0 \in Y$ is said to be an optimal [resp. weak optimal] response strategy for player 2 against a strategy $x \in X$ of player 1 if $f(x, y_0) \in \mathrm{Max} f(x, Y)$. [resp. $f(x, y_0) \in \mathrm{Max}_{\mathrm{w}} f(x, Y)$]. The sets of all optimal response strategies for each player against an opponent's given strategy ($y \in Y$ and $x \in X$) are defined by $R_1(y)$ and $R_2(x)$ [resp. $R_1^{\mathrm{w}}(y)$ and $R_2^{\mathrm{w}}(x)$], respectively.

For the convenience, we denote each optimal response set as follows:

$$D_1 := \{(x, y) \in X \times Y \mid y \in R_2(x), x \in X\}, \tag{2.2}$$

$$D_2 := \{(x, y) \in X \times Y \mid x \in R_1(y), y \in Y\}. \tag{2.3}$$

Similarly, weak optimal response sets are denoted by D_1^{w} and D_2^{w}. It should be remarked that for any $y \in Y$ and $x \in X$, $R_1(y) \subset R_1^{\mathrm{w}}(y)$, $R_2(x) \subset R_2^{\mathrm{w}}(x)$ and that $D_1 \subset D_1^{\mathrm{w}}$, $D_2 \subset D_2^{\mathrm{w}}$. Then, we find the following idea of equilibrium strategies.

Definition 2.2 A point (x_0, y_0) is said to be an equilibrium optimal response strategy pair [resp. equilibrium weak optimal response strategy pair] of the game if $x_0 \in R_1(y_0)$ and $y_0 \in R_2(x_0)$ [resp. $x_0 \in R_1^{\mathrm{w}}(y_0)$ and $y_0 \in R_2^{\mathrm{w}}(x_0)$]. The set of all equilibrium optimal response strategy pairs (resp. equilibrium optimal response strategy pairs) is given by $D_1 \cap D_2$ [resp. $D_1^{\mathrm{w}} \cap D_2^{\mathrm{w}}$].

These concepts coincide with those of C-saddle point and weak C-saddle point of f, respectively; $(x_0, y_0) \in D_1 \cap D_2$ [resp. $(x_0, y_0) \in D_1^{\mathrm{w}} \cap D_2^{\mathrm{w}}$] if and only if $f(x_0, y_0) \in \mathrm{Max} f(x_0, Y) \cap \mathrm{Min} f(X, y_0)$,[resp. $f(x_0, y_0) \in \mathrm{Max_w} f(x_0, Y) \cap \mathrm{Min_w} f(X, y_0)$]. Such concepts of generalized saddle point are defined in Section 3.

We called the following subsets of Z

$$\mathrm{Min} \bigcup_{x \in X} \mathrm{Max_w} f(x, Y), \qquad \mathrm{Max} \bigcup_{y \in Y} \mathrm{Min_w} f(X, y) \tag{2.4}$$

the set of all minimax values of f and the set of all maximin values of f, respectively (see [16] and [15]). Also, we call a strategy x^* [resp. y^*] attaining a minimax value [resp. a maximin value] above, i.e., satisfying $f(x^*, y) \in \mathrm{Min} f(D_1^{\mathrm{w}})$ for some $y \in R_2^{\mathrm{w}}(x^*)$ [resp. $f(x, y^*) \in \mathrm{Max} f(D_2^{\mathrm{w}})$ for some $x \in R_1^{\mathrm{w}}(y^*)$] a minimax strategy [resp. a maximin strategy] of player 1 [resp. player 2], written $\underset{x \in X}{\mathrm{mini}}\underset{y \in Y}{\mathrm{max}}\ f(x, y)$ [resp. $\underset{y \in Y}{\mathrm{maxi}}\underset{x \in X}{\mathrm{min}}\ f(x, y)$].

Theorem 2.1 (See Lemma 5.5 in [16].) Let X and Y be nonempty compact sets in two topological spaces, respectively. If a vector-valued function $f : X \times Y \to Z$ is continuous, and if C satisfies condition (2.1), then

$$\left[\mathrm{Min} \bigcup_{x \in X} \mathrm{Max_w} f(x, Y)\right] + C \supset \mathrm{Max_w} f(x', Y) \neq \emptyset, \tag{2.5}$$

$$\left[\mathrm{Max} \bigcup_{y \in Y} \mathrm{Min_w} f(X, y)\right] - C \supset \mathrm{Min_w} f(X, y') \neq \emptyset \tag{2.6}$$

for each $x' \in X$ and $y' \in Y$.

The theorem above shows that under the conditions, there exists at least one optimal response strategy for each player against an opponent's given strategy, and that there exist a minimax strategy and a maximin strategy, too.

3. Existence of Cone Saddle Points

Under notation given in Section 2, we will give reasonable definitions for saddle point and saddle value of a vector-valued function. Let $f : X \times Y \to Z$ be a vector-valued function.

Definition 3.1 (i) A point (x_0, y_0) is said to be a C-saddle point of f with respect to $X \times Y$, if $f(x_0, y_0) \in \mathrm{Max} f(x_0, Y) \cap \mathrm{Min} f(X, y_0)$;

(ii) A point (x_0, y_0) is said to be a weak C-saddle point of f with respect to $X \times Y$, if $f(x_0, y_0) \in \mathrm{Max_w} f(x_0, Y) \cap \mathrm{Min_w} f(X, y_0)$.

We note that any C-saddle point of f is a weak C-saddle point of f obviously. Also, in the case $C^0 = C$, the two concepts are coincident. For the convenience, we denote the set of all C-saddle values [resp. weak C-saddle values] of f by $\mathrm{SV}(f)$ [resp. $\mathrm{SV_w}(f)$].

Now, we give the definition of lower level-closedness (i.e., C-semicontinuous in Definition 2.4 in [2]).

Definition 3.2 Let X be a topological space. A vector-valued function $f : X \to Z$ is said to be lower level-closed (or C-semicontinuous) if $f^{-1}(z - \mathrm{cl}\, C)$ is closed in X for each $z \in Z$.

First of all, we state the first existence theorem for generalized saddle points, which is a generalization of Theorem 3.2.1 in [15].

Theorem 3.1 (See Theorem 3.1 in [17].) Let X and Y be nonempty compact sets in two topological spaces, respectively. A vector-valued function $f : X \times Y \to Z$ has at least one weak C-saddle point if one of the following conditions holds:

(i) f is of the type $f(x,y) = u(x) + v(y)$ where u and $-v$ are lower level-closed;

(ii) f is of the type $f(x,y) = u(x) + \beta(x)v(y)$ where u is continuous, $-v$ is lower level-closed, and $\beta . X \to \boldsymbol{R}_+$ is continuous.

If, in addition, C satisfies condition (2.1), then f has at least one C-saddle point.

Remark 3.1 Since every continuous function is lower level-closed obviously, the theorem above holds for continuous functions u and v. Hence, the theorem is a generalization of Lemma 3.4 in [14].

To obtain another generalized version of Lemma 3.4 in [14], we introduce a vector-valued version of lower semicontinuity, which is a generalization of the ordinary lower semicontinuity on real-valued functions.

Definition 3.3 Let X be a topological space. A vector-valued function $f : X \to Z$ is said to be C-lower semicontinuous on X if it satisfies the three equivalent conditions (See Proposition 2.1 in [17]):

(i) For all $a \in Z$, $f^{-1}(a + \mathrm{int}\, C)$ is open;

(ii) For each $x_0 \in X$ and any open neighborhood V of $f(x_0)$, there exists an open neighborhood U of x_0 such that $f(x) \in V + C$ for all $x \in U$;

(iii) For each $x_0 \in X$ and any $d \in \mathrm{int}\, C$, there exists an open neighborhood U of x_0 such that $f(x) \in f(x_0) - d + \mathrm{int}\, C$ for all $x \in U$.

Also, it is said to be C-upper semicontinuous on X if $-f$ is C-lower semicontinuous on X.

Then, we can easily prove the following lemma.

Lemma 3.1 Let X be a topological space. If f and g are C-lower semicontinuous functions from X to Z, then

(i) $f+g$ is C-lower semicontinuous;

(ii) αf is C-lower semicontinuous for each $\alpha > 0$;

Moreover, if $\beta : X \longrightarrow \boldsymbol{R}_+$ is lower semicontinuous, then for each $v \in C$,

(i) $x \longmapsto \beta(x)v$ is C-lower semicontinuous;

(ii) $x \longmapsto f(x) + \beta(x)v$ is C-lower semicontinuous.

Theorem 3.2 (See Theorem 3.2 in [17].) Let X and Y be nonempty compact sets in two topological spaces, respectively. A vector-valued function $f : X \times Y \longrightarrow Z$ has at least one weak C-saddle point if f is of the type $f(x,y) = u(x) + \beta(x)v(y)$ where u is C-lower semicontinuous, v is C-upper semicontinuous, and $\beta : X \longrightarrow \boldsymbol{R}_+$ is lower semicontinuous, $C \cap \mathrm{Max_w} v(Y) \neq \emptyset$. If, in addition, C satisfies condition (2.1), then f has at least one C-saddle point.

Second, to present the second existence theorem for generalized saddle points, we recall a kind of convexity of vector-valued functions.

Definition 3.4 Let X be a convex set in a real vector space. A vector-valued function $f : X \longrightarrow Z$ is said to be naturally quasi C-convex on X if

$$f(\lambda x_1 + (1-\lambda)x_2) \in \mathrm{co}\,\{f(x_1), f(x_2)\} - C \tag{3.1}$$

for every $x_1, x_2 \in X$ and $\lambda \in [0,1]$, where co$A$ denotes the convex hull of the set A. The condition (3.1) is equivalent to the following condition: there exists $\mu \in [0,1]$ such that $f(\lambda x_1 + (1-\lambda)x_2) \leq_C \mu f(x_1) + (1-\mu) f(x_2)$. Also, it is said to be naturally quasi C-concave on X if $-f$ is naturally quasi C-convex on X.

Remark 3.2 We note that every C-convex function is naturally quasi C-convex, and that every properly quasi C-convex function is naturally quasi C-convex (See [16]). Moreover, if a mapping $f : X \longrightarrow Z$ is naturally quasi C-convex on X then for each $\varphi \in Z^*$, the composite mapping $\varphi \circ f$ is a (ordinary) quasi convex function. If a mapping $f : X \longrightarrow Z$ is C-lower semicontinuous on X, then for each $\varphi \in C^*$, the composite mapping $\varphi \circ f$ is lower semicontinuous. Similarly, if $f : X \longrightarrow Z$ is C-upper semicontinuous on X, then for each $\varphi \in C^*$, the composite mapping $\varphi \circ f$ is upper semicontinuous.

Then, we have the second existence theorem of weak C-saddle points, which generalizes Lemma 3.3 in [14] and Theorem 3.1 in [16].

Theorem 3.3 (See Theorem 3.3 in [17].) Let X and Y be nonempty compact convex sets in two t.v.s.'s, respectively. If a vector-valued function $f : X \times Y \longrightarrow Z$ satisfies that

(i) $x \longmapsto f(x,y)$ is C-lower semicontinuous and naturally quasi C-convex on X for every $y \in Y$;

(ii) $y \longmapsto f(x,y)$ is C-upper semicontinuous and naturally quasi C-concave on Y for every $x \in X$, then the vector-valued function f has at least one weak C-saddle point.

At last, we shall give the third existence theorem for generalized saddle points.

Theorem 3.4 (See Theorems 4.1 in [11] and 3.1 in [12].) Let X and Y be nonempty compact convex sets in two l.c.s.'s, respectively. If a vector-valued function $f : X \times Y \to Z$ is continuous and if sets $T(y) := \{x \in X | f(x, y) \in \mathrm{Min_w} f(X, y)\}$ and $U(x) := \{y \in Y | f(x, y) \in \mathrm{Max_w} f(x, Y)\}$ are convex for every $y \in Y$ and $x \in X$, respectively, then the vector-valued function f has at least one weak C-saddle point.

4. Minimax Theorems for Vector-Valued Functions

In a few of the author's papers, he has proposed some minimax theorems for vector-valued functions. In this section, we shall present some of most general versions of such minimax theorems. To this end, we need the following saddle point theorem of vector-valued function, which is a corollary of Theorem 2.1.

Theorem 4.1 (Saddle Point Theorem) Let X and Y be nonempty compact sets in two topological spaces, respectively. If a vector-valued function $f : X \times Y \to Z$ is continuous and if C satisfies condition (2.1), then

$$\left[\mathrm{Min} \bigcup_{x \in X} \mathrm{Max_w} f(x, Y)\right] + C \supset \mathrm{SV_w}(f), \quad \left[\mathrm{Max} \bigcup_{y \in Y} \mathrm{Min_w} f(X, y)\right] - C \supset \mathrm{SV_w}(f).$$

Hence, if f has a weak C-saddle point $(x_0, y_0) \in X \times Y$, then there exist

$$z_1 \in \mathrm{Min} \bigcup_{x \in X} \mathrm{Max_w} f(x, Y) \quad \text{and} \quad z_2 \in \mathrm{Max} \bigcup_{y \in Y} \mathrm{Min_w} f(X, y)$$

such that $z_1 \leq_C f(x_0, y_0)$ and $f(x_0, y_0) \leq_C z_2$.

We will say that the minimax inequality holds if there exist

$$z_1 \in \mathrm{Min} \bigcup_{x \in X} \mathrm{Max_w} f(x, Y) \quad \text{and} \quad z_2 \in \mathrm{Max} \bigcup_{y \in Y} \mathrm{Min_w} f(X, y)$$

such that $z_1 \leq_C z_2$.

Theorem 4.2 (Minimax Theorem I) Let X and Y be nonempty compact sets in two topological spaces, respectively. If a vector-valued function $f : X \times Y \to Z$ is of the type $f(x, y) = u(x) + \beta(x) v(y)$ where u and v are continuous and β is continuous into $\boldsymbol{R}_+$, and if C satisfies condition (2.1), then the minimax inequality holds. If, in particular, the vector-valued function f is of the type $f(x, y) = u(x) + v(y)$ where u and v are continuous, then

$$\mathrm{Max} \bigcup_{y \in Y} \mathrm{Min_w} f(X, y) \subset \left[\mathrm{Min} \bigcup_{x \in X} \mathrm{Max_w} f(x, Y)\right] + C$$

and

$$\mathrm{Min} \bigcup_{x \in X} \mathrm{Max_w} f(x, Y) \subset \left[\mathrm{Max} \bigcup_{y \in Y} \mathrm{Min_w} f(X, y)\right] - C.$$

Proof. The first part of the proof follows immediately from Theorems 3.1 and 4.1. The second part of the proof can be done in the same way as in the proof of Theorem 3.2 in [11]. □

Theorem 4.3 (Minimax Theorem II) Let X and Y be nonempty compact convex sets in two t.v.s.'s, respectively. If a vector-valued function $f : X \times Y \to Z$ is continuous and satisfies:

(i) $x \longmapsto f(x, y)$ is naturally quasi C-convex on X for every $y \in Y$;

(ii) $y \longmapsto f(x, y)$ is naturally quasi C-concave on Y for every $x \in X$,

and if C satisfies condition (2.1), then the minimax inequality holds.

Proof. The proof follows immediately from Theorems 3.3 and 4.1. □

Theorem 4.4 (Minimax Theorem III) Let X and Y be nonempty compact convex sets in two l.c.s.'s, respectively. Assume that a vector-valued function $f : X \times Y \to Z$ is continuous and that C satisfies condition (2.1). If the sets $T(y)$ and $U(x)$ in Theorem 3.4 are convex for every $y \in Y$ and $x \in X$, respectively, then the minimax inequality holds.

Proof. The proof follows immediately from Theorems 3.4 and 4.1. □

5. Conclusions

Minimax theorems for vector-valued functions have a similar statement to the ordinary minimax theorems for real-valued functions. For a real-valued function F, the following inequality

$$\min_{x \in A} \sup_{y \in B} F(x, y) \leq F(x_0, y_0) \leq \max_{y \in B} \inf_{x \in A} F(x, y) \tag{5.1}$$

holds under suitable conditions which are sufficient for the function F to possess a saddle point $(x_0, y_0) \in A \times B$. Replacing the usual total ordering $\leq$ by a general partial ordering $\leq_C$, defined by a pointed convex cone C, we may expect to get a similar minimax inequality to (5.1).

This has come true in Section 4, and minimax theorems presented in Section 4 tell us that there exist some minimax strategy and maximin strategy of f such that their values are ordered by $\leq_C$ and dominated each other whenever f has a weak C-saddle point. First type minimax theorem means that minimax values and maximin values of f are entirely contained in the set of maximin values of f minus the pointed convex cone C and in the set of minimax values of f plus the pointed convex cone C, respectively. Also, second and third type minimax theorems mean that there exist some minimax values and maximin values of f such that both vectors are ordered by $\leq_C$ and dominated each other.

Unfortunately, minimax strategies [resp. maximin strategies] do not always give equilibrium values, but may give equilibrium values in the sense of security levels in the case that minimax strategies and maximin strategies satisfy some conditions. Thus, we note that the minimax values and maximin values are lower bounds and upper bounds of such equilibrium values, respectively.

Acknowledgements

The author would like to thank Professor R. Caballero of University of Malaga and Professor M. Tamiz of University of Portsmouth for this opportunity of giving a talk about this work.

References

[1] Chen GY (1991) A generalized section theorem and a minimax inequality for a vector-valued mapping. Optimization 22:745–754.

[2] Corley HW (1980) An existence result for maximizations with respect to cones. J Optim Theory Appl 31:277–281.

[3] Corley HW (1985) Games with vector payoffs. J Optim Theory Appl 47:491–498.

[4] Ferro F (1982) Minimax type theorems for n-valued functions. Annali di Matematica Pura ed Applicata, 32:113–130.

[5] Ferro F (1989) A minimax theorem for vector-Valued functions. J Optim Theory Appl 60:19–31.

[6] Ferro F (1991) A minimax theorem for vector-valued functions, part 2. J Optim Theory Appl 68:35–48.

[7] Ghose D, Prasad UR (1989) Solution concepts in two-person multicriteria games. J Optim Theory Appl 63:167–189.

[8] Luc DT (1989) An existence theorem in vector optimization. Math Oper Res 14:693–699.

[9] Luc DT (1989) Theory of Vector Optimization. Lecture Notes in Economics and Mathematical Systems, Vol.319, Springer-Verlag, Berlin.

[10] Nieuwenhuis JW (1983) Some minimax theorems in vector-valued functions. J Optim Theory Appl 40:463–475.

[11] Tanaka T (1988) Some minimax problems of vector-valued functions. J Optim Theory Appl 59:505–524.

[12] Tanaka T (1989) Existence theorems for cone saddle points of vector-valued functions in infinite-dimensional spaces. J Optim Theory Appl 62:127–138.

[13] Tanaka T (1990) A characterization of generalized saddle points for vector-valued functions via scalarization. Nihonkai Mathematical Journal 1:209–227.

[14] Tanaka T (1991) Two types of minimax theorems for vector-valued functions. J Optim Theory Appl 68:321–334.

[15] Tanaka T (1992) Minimax Theorems in Vector Optimization. Ph.D. thesis, Niigata University, Niigata.

[16] Tanaka T (1994) Generalized quasiconvexities, cone saddle points, and a minimax theorem for vector-valued functions. J Optim Theory Appl 81:355–377.

[17] Tanaka T (1997) Generalized semicontinuity and existence theorems for cone saddle points. J Appl Math Optim, printed.

[18] Yu PL (1974) Cone convexity, cone extreme points, and nondominated solutions in decision problems with multiobjectives. J Optim Theory Appl 14:319–377.

[19] Zhukovskiy VI, Salukvadze ME (1994) The Vector-Valued Maximin. Academic Press, London.

A New Approach to Second Order Optimality Conditions in Vector Optimization

A. CAMBINI - L. MARTEIN - R. CAMBINI
Department of Statistics and Applied Mathematics,
University of Pisa, via Ridolfi 10, 56124 Pisa, ITALY

Abstract The aim of this paper is to establish some second order necessary and sufficient optimality conditions for a multiobjective problem where optimality is studied with respect to arbitrary closed convex cones. The proposed approach is an extension to the one recently given by the same authors.

1. Introduction

In these last years, the image space has been frequently used [5-8, 12, 13] as a general framework within which some topics related to optimization have been studied; recently, for a scalar problem, Cambini and Martein [9] have suggested a new approach for studying necessary second order optimality conditions based on a characterization of a suitable tangent cone in the image space [9]; such an approach has been deepened by R. Cambini in [10, 11], where several second order necessary and sufficient optimality conditions are obtained for a multiobjective Pareto problem and where a new look inside second order constraints qualification is given. The aim of this paper is to extend and to collect all of these recent results for a vector optimization problem where optimality is studied with respect to arbitrary closed convex cones.

2. Statement of the problem

In this paper we will consider the following multiobjective problem

$$P: \ U\text{-}\max\ \varphi(x)\ ,\ \ x\in S=\{x\in X:\ g(x)\in V\}\ ,$$

where $X\subset\mathbf{R}^n$ is an open set, $\varphi=(\varphi_1,\dots,\varphi_s):X\to\mathbf{R}^s$, $g=(g_1,\dots,g_m):X\to\mathbf{R}^m$, $s\geq 1$, $m\geq 1$, φ_j , g_i , $j=1,..,s$, $i=1,..m$ are twice differentiable functions, and $U\subset\mathbf{R}^s$, $V\subset\mathbf{R}^m$ are closed convex cones with vertices at the origin such that $\mathrm{int}U\neq\emptyset$, $\mathrm{int}V\neq\emptyset$. A point $x_0\in S$ is said to be a *local efficient point* for problem P if there is no a feasible x belonging to a suitable neighbourhood of x_0 such that:

$$\varphi(x)\in\varphi(x_0)+U^0 \tag{2.1}$$

where $U^0=U\setminus\{0\}$.

We say that x_0 is an *efficient point* for P if (2.1) holds for every $x\in S$.

Let us note that when $s=1$, $U=\mathbf{R}_+$, $V=\mathbf{R}^m_+$, problem P reduces to a scalar optimization problem and (2.1) collapses to the ordinary definition of a local

maximum point, while when $U=\mathbf{R}^s_+$, $V=\mathbf{R}^m_+$, (2.1) collapses to the ordinary definition of a local Pareto point.
Let x_0 be a feasible point and assume, without lost of generality, that $g(x_0)=0$.
Set: $f(x)=\varphi(x)-\varphi(x_0)$, $F(x)=(f(x),g(x))$, $K=F(X)$, $H=U^0\times V$.
We will refer to $\mathbf{R}^n$ as the *decision space* and to $\mathbf{R}^{s+m}$ as the *image space* .
It is easy to prove that x_0 is an efficient point if and only if:

$$K\cap H=\emptyset. \tag{2.2}$$

In such a way, the efficiency of x_0 is reduced to study the disjunction of these two sets. Since K does not have in general particular properties, the idea underlying our suggested approach is to substitute K with the following tangent cone, which is a subcone of the Bouligand tangent cone to K at $F(x_0)$:

$$T_1=\{\ t:\ \exists\alpha_n\to+\infty\ ,\ x_n\to x_0\ \text{ with }\ \alpha_nF(x_n)\to t\ \}.$$

By means of T_1 it can be proven [7] the following optimality conditions.

Theorem 2.1
i) Let x_0 be a local efficient point for problem P. Then $T_1\cap \mathrm{int}H=\emptyset$.
ii) If $T_1\cap \mathrm{cl}H=\{0\}$, then x_0 is a local efficient point for P.
iii) The feasible point x_0 is a local efficient point for P if and only if the following condition holds:
$\forall\ t\in T_1\cap \mathrm{cl}H$, $t\neq0$, and for any sequence $x_n\to x_0$ such that there exists $\alpha_n\to+\infty$ with $\alpha_nF(x_n)\to t$, we have $F(x_n)\notin H\ \ \forall n$.

Some necessary and sufficient first order optimality conditions have been established in [7] by means of a first order characterization of T_1.
The following second order characterization of T_1 has been suggested and utilized in [9] in establishing second order necessary optimality conditions for a scalar problem and in [10] in establishing second order necessary and sufficient optimality conditions for a vector Pareto problem:

$$T_1=(K_L+\bar{K}_Q)\cup A_2 \tag{2.3}$$

with: $K_L=F'(x_0)(\mathbf{R}^n)$, where $F'(x_0)$ is the Jacobian matrix of F at x_0,
$K_Q=\{F''(x_0)(w,w):\ w\in W\}$, $\bar{K}_Q=K_Q\cup\{0\}$, $W=\{w\in\mathbf{R}^n\text{: } F'(x_0)(w)=0\}$,
$F''(x_0)(w,w)=(w^TH_{f_1(x_0)}w,\ldots,w^TH_{f_s(x_0)}w,\ w^TH_{g_1(x_0)}w,\ldots,w^TH_{g_m(x_0)}w)$,
where $H_{f_j(x_0)}$, $H_{g_i(x_0)}$, are the Hessian matrices of the functions f_j, g_i, $j=1,\ldots,s$, $i=1,\ldots,m$ at the point x_0 respectively,

$$A_2=\{t\in T_1:\ t\neq0,\ \exists x_n\to x_0,\ \exists\alpha_n\to+\infty,\ \text{with } \alpha_n(F(x_n)-F(x_0))\to t,\ \frac{F(x_n)-F(x_0)}{\|x_n-x_0\|^2}\to0\}$$

The following Theorem [10] states a sufficient condition for A_2 to be empty.

Theorem 2.2 If $0\notin(K_L+K_Q)$ then $A_2=\emptyset$.

In the following sections we will extend the results given in [10] to arbitrary convex cones.

3. Second order necessary optimality conditions and regularity conditions

Taking into account (3.1) and i) of Theorem 2.1, we have the following second order necessary optimality condition for problem P.

Theorem 3.1 Let x_0 be a local efficient point for problem P. Then:

$$(K_L+\bar{K}_Q)\cap \text{int}H=\emptyset. \tag{3.1}$$

Let us note that (3.1) does not involve either Lagrange multipliers or second order constraints qualification unlike the classical second order necessary optimality condition of a scalar problem ($s=1$, $U=\mathbf{R}_+$, $V=\mathbf{R}^m_+$); the suggested approach in the image space allows to give a new look inside this aspect.
In order to find second order optimality conditions involving Lagrange multipliers, the problem of separation between the two cones $K_L+\bar{K}_Q$ and intH arises, that is the problem of the existence of $\alpha\in H^+=\{\alpha\colon \alpha^T h\geq 0\ \forall h\in H\}$ such that:

$$K_L+\bar{K}_Q\subseteq \alpha^{\perp}_{-}\text{, where } \alpha^{\perp}=\{z\in \mathbf{R}^{s+m}\colon \alpha^T z=0\},\ \alpha^{\perp}_{-}=\{z\in \mathbf{R}^{s+m}\colon \alpha^T z\leq 0\}.$$

With this regards we have the following Theorem holds, where co(•) denotes the convex hull of set (•).

Theorem 3.2 The following conditions

i) $\exists\alpha\in H^+$, $\alpha\neq 0$, such that $K_L\subseteq\alpha^{\perp}$ and $K_Q\subseteq\alpha^{\perp}_{-}$,

ii) $$\text{Co}(K_L+\bar{K}_Q)\cap\text{Int}(H)=\emptyset. \tag{3.2}$$

are equivalent. Furthermore the two cones $K_L+\bar{K}_Q$ and intH are separate if and only if i) or ii) holds.

Proof i)⇒ii) Suppose ab absurdo that $z\in \text{Co}(K_L+\bar{K}_Q)\cap\text{Int}H\neq\emptyset$; since $z\in\text{Co}(K_L+\bar{K}_Q)$, z is a convex combination of k elements of $(K_L+\bar{K}_Q)$, that is $\exists\ l_1,\dots,l_k\in K_L$, $\exists\ q_1,\dots,q_k\in K_Q$, $\exists\ \lambda_1,\dots,\lambda_k>0$ such that $\sum_{i=1}^{k}\lambda_i=1$ and $z=\sum_{i=1}^{k}\lambda_i(l_i+q_i)$; then we have $\alpha^T z=\sum_{i=1}^{k}\lambda_i\alpha^T l_i+\sum_{i=1}^{k}\lambda_i\alpha^T q_i=\sum_{i=1}^{k}\lambda_i\alpha^T q_i\leq 0$, and this is absurd since $z\in\text{Int}H$, $\alpha\in H^+$, $\alpha\neq 0$, implies $\alpha^T z>0$.

ii)⇒i) For a known separation theorem, $\exists\alpha\in H^+$, $\alpha\neq 0$, such that $\alpha^T z\leq 0$ $\forall z\in\text{Co}(K_L+\bar{K}_Q)$; since $K_L\subseteq\text{Co}(K_L+\bar{K}_Q)$ it results $\alpha^T F'(x_0)v\leq 0$ $\forall v\in\mathbf{R}^n$ and this implies $\alpha^T F'(x_0)=0$, that is $K_L\subseteq\alpha^{\perp}$; as a consequence $\alpha^T(l+q)=\alpha^T l+\alpha^T q=\alpha^T q\leq 0$ $\forall q\in K_Q$ so that $K_Q\subseteq\alpha^{\perp}_{-}$.
The last statement of the Theorem is obvious. ♦

Since (3.2) is equivalent to the existence of a vector of "multipliers" $\alpha \in H^+$, any condition which ensure the validity of (3.2) can be interpreted as a second order regularity condition (let us note that we use the term "regularity conditions" instead of "constraints qualification" since in the image space both the objective functions and the constraints are involved).
In such a way the suggested approach allows us to define two different kinds of regularity conditions [10]:

weak regularity Assume that $(K_L+\bar{K}_Q)\cap \mathrm{Int}H=\emptyset$. We will say that a condition $\mathcal{R}_w$ is a weak second order regularity condition if it implies:

$$\mathrm{Co}(K_L+\bar{K}_Q)\cap \mathrm{Int}H=\emptyset$$

strong regularity Assume that $(K_L+\bar{K}_Q)\cap \mathrm{Int}H=\emptyset$. We will say that a condition $\mathcal{R}_s$ is a strong second order regularity condition if it implies:

$$K_Q \subseteq K_L - \mathrm{Cl}H$$

Some second order regularity conditions are given in the following Theorems [10]:

Theorem 3.3 Consider problem P; then the following conditions are weak second order regularity conditions:

i) $K_L+\bar{K}_Q$ is a convex cone,

ii) $\bar{K}_Q$ is a convex cone,

iii) $\dim(K_L)= n-1$.

Theorem 3.4 Consider problem P; then the following conditions are strong second order strict regularity conditions:

i) $K_Q \subseteq K_L$,

ii) $K_Q = \emptyset$,

iii) $\dim(K_L)= n$,

iv) $\dim(K_L)= s+m-1$.

Remark 3.1 When s=1, the linearly independence of the gradients of the constraints binding at x_0 is equivalent to iv) of Theorem 3.4.
Another second order strong regularity condition in the decision space which generalizes, in the scalar case, the McCormick constraint qualification can be found in [9,10].

Let us note that strong regularity implies weak regularity but the converse is not true as is shown in the following example:

Example 3.1 Consider problem P where s=1, m=1, $\varphi(x,y)=x^2+y^3$, $g(x,y)=-x^2-y^2$ $U=\mathbf{R}_+$, $V=\mathbf{R}_+$, and the feasible point (0,0). Since (0,0) is the only feasible point then it is also an efficient point.
It is easy to verify that $K_L=\{(0,0)\}$, $K_Q=\{q: q=\lambda(1,-1)+\mu(0,-1), \lambda,\mu\geq 0, \lambda+\mu\neq 0\}$, so that $co(K_L+\bar{K}_Q)=(K_L+\bar{K}_Q)$ and the weak regularity condition i) of Theorem 3.3 holds; on the other hand the vector $\alpha=(2,1)^T$ is such that $K_L\subseteq\alpha^{\perp}$ but $K_Q\not\subseteq\alpha^{\perp}_{-}$ so that strong regularity cannot hold.

Now we are able to state some second order necessary optimality conditions in the image space and in the decision space.

Theorem 3.5
Let x_0 be a local efficient point for problem P. Then:
i) if a second order weak regularity condition $\mathcal{R}_w$ holds then:

$$\exists\ \alpha\in H^+,\ \alpha\neq 0,\ \text{such that}\ \ K_L\subseteq\alpha^{\perp}\ \text{and}\ \ K_Q\subseteq\alpha^{\perp}_{-},$$

ii) if a second order strong regularity condition $\mathcal{R}_s$ holds then:

$$\forall\ \alpha\in H^+,\ \alpha\neq 0,\ \text{such that}\ \ K_L\subseteq\alpha^{\perp},\ \text{we have}\ \ K_Q\subseteq\alpha^{\perp}_{-}.$$

Proof It follows immediately from Theorems 3.1 and 3.2 and by the given definitions of regularity. ♦

Theorem 3.5 is equivalent to the following one stated in the decision space:

Theorem 3.6
Let x_0 be a local efficient point for problem P.
i) if a second order weak regularity condition $\mathcal{R}_w$ holds then:

$$\exists\alpha=(\bar{\alpha},\alpha^*)\in H^+,\ \bar{\alpha}\in\mathbf{R}^s,\ \alpha^*\in\mathbf{R}^m,\ \text{such that}\ \sum_{i=1}^{s}\bar{\alpha}_i\nabla f_i(x_0)+\sum_{j=1}^{m}\alpha^*_j\nabla g_j(x_0)=0$$

$$\text{and}\ \ w^T\Big[\sum_{i=1}^{s}\bar{\alpha}_i H_{f_i}(x_0)+\sum_{j=1}^{m}\alpha^*_j H_{g_j}(x_0)\Big]w\leq 0\quad\forall w\in W$$

ii) if a second order strong regularity condition $\mathcal{R}_s$ holds then :

$$\forall\alpha=(\bar{\alpha},\alpha^*)\in H^+,\ \bar{\alpha}\in\mathbf{R}^s,\ \alpha^*\in\mathbf{R}^m,\ \text{such that}\ \sum_{i=1}^{s}\bar{\alpha}_i\nabla f_i(x_0)+\sum_{j=1}^{m}\alpha^*_j\nabla g_j(x_0)=0$$

$$\text{it results:}\ \ w^T\Big[\sum_{i=1}^{s}\bar{\alpha}_i H_{f_i}(x_0)+\sum_{j=1}^{m}\alpha^*_j H_{g_j}(x_0)\Big]w\leq 0\quad\forall w\in W$$

Remark 3.2 When s=1, ii) of Theorem 3.6 reduces to the well known and classic second order necessary optimality condition.

4. Second order sufficient optimality conditions in the image space

In this section we will derive some second order sufficient optimality conditions by means of the second order characterization of the tangent cone T_1.

As a direct consequence of ii) of Theorem 2.1, taking into account (2.3) and Theorem 2.2 we have:

Theorem 4.1 If (4.1) holds

$$(K_L+\bar{K}_Q)\cap \mathrm{Cl}H=\{0\} \text{ and } 0\notin(K_L+K_Q)\ , \tag{4.1}$$

then x_0 is a local efficient point for P.

Let us note that Theorem 4.1 is stated without use of multipliers; obviously any condition which ensures (4.1) becomes a second order sufficient optimality condition. In this order of ideas, set $\alpha^{\perp}_{\leqq}=\{z\in\mathbf{R}^{s+m}:\ \alpha^Tz<0\}$, we have the following corollaries.

Corollary 4.1 If i) and ii) hold then $x_0\in S$ is a local efficient point for P:

i) $0\notin(K_L+K_Q)$,

ii) $\exists\ \alpha\in \mathrm{int}H^+$ such that $K_L\subseteq\alpha^{\perp}$ and $K_Q\subseteq\alpha^{\perp}_{-}$.

Proof. We will prove that i) and ii) imply (4.1).

Assume that $(K_L+\bar{K}_Q)\cap \mathrm{Cl}H\neq\{0\}$; then there exist $l\in K_L$, $q\in K_Q$ such that $l+q\in \mathrm{cl}H\setminus\{0\}$, so that $\alpha^T(l+q)>0$; on the other hand for ii) we have $\alpha^T(l+q)=\alpha^Tl+\alpha^Tq=\alpha^Tq\leq 0$ and this is absurd. ♦

Corollary 4.2 If the following condition (4.2) holds:

$$\exists\alpha\in \mathrm{int}H^+ \text{ such that } K_L\subseteq\alpha^{\perp} \text{ and } K_Q\subseteq\alpha^{\perp}_{\leqq}\ , \tag{4.2}$$

then $x_0\in S$ is a local efficient point for problem P.

Corollaries 4.1 and 4.2 are equivalent in the decision space to Theorems 4.2 and 4.3, respectively:

Theorem 4.2 Consider problem P and assume that $x_0\in S$ verifies the following conditions:

i) $\exists\ \alpha=(\bar{\alpha},\alpha^*)\in \mathrm{int}H^+$, such that $\sum_{i=1}^{s}\bar{\alpha}_i\nabla f_i(x_0)+\sum_{j=1}^{m}\alpha^*_j\nabla g_j(x_0)=0$

ii) $w^T\left[\sum_{i=1}^{s}\bar{\alpha}_iH_{f_i}(x_0)+\sum_{j=1}^{m}\alpha^*_jH_{g_j}(x_0)\right]w\leq 0\ \ \forall w\in W$

iii) $0\notin(K_L+K_Q)$

Then x_0 is a local efficient point for P.

Theorem 4.3 Consider problem P and assume $x_0 \in S$ verifies conditions i) and ii):

i) $\exists\, \alpha=(\bar{\alpha},\alpha^*)\in \mathrm{int}H^+$, $\bar{\alpha}\in \mathbf{R}^s$, $\alpha^*\in \mathbf{R}^m$, such that $\sum_{i=1}^{s} \bar{\alpha}_i \nabla f_i(x_0) + \sum_{j=1}^{m} \alpha^*_j\, g_j(x_0)=0$

ii) $w^T [\, \sum_{i=1}^{s} \bar{\alpha}_i H_{f_i}(x_0) + \sum_{j=1}^{m} \alpha^*_j H_{g_j}(x_0)\,]\, w < 0 \;\; \forall w\in W$

Then x_0 is a local efficient point for P.

Let us note that conditions i) and ii) of Theorem 4.3 imply i), ii), iii) of Theorem 4.2, but the converse is not true as is shown by the following example.

Example 4.1 Consider problem P where s=1, m=1, $\varphi(x)=x^2$, $g(x)= -x^2$, $U=\mathbf{R}_+$, $V=\mathbf{R}_+$, and consider the feasible point $x_0=0$.
It is easy to verify that $f'(x_0)+g'(x_0)=0$, $W=\mathbf{R}$. $K_L = \{0\}$, $K_Q = \{(x^2,-x^2),\, x\neq 0\}$, $0\notin (K_L+K_Q)$. Conditions i), ii), iii) of Theorem 4.2 are verified so that x_0 is a local efficient point for the problem, while ii) of Theorem 4.3 is not verified.

The previous sufficient optimality conditions have been obtained starting from the sufficient condition $T_1 \cap \mathrm{cl}H =\{0\}$. In order to study the case $T_1\cap \mathrm{cl}H\neq\emptyset$, we define the following sets:
$K^*_Q=\{q\in \mathbf{R}^{s+m}\colon q=w^T H_F(x_0)w,\ w\in W^*\}$, $W^*=\{w\in \mathbf{R}^n\colon F'(x_0)w\in \mathrm{Fr}(H),\ w\neq 0\}$.
$K^*_\alpha=\{ q\in \mathbf{R}^{s+m}\colon q=v^T H_F(x_0)v,\ v\in W\alpha\}$. $W\alpha=\{w\in \mathbf{R}^n\colon F'(x_0)w\in \alpha^\perp,\ w\neq 0\}$

The following Theorem holds.

Theorem 4.4 If the following condition (4.3) holds:

$$\exists \alpha\in H^+,\ \alpha\neq 0, \text{ such that } K_L \subseteq \alpha^\perp \text{ and } K^*_Q \subseteq \alpha^\perp_{\leqq}, \tag{4.3}$$

then $x_0\in S$ is a local efficient point for problem P.
Proof Suppose ab absurdo that $x_0\in S$ is not a local efficient point; then there exists a sequence $\{x_k\} \subset S\backslash\{x_0\}$, $x_k\to x_0$, such that $F(x_k)\in H\ \forall k$.
The first order Taylor expansion gives $F(x_k) - F(x_0) = F'(x_0)(x_k - x_0) + \sigma(x_k, x_0)$ with $\lim_{k\to+\infty} \frac{\sigma(x_k, x_0)}{\|x_k - x_0\|} = 0$, so that there exists a subsequence of $\{x_k\}$ (without loss of generality we can suppose to be the same sequence) such that $\lim_{k\to+\infty} \frac{F(x_k) - F(x_0)}{\|x_k - x_0\|} = F'(x_0)w$, with $w= \lim_{k\to+\infty} \frac{x_k - x_0}{\|x_k - x_0\|}$. Since $F(x_k) - F(x_0) \in H$ $\forall k$, it results $\frac{F(x_k) - F(x_0)}{\|x_k - x_0\|} \in H\ \forall k$, so that $F'(x_0)w \in \mathrm{cl}H$ and consequently, taking into account that $K_L \subseteq \alpha^\perp$, $F'(x_0)w \in \mathrm{Fr}H$, that is $w \in W^*$.
Consider now the second order Taylor expansion:
$F(x_k) - F(x_0) = F'(x_0)(x_k - x_0) + (1/2)\, F''(x_0)((x_k - x_0),(x_k - x_0)) + \sigma(x_k, x_0)$ with $\lim_{k\to+\infty} \frac{\sigma(x_k, x_0)}{\|x_k - x_0\|^2} = 0$. Taking into account that $\alpha^T F'(x_0)=0$, we have:

$\alpha^T(F(x_k) - F(x_0)) = (1/2)\,(x_k - x_0)^T(\alpha^T F''(x_0))(x_k - x_0) + \alpha^T \sigma(x_k, x_0)$, so that $\lim_{k\to+\infty} \frac{\alpha^T[F(x_k) - F(x_0)]}{\|x_k - x_0\|^2} = (1/2)\, w^T(\alpha^T F''(x_0))w \geq 0$; on the other hand $w \in W^*$ implies $w^T(\alpha^T F''(x_0))w < 0$ and this is absurd. ♦

By means of the previous theorem, taking into account that $K_Q^* \subseteq K_\alpha^*$, we have the following corollary.

Corollary 4.3 If the following condition (4.4) holds:

$$\exists \alpha \in H^+,\ \alpha \neq 0, \text{ such that } K_L \subseteq \alpha^\perp \text{ and } K_\alpha^* \subseteq \alpha_=^\perp, \tag{4.4}$$

then $x_0 \in S$ is a local efficient point for problem P.

Theorem 4.4 and Corollary 4.3 are equivalent in the decision space to Theorem 4.5 and Theorem 4.6, respectively.

Theorem 4.5 Consider problem P and assume $x_0 \in S$ verifies conditions i) and ii):

i) $\exists \alpha = (\bar{\alpha}, \alpha^*) \in H^+$, $\bar{\alpha} \in \mathbf{R}^s$, $\alpha^* \in \mathbf{R}^m$, such that $\sum_{i=1}^{s} \bar{\alpha}_i \nabla f_i(x_0) + \sum_{j=1}^{m} \alpha^*_j \nabla g_j(x_0) = 0$

ii) $w^T [\sum_{i=1}^{s} \bar{\alpha}_i H_{f_i}(x_0) + \sum_{j=1}^{m} \alpha^*_j H_{g_j}(x_0)]\, w < 0 \quad \forall w \in W^*$

Then x_0 is a local efficient point for P.

Theorem 4.6 Consider problem P and assume $x_0 \in S$ verifies conditions i) and ii):

i) $\exists \alpha = (\bar{\alpha}, \alpha^*) \in H^+$, $\bar{\alpha} \in \mathbf{R}^s$, $\alpha^* \in \mathbf{R}^m$, such that $\sum_{i=1}^{s} \bar{\alpha}_i \nabla f_i(x_0) + \sum_{j=1}^{m} \alpha^*_j \nabla g_j(x_0) = 0$

ii) $w^T [\sum_{i=1}^{s} \bar{\alpha}_i H_{f_i}(x_0) + \sum_{j=1}^{m} \alpha^*_j H_{g_j}(x_0)]\, w < 0 \quad \forall w \in W\alpha$

Then x_0 is a local efficient point for P.

Let us note that in the scalar case condition ii) of Theorem 4.6 states the well known second order sufficient optimality condition; the following example points out that the optimality condition stated in Theorem 4.5 is more general than the previous one.

Example 4.2 Consider problem P where s=1, m=3, $U = \mathbf{R}_+$, $V = \mathbf{R}_+^3$, $\varphi(x,y,z) = x+y-x^2-y^2+z^2$, $g_1(x,y,z) = -x-y$, $g_2(x,y,z) = z$, $g_3(x,y,z) = -z$ and consider the feasible point $x_0 = (0,0,0)$. By means of simple calculations we obtain: $\nabla f(x_0) + \nabla g_1(x_0) = 0$, so that i) of Theorem 4.5 and i) of Theorem 4.6 hold; $W\alpha = \{(w_1, -w_1, w_3),\ w_1, w_3) \in \mathbf{R}\}$, so that $w^T(Hf(x_0) + Hg_1(x_0))w = -2w_1^2 + w_3 \in \mathbf{R}$ and this implies that ii) of Theorem 4.6 is not verified. On the other hand

$W^* = \{(w_1, -w_1, 0), w_1 \in \mathbf{R}\}$ and it results $w^T(Hf(x_0)+Hg_1(x_0))w = -2w_1^2$; condition ii) of Theorem 4.5 is verified and the origin is a local efficient point for P.

References

[1] M.S. Bazaraa and C.M. Shetty, Foundations of optimization, Lecture Notes in Economics and Mathematical Systems 122, Springer-Verlag, 1976.

[2] A. Ben-Tal, Second-order and related extremelity conditions in nonlinear programming, Journal of Opt. Theory and Applications 31, n°2 (1980) 143-165.

[3] A. Ben-Tal and J. Zowe, A unified theory of first and second order conditions for extremum problems in topological vector spaces, Mathematical Programming Study 19 (1982) 39-76.

[4] J.M. Borwein, Proper efficient points for maximizations with respect to cones, SIAM Journal Control Opt. 15 (1977) 57-63.

[5] A. Cambini and L. Martein, Some optimality conditions in vector optimization, Journal of Informations and Opt. Sciences 10, n°1 (1989) 141-151.

[6] A. Cambini and L. Martein, Tangent cones in optimization, in "Generalized Concavity for Economic Applications, Proceedings of the Workshop held in Pisa, April 2, 1992", edited by P. Mazzoleni, Tecnoprint, Bologna (1992) 29-39.

[7] A. Cambini and L. Martein, Generalized concavity and optimality conditions in vector and scalar optimization, in "Generalized Convexity", edited by S. Komlósi, T. Rapcsák, and S. Schaible, Springer-Verlag (1994) 337-357.

[8] A. Cambini and L. Martein, A survey of recent results in vector optimization, in "Optimization of Generalized Convex Problems", edited by P. Mazzoleni (1994) 39-55.

[9] A. Cambini and L. Martein, Second order necessary optimality conditions in the image space: preliminary results, in "Scalar and Vector Optimization in Economic and Financial Problems, Proceedings of the Workshop held in Milan, March 28, 1995", edited by E. Castagnoli and G. Giorgi, (1995) 27-38.

[10] R. Cambini, Second order optimality conditions in the image space, Report n.99, Dept. Statistics and Applied Mathematics, University of Pisa, 1996.

[11] R. Cambini, Generalized Concavity and Optimality Conditions in Vector Optimization, in "Operations Research and its Applications", Proceedings of ISORA'96, World Publishing Corporation, Beijing (1996) pp.172-180.

[12] L. Martein, Some results on regularity in vector optimization, Optimization 20 (1989) 787-798.

[13] L. Martein, Stationary points and necessary conditions in vector extremum problems, Journal of Informations and Optimiz. Sciences 10, n°1 (1989) 105-128.

[14] L. Martein, Soluzioni efficienti e condizioni di ottimalità nell'ottimizzazione vettoriale, in "Metodi di Ottimizzazione per le Decisioni", Masson, Milano (1994) 215-241.

[15] G.P. McCormick, Second order conditions for constrained minima, SIAM Journal of Applied Mathematics 15, n°3 (1967) 641-652.

Invex and Pseudoinvex Functions in Multiobjective Programming

Osuna-Gómez R.[1],Beato-Moreno A.[1], Luque-Calvo, P.[1], Rufián-Lizana, A.[1]

[1] Departamento de Estadística e Investigación Operativa, Fac. de Matemáticas, 41012 Sevilla, España.

Abstract. D. H. Martin studied the optimality conditions of invex functions in the scalar case. In this work we will generalize his results making them applicable to the vectorial case. We will prove that equivalences between minima and stationary points are still true if we have to optimize p-objective functions instead of one objective function.

Key words: Invex and Pseudoinvex functions, weakly efficient solutions, weighting problems.

1 Introduction

The following convention for equalities and inequalities will be used. If $x, y \in \mathbb{R}^n$, then

- $x = y$ iff $x_i = y_i \quad i = 1, \cdots, n$
- $x \leqq y$ iff $x_i \leq y_i \quad i = 1, \cdots, n$
- $x \leq y$ iff $x_i \leq y_i \quad i = 1, \cdots, n$ with strict inequality holding for at least one i
- $x < y$ iff $x_i < y_i \quad i = 1, \cdots, n$

Hanson [5] in 1981, introduced a certain generalization of convexity for constrained scalar programming problems of the following form:

$$\textbf{(CP)} \quad \begin{array}{ll} Min & \theta(x) \\ s.t: & g(x) \leqq 0 \\ & x \in S \subseteq \mathbb{R}^n \end{array}$$

where $\theta : S \longrightarrow \mathbb{R}$ and $g : S \longrightarrow \mathbb{R}^m$ are differentiable functions on an open set $S \subseteq \mathbb{R}^n$.

If θ and g are convex functions we have the following well-known inequalities,

$$\theta(x) - \theta(\bar{x}) \geq (x - \bar{x})^T \nabla\theta(\bar{x})$$
$$g(x) - g(\bar{x}) \geq (x - \bar{x})^T \nabla g(\bar{x})$$
$$\forall x, \bar{x} \in S$$

Hanson noticed that the functional form of the factor $(x - \bar{x})$ here plays no role in establishing the following important property of convex problems:

- Every feasible Kuhn-Tucker point is a global minimizer.

So, he considered functions for which exists a vectorial function

$$\eta : S * S \longrightarrow \mathbb{R}^n \text{ such that } \forall x, \bar{x} \in S$$

$$\theta(x) - \theta(\bar{x}) \geq \eta(x, \bar{x})^T \nabla\theta(\bar{x})$$
$$g(x) - g(\bar{x}) \geq \eta(x, \bar{x})^T \nabla g(\bar{x})$$

Craven [4] called invex to these functions. And later, Martin [9] pointed out for unconstrained scalar programming problems,

THEOREM 1.1 *A function, θ, is invex on S if and only if every critical point of θ is a global minimizer of θ on S.*

Israel and Mond [1] defined other types of functions, which generalized pseudoconvex functions in the same way that invex functions generalized the convex ones.

DEFINITION 1.1 *Let $\theta : S \subseteq \mathbb{R}^n \rightarrow \mathbb{R}$ be a differentiable function on the open set S. Then θ is Pseudoinvex on S with respect to η if for all $x_1, x_2 \in S$ there exists $\eta(x_1, x_2) \in \mathbb{R}^n$ such that:*

$$\eta(x_1, x_2)^T \nabla\theta(x_2) \geq 0 \;\Rightarrow\; \theta(x_1) - \theta(x_2) \geq 0$$

Or equally,

$$\theta(x_1) - \theta(x_2) < 0 \;\Rightarrow\; \eta(x_1, x_2)^T \nabla\theta(x_2) < 0$$

It is clear that:

$$\text{convex} \Longrightarrow \text{pseudoconvex} \Longrightarrow \text{invex} \Longrightarrow \text{pseudoinvex}$$

2 The Multiobjective Problem

In this section we intend to lay out Martin's results, compiled in the above section from the scalar programming problem to the multiobjective one.

In general, the multiobjective optimization problem (VOP) is represented as the following vector-minimization problem:

$$\textbf{(VOP)} \quad Min \quad f(x) = (f_1(x), \cdots, f_p(x))$$
$$s.t: \qquad x \in S \subseteq \mathbb{R}^n$$

where $f : S \longrightarrow \mathbb{R}^p$ is a differentiable function on the open set S. We will represent a matrix of dimensions $p * n$ whose rows are the gradient vectors of each component valued at the point $\bar{x}$ for $\nabla f(\bar{x})$.

In multiobjective optimization problems, (VOP), multiple objectives are usually noncommensurable and cannot be combined into a single objective. Moreover, the objectives often conflict with each other. Consequently, the concept of optimality for single-objective optimization problems cannot be directly applied for (VOP). The concept of Pareto optimality, also know as an efficient solution, in turn has been introduced for (VOP). Mathematically, we can define slightly different notions of Pareto optimality.

DEFINITION 2.1 *A feasible point, $\bar{x}$, is said to be an Efficient Solution if and only if there does not exist another $x \in S$ such that $f_i(x) \leq f_i(\bar{x})$ $i = 1, \cdots, p$.*

DEFINITION 2.2 *A feasible point, $\bar{x}$, is said to be a Weak Efficient Solution, (W.E.P.), if and only if there does not exist another $x \in S$ such that $f_i(x) < f_i(\bar{x})$ $i = 1, \cdots, p$.*

From the definitions it is clear that every efficient solution is also a weak efficient solution.

Either efficient solutions or the weakly efficient solutions of the vectorial problem coincide with the concept of minimum. In our case we will take into account the weakly efficient points. Characterizations of weakly efficient solutions has been achieved, in the cases of no differentiable functions, by Ruiz and Rufian [10]. In this case we will study them when the functions are differentiable.

To generalize the results from the first section we have to define similar stationary points in the multiobjective case.

DEFINITION 2.3 *A feasible point, $\bar{x} \in S$, is said to be a Vectorial Critical Point, (VCP), to (VOP) if there exists a vector $\lambda \in R^p$ with $\lambda \geq 0$ such that $\lambda^T \nabla f(\bar{x}) = 0$.*

Craven [4] established the following theorem for (VOP).

THEOREM 2.1 *Let $\bar{x}$ be a weakly efficient solution for (VOP). There exists $\bar{\lambda} \geq 0$ such that $\bar{\lambda}^T \nabla f(\bar{x}) = 0$.*

We will consider the following invexity and pseudoinvexity definitions for vectorial functions, which coincide with those given in the above section when the function is scalar.

DEFINITION 2.4 *Let $f : S \subseteq \mathbb{R}^n \longrightarrow \mathbb{R}^p$ be a differentiable function on the open set S. Then f is invex on S with respect to η if for all $x_1, x_2 \in S$ there exists $\eta(x_1, x_2) \in \mathbb{R}^n$ such that,*

$$f(x_1) - f(x_2) \geq \nabla f(x_2)\eta(x_1, x_2)$$

DEFINITION 2.5 *Let $f : S \subseteq \mathbb{R}^n \longrightarrow \mathbb{R}^p$ be a differentiable function on the open set S. Then f is pseudoinvex on S with respect to η if for all $x_1, x_2 \in S$ there exists $\eta(x_1, x_2) \in \mathbb{R}^n$ such that,*

$$f(x_1) - f(x_2) < 0 \Rightarrow \nabla f(x_2)\eta(x_1, x_2) < 0$$

For scalar functions, Israel and Mond [1] proved that concepts of invexity and pseudoinvexity coincide. But for vectorial functions this equivalence is not proved. So, in general

$$\text{Invex} \Longrightarrow \text{Pseudoinvex}$$

To stablish the reciprocal of theorem 2.1 we need to impose the pseudoinvexity hyphotesis.

PROPOSITION 2.1 *Let $\bar{x}$ be a vectorial critical point for (VOP) and f is a pseudoinvex function at $\bar{x}$ with respect to η, then $\bar{x}$ is a weakly efficient solution.*

PROOF Let $\bar{x}$ a vectorial critical point for (VOP), then $\exists \bar{\lambda} \geq 0$ such that $\bar{\lambda}^T \nabla f(\bar{x}) = 0$.
On the other hand, if there exists another $x \in S$ such that $f(x) < f(\bar{x})$ then

$\exists \eta(x, \bar{x}) \in \mathbb{R}^n$ such that $\nabla f(\bar{x})\eta(x, \bar{x}) < 0$.
By the Gordan's theorem, the system

$$\lambda^T \nabla f(\bar{x}) = 0 \quad \lambda \geq 0 \quad \lambda \in \mathbb{R}^p$$

has no solution at λ. □

The next theorem proved that the pseudoinvexity hyphotesis is a necessary condition also.

THEOREM 2.2 *Let (VOP) be the unconstrained multiobjective programming problem, then:*

Every (V.C.P.) *is* (W.E.P.) *if and only if f is a pseudoinvex function on S.*

PROOF
The sufficient condition has already been proved.
Then we should prove that if every vectorial critical point is a weakly efficient point, then vectorial function f, fulfils the pseudoinvexity condition.
Be $\bar{x}, x \in S$.
If $\bar{x}$ is a weakly efficient point, then the system

$$\{f_i(x) - f_i(\bar{x}) < 0\}_{i=1,\ldots,p}$$

has no solution at $x \in S$. Then f is a pseudoinvex function on S.
If $\bar{x}$ is not a weakly efficient point for (VOP), then by hyphotesis, the system

$$\lambda^T \nabla f(\bar{x}) = 0$$

has no solution at $\lambda \in \mathbb{R}^p, \lambda \geq 0$.
Therefore, by Gordan's theorem the next system has solution at $u \in \mathbb{R}^n$

$$\nabla f(\bar{x})u < 0$$

So, if $f(x) - f(\bar{x}) < 0 \Longrightarrow \exists \eta(x, \bar{x}) \in \mathbb{R}^n$ such that $\nabla f(\bar{x})\eta(x, \bar{x}) < 0$.

□

Above result shows how pseudoinvexity hypothesis is a sufficient and necessary condition for every vectorial critical point to be a weakly efficient solution. We conclude with this the generalization of Martin's results, and we establish that it is still true when there are more than one function to optimize, because pseudoinvex and invex scalar functions coincide.

On the other hand, a matter of great interest, both computationally and theoretically is the relation of the Vector Optimization Problem, (VOP), to the following scalar optimization problem, called the weighting problem associated,

$$(\mathbf{P}_\lambda) \quad Min \quad \lambda^T f(x)$$
$$s.t: \quad x \in S \subseteq \mathbb{R}^n$$

where $\lambda \in \mathbb{R}^p$, $\lambda_j \geq 0$ and $\sum_{j=1}^{p} \lambda_j = 1$.

THEOREM 2.3 *Let $\bar{x}$ be an optimal solution for (P_λ) with $\lambda \geq 0$, then $\bar{x}$ is a weakly efficient solution for (VOP)*

PROOF .

Let's suppose that $\bar{x}$ is an optimal solution for (P_λ) with $\lambda \geq 0$, then there exists at least a $\lambda_j > 0$.

Let be $J = \{j = 1, ..., p$ such that $\lambda_j > 0\}$.

If $\bar{x}$ is not a weakly efficient solution for (VOP) then there exists another feasible point, $x \in S$, such that $f(x) < f(\bar{x})$.

In particular,

$$f_j(x) < f_j(\bar{x}) \quad \forall j \in J \Longrightarrow \lambda^T f(x) < \lambda^T f(\bar{x})$$

And this is a contradiction with the hyphotesis.

□

In order to establish the reciprocal of previous theorem; and to guarantee that all weakly efficient solutions for the problem of multiobjective programming problem, (VOP), can be found as solutions of a certain weighting programming problem, we need to impose that the objective function shows some property of convexity.

Next, we will prove that this hypothesis of convexity can be replaced by the invexity of the objective function. Therefore, it is not necessary to demand any condition of convexity to the feasible set.

THEOREM 2.4 *Let $\bar{x}$ be a weakly efficient solution for (VOP). If f is an invex function on S then there exists a $\lambda \geq 0$ such that $\bar{x}$ is an optimal solution for (P_λ).*

PROOF .

If f is invex on S with respect to η and $\lambda \geq 0$ then $\lambda^T f$ is invex on S with respect to η [2].

On the other hand, if $\bar{x}$ is a (W.E.P.) then $\exists \lambda \geq 0$ such that $\lambda^T \nabla f(\bar{x}) = 0$. Then $\bar{x}$ is a stationay point for the scalar programming problem (P_λ), therefore $\bar{x}$ is an optimal solution for (P_λ) [8].

If it is necessary we can normalize the vector λ, $\lambda' = \dfrac{\lambda}{\sum_{j=1}^{p}}$.

□

References

[1] BEN-ISRAEL AND MOND B. *What is the Invexity?*
Journal Australian Mathematical Society, Serie B, Vol 28, 1-9 (1986)

[2] CHANKONG V. AND HAIMES Y.Y. *Multiobjective Decision Making: Theory and Methodology.*
North-Holland.

[3] CRAVEN. B. D. *Lagrangian conditions and quasiduality*
Bulletin Australian Mathematical Society, Vol. 16, 325-339 (1977)

[4] CRAVEN B. D. *Invex Functions and constrained local minima*
Bulletin Australian Mathematical Society, Vol 24, 357-366 (1981)

[5] HANSON M. A. *On sufficiency of Kuhn-Tucker conditions*
Jour. Math. Ana. Appl. 30, (1981) 545-550

[6] HANSON M.A. AND MOND B. *Further generalizations of convexity in Mathematical Programming.*
J.Inform.Optim.Sci 3 (1982) 25-32

[7] JEYAKUMAR *Strong and Weak invexity in mathematical programming*
Collection: Methods of Operations Research 55, (1980), 109-125

[8] MANGASARIAN O. *Nonlinear Programming.*
McGraw-Hill.

[9] MARTIN D.H. *The essence of invexity*
Jour. Opt. Th. Appl. 47 No. 1 (1985) 65-76

[10] RUIZ P. AND RUFIAN A. *A characterization of weakly efficient points*
Mathematical Programming, 68, (1995), 205-212.

Vector-Values Risk in Multicriteria Problems

M. Salukvadze[(1)], A. E. Bardin[(2)]

(1) Professor. Institute of Control Systems. Tbilisi. Georgia.
(2).Assistant professor. Orekovo-Zuevo Pedagogical Institute. Orekovo-Zuevo. Russia.

1 Introduction

Most of the social-economical, technical, organizational situations and processes are characterized by following two factors:

- functioning quality of controlled systems is estimated by a set of criteria, and not by a single one;
- it is necessary to take into account the influence of perturbations, interferences and other kind of uncertainties.

The first attempt to consider these peculiarities from "maximin positions" has been made in [1,2]. But the "maximin position" is orientated on:

1° the possibility to realise the worst variant for a decision maker, the probability of which is, as a rule, small;

2° possible "losses" in different criteria when deviating from the maximun solution might be incomparably smaller than possible "acquisitions" (even by the same criteria) when choosing some other solution.

The proposed paper aims at finding a decision-making method for multicriteria problems under uncertainties based on the Sevidge's principle of "minimax risk" [3]. It allows to take into account peculiarities mentioned in 1° and 2°. It should be noted that the "minimax risk" principle has been earlier applied only to single criteria problems.

2 r-Guaranteeing Solution

Consider the multicriteria problem under uncertainty

$$\langle X, Y, f(x,y)\rangle, \tag{2.1}$$

where solution $x \in X \in \text{comp}\mathbf{R}^n$, uncertainty $y \in Y \in \text{comp}\mathbf{R}^m$, and the components $f_i(x, y)$, $i \in \mathbf{N} = \{1, ..., N\}$, of the vector $f(x, y) = (f_1(x, y),....., f_N(x, y))$ of criteria are continuous on the product of the compacts $X \times Y$.

In the problem (2.1) the decision maker tends to choose the solution $x \in X$ in such a way that he could achieve possibly greater values for all the components of the objective vector-function $f(x, y)$. For this, the decision maker takes into account the possibility of the realization of any uncertainty $y \in Y$.

We correspond to each uncertainty $y \in Y$ the vector of the components maximum values

$$f(y) = (\max_{x \in X} f_1(x, y), \ldots, \max_{x \in X} f_N(x, y)),$$

which determines the "utopia" point [4], and the difference

$$\varphi(x, y) = f(y) - f(x, y) =$$
$$= (\max_{z \in X} f_1(z, y) - f_1(x, y), \ldots, \max_{z \in X} f_N(z, y) - f_N(x, y)) \tag{2.2}$$

further will be referred to as **Vector Risk.**

Lemma 2.1.

Components $\varphi_i(x, y)$, $i \in \mathbf{N}$, of the function $\varphi(x, y) = (\varphi_1(x, y), \ldots, \varphi_N(x, y))$ are continuous in X x Y.

Actually, due to the continuity of functions $f_i(x, y)$ in X for each $y \in Y$, maximums in (2.2) are achieved, and, corresponding to [5], the functions $\max_z f_i(z, y)$ are continuos by $y \in Y$. Then the difference of the continuos functions is (2.2) should also be continuos in X x Y.

Let's consider the problem

$$<X, Y, \varphi(x, y)> \tag{2.3}$$

where the risk $\varphi(x, y)$ is determined in (2.2) and the decision maker chooses the solution $x \in X$ in order to achieve possibly small values for all components of the Risk Vector.

Definition 2.1 .

The solution $x^* \in X$ is called r-guaranteeing for the problem (2.1) if there exists uncertainty $y^* \in Y$, such that

$$\varphi(x^*, y) \not> \varphi(x^*, y^*) \not> \varphi(x, y^*) \quad \forall x \in X, \ y \in Y \tag{2.4}$$

is valid.

The vector $\varphi(x^*, y^*)$ will further be referred to as the Optimal Risk.

Correlations (2.4) represent the unification of the saddle point notion from the game theory and the Slater optimum from the multicriteria problems theory. This notion was actively used in [1,2], when dealing with the decision making in multicriteria problems under uncertainty and was called Slater saddle point.

Let's consider geometric interpretation of the r-guaranteeing solution, for a bicriterium problem (2.1) of the type

$$\Gamma = <X, Y, f(x, y) = (f_1(x, y), f_2(x, y))>.$$

Suppose that x^* is the r-guaranteeing solution for the problem Γ, and y^* is the corresponding uncertainty from (2.1). Then, if the decision maker has chosen and uses x^*, he provides himself with any uncertainty $y \in Y$ the risk vector components values $\varphi_i(x^*, y)$ (simultaneously for all $i \in \mathbf{N}$), which are not greater than the Optimal Risk $\varphi(x^*, y^*)$.

So, using x^*, the decision maker guarantees "himself" the value $\varphi(x^*, y^*)$ for any uncertainties $y \in Y$.

Property 1.1. For $N = 1$ the r-guaranteeing solution of the problem (2.1) is a minimax strategy in an antagonistic game

$$<X, Y, \max_{z \in X} f_1(z,y) - f_1(x,y)> \tag{2.5}$$

and, consequently, realizes the Sevidge's minimax risk principle [3] for a scalar criterium $\max_{z \in X} f_1(z,y) - f_1(x,y)$.

Property 1.2. When there are no uncertainties in (2.1), i.e. for $Y = \varnothing$, the r-guaranteeing solution coincides with the weakly efficient one [6, p. 32] (Slater maximum) for the problem $<X, f(x)>$.

The given two properties have shown that the definition of the r-guaranteeing solution is sufficiently complete, because it includes generally accepted solutions from games theory (a saddle point) and multicriteria problems theory (weakly efficient solutions).

Let's consider the multicriterium problem under uncertainty with the "separated" vector criterium:

$$<X, Y, h(x) + g(y)>. \tag{2.6}$$

We'll assume that the vectors $h(x) = (h_1(x), \ldots, h_N(x))$ and $g(y) = (g_1(y), \ldots, g_N(y))$ have continuous components on $X \times Y$.

Property 2.3. The r-guaranteeing solution x* of the problem (2.6) coincides with Slater maximum solution of the multicriterium problem

$$<X, h(x)> \tag{2.7}$$

i.e. $h(x) \not> h(x^*)$, $\forall x \in X$. For this uncertainty $y^* \in Y$ (definition 2.1) might be any vector from Y.

Property 1.3 makes it possible for multicriteria problems under uncertainty with "separated" vector criterium to formulate the following method of constructing the r-guaranteeing solutions:

-At least one Slater maximum solution x^* of the problem (2.7) should be found. According to the property 2.3 this x^* should be the r-guaranteeing solution for (2.6).

Property 2.4. The set of r-guaranteeing solutions x^* of the problem (2.1) is a closed and bounded subset (may be an empty one) of the set X.

Construction of the r-guaranteeing solution x^* for (2.1) reduces to the determination of the Slater saddle point (x^S, y^S) of the problem (2.3). Therefore, we use already known sufficient conditions of the vector saddle points existence.

Statement 2.1. Suppose there exist nonnegative constants α_i, $i \in \mathbf{N}$, $\sum_{i \in \mathbf{N}} \alpha_i > 0$, such that the antagonistic game

$$<X, Y, \sum_{i \in N} \alpha_i \, \varphi_i(x, y)>$$

has a saddle point (x^o, y^o). Then x^o is the r-guaranteeing solution.

The given statement permits to establish existence of the r-guaranteeing solutions in mixed strategies.

Statement 2.2. A set of the r-guaranteeing mixed solutions of the problem (2.1) is no empty.

Note that statement 2.2. "provides" existence of the r-guaranteeing solutions only in the class of probability measures. If there are used "pure" solutions the r-guaranteeing solution of the problem (2.1) might not exist.

3 Minimax Risk

Consider the multicriteria problem under uncertainty (2.1), to which (as it was made in the previous paragraph) we correspond the problem (2.3):

$$<X, Y, \varphi(x, y)>, \tag{3.1}$$

where $\varphi(x, y)$ is the Vector Risk. In (3.1) a decision maker tends to choose a solution $x \in X$, such that all of the components of the vector-function $\varphi(x,y) = (\varphi_1(x,y), \ldots, \varphi_N(x,y))$ simultaneously achieve possibly smaller values. For this, the decision maker is to take into account the possibility of realization of such uncertainties which give (with the chosen solution x^*) Slater (Pareto) maximum to the Vector Risk. These requirements can be formalized as notions of Slater (Pareto) r-minimax solution of the problem (2.1).

Fix an arbitrary solution $x^* \in X$ and consider the multicriterium problem

$$<x^*, Y, \varphi(x^*, y)>. \tag{3.2}$$

Define by $Y^S(x^*)$ a set of Slater maximum solutions $y^S(x^*) \in Y$ of the problem (3.2), i.e.

$$Y^S(x^*) = \{ y^S(x^*) \in Y \,/\, \varphi(x^*, y^S(x^*)) \nless \varphi(x^*, y), \ \forall y \in Y \}.$$

Definition 3.1.

The solution $x_s^r \in X$ is called Slater r-minimax for (2.1) if there exists an uncertainty $\hat{y}(x_s^r) \in Y^s(x_s^r)$ such that

$$\varphi(x_s^r, \hat{y}(x_s^r)) \ngtr \varphi(x, y(x)), \forall x \in X, y(x) \in Y^s(x) \tag{3.3}$$

Call the vector $\varphi(x_s^r, \hat{y}(x_s^r))$ Slater r-minimax.

Analogously, let's define Pareto r-minimax. Exactly, we'll correspond to each fixed solution $x^* \in X$ the set

$$Y^P(x^*) = \left\{ y^P(x^*) \in Y \,/\, \varphi(x^*, y^P(x^*)) \nleq \varphi(x^*, y), \ \forall y \in Y \right\}$$

of Pareto maximum solutions $y^P(x^*)$ of the problem (3.2).

Definition 3.2.

The solution $x_p^r \in X$ is called Pareto r-minimax for (2.1) if there exists an uncertainty $\hat{y}(x_p^r) \in Y^P(x_p^r)$, such that

$$\varphi(x_p^r, \hat{y}(x_p^r)) \not\leq \varphi(x, y(x)),\ \forall x \in X, y(x) \in Y^P(x) \qquad (3.4)$$

Analogously, the vector $\varphi(x_p^r, \hat{y}(x_p^r))$ is Pareto r-minimax.

Now we'll give geometric interpretation of Slater (Pareto) r-minimax for the bicriterium problem

$$<X, Y, (f_1(x, y), f_2(x, y))>, \qquad (3.5)$$

which we get from the problem (2.1) supposing that $\mathbf{N} = \{1, 2\}$. Let $x_s^r \in X$ be the Slater r-minimax solution of the problem (3.5) and $\hat{y}(x_s^r)$ the corresponding value of uncertainty from (3.3). The set $\varphi(x_s^r, Y^s(x_s^r))$ of the Slater maximum values of the risk vector-function with the chosen r-minimax solution $x = x_s^r$ is denoted by a thick line on the fig. 3.1. From (3.3) it follows that for any uncertainty $y \in Y^s$ points $\varphi(x_s^r, y)$ of the set $\varphi(x_s^r, Y)$ can not get inside the right angle G with the vertex $\varphi(x_s^r, \hat{y}(x_s^r))$ (where $\hat{y}(x_s^r)$ is the uncertainty from (2.3)).

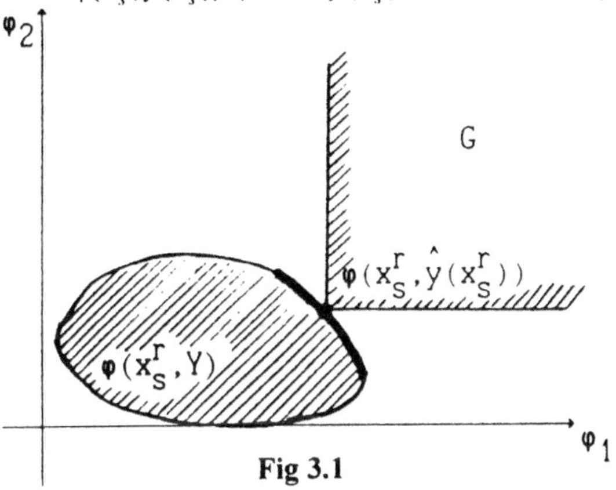

Fig 3.1

Thus, $\varphi(x_s^r, \hat{y}(x_s^r))$ is the guaranteed Vector Risk which can not be exceeded (by all components simultaneously) by $\varphi(x_s^r, y)$ for any uncertainties $y \in Y$.

On the other hand, the vector criteria value $\varphi(x_s^r, \hat{y}(x_s^r))$ is compared with the points of the set $\varphi(x, Y^s(x))$ for every solution $x \in X$. These sets are Slater maximum values of the Risk Vector for fixed $x \in X$, i.e. for the chosen solution x the vector $\varphi(x, y^s(x))$ (where $y^s(x) \in Y^S(x)$) can not be increased (at once by all components) by the choice of $y \in Y$.

Then the problem of finding the smallest risk out of all guaranteed ones (in the multicriterium sence) is stated. Just this "minimum" is achieved on the r-minimax solution.

We'll suppose in the given below properties that the sets X and Y are compacts and the components $f_i(x, y)$ of the vector criterium $f(x, y) = (f_1(x, y),...., f_N(x, y))$ are continuous on $X \times Y$. The existence of the r-minimax with the pointed out restrictions establishes.

Theorem 3.1.

A set of Slater r-minimax solutions of the problem (2.1) is a nonempty compact.

Proof. By lemma 1.1 components $\varphi_i(x, y)$ of the Vector Risk are continuos on the product of compacts X x Y and the r-minimax solution of the problem (2.1) is Slater minimax for the multicriterium problem (3.1). In this case, from the remark 2.4 [2, p.21] it follows the validity of theorem 3.1.
Remark 3.1. From [2, p. 20] we also get that the set

$$\mathrm{F}_s = \varphi(X_s^r) = \bigcup_{x \in X_s^r} \bigcup_{\hat{y}(x) \in \hat{Y}^s(x)} \varphi(x, \hat{y}(x)) \tag{3.6}$$

will be compact in $\mathbf{R}^N$ as well. Here X_s^r is a set (a nonempty compact by theorem (3.1) of all Slater r-minimax solutions of the problem (2.1) and $\hat{Y}(x_s^r) = \{\hat{y}(x_s^r)\}$ is a set of the corresponding uncertainties (by definition 3.1). Consequently, F_S is a set of Slater r-minimaxes of the problem (2.1).
Remark 3.2. In [2, p. 18] it was also found out that the set

$$\varphi_s(X) = \bigcup_{x \in X} \bigcup_{y \in Y^s(x)} \varphi(x, y) \tag{3.7}$$

is a compact subset in $\mathbf{R}^N$.

So, theorem 3.1 and remarks 3.1 and 3.2 show that the r- minimax solution of the problem (2.1) pocesses "good topological Property-a set of such solutions is nonempty, closed and bounded. Now we'll investigate the properties of the set X_s^r, connected with external and internal stability.
Statement 3.1. A set of Slater r-minimaxes F_S from (3.6) is internally stable, i.e. for any two r-minimax solutions of the problem (2.1) $x_s^{(1)}$ and $x_s^{(2)}$ and corresponding to them uncertainties $\hat{y}(x_s^{(1)})$ and $\hat{y}(x_s^{(2)})$ the following inequality

$$\varphi(x_s^{(1)}, \hat{y}(x_s^{(1)})) \nless \varphi(x_s^{(2)}, \hat{y}(x_s^{(2)})) \tag{3.8}$$

is valid.

Actually, the relation (3.8) follows from (3.3) if we put $x = x_s^{(1)}$ and $x_s^r = x_s^{(2)}$.
Remark 3.3. Statement 3.1 shows that in definition 3.1 none of the Slater r-minimaxes is not "better" (by Slater) than the other. Therefore, for the extraction of any minimax from the ser of Slater r-minimaxes it is necessary to give additional criteria on the set of pairs

$$\bigcup_{x \in X_s^r} \bigcup_{y = \hat{y}(x) \in \hat{Y}^s(x)} (x, y) \tag{3.9}$$

Notice that according to [2, p. 20] the set (3.9) is a nonempty compact and in the case of continuity of these additional criteria the compactness (3.9) provides existence of "extracted" solutions.

Statement 3.2. The set of Slater r-minimaxes $\mathbf{F}_S$ of the problem (2.1) is externally stable with the respect to the compact (3.7), i.e. for each $\varphi(x, y(x)) \in \varphi_S(X)$ there exists "its own" Slater r-minimax $\varphi_S \in \mathbf{F}_S$, such that

$$\varphi(x, y(x)) \geq \varphi_S. \tag{3.10}$$

Actually, by remark 3.2, the set $\varphi_S(X)$ is a compact, and $\mathbf{F}_S$ is a set of Slater minimums for this compact. Then, owing to [6, p. 158] the set $\mathbf{F}_S$ is externally stable, i.e. (3.10) takes place.

Reveal peculiarities of r-minimax solutions by the example of bicriterium problem

$$<X, Y, (f_1(x, y), f_2(x, y))>, \tag{3.11}$$

where the sets X and Y are represented as

$$X = \{x = (x_1, x_2) \in \mathbf{R}^2 \,/\, 0 \leq x_1 \leq a, \quad 0 \leq x_2 \leq a,\ a = \text{const} > 0\}$$

$$Y = \{y = (y_1, y_2) \in \mathbf{R}^2 \,/\, -b \leq y_1 \leq 1,\ -b \leq y_2 \leq 1,\ b = \text{const} \geq 0\}$$

and criteria

$$f_1(x, y) = x_1 y_1,\ f_2(x, y) = x_2 y_2.$$

Find components of the Risk Vector for the problem (3.11).

$$\varphi_1(x, y) = \max_{x \in X} x_1 y_1 - x_1 y_1 = \begin{cases} a y_1 - x_1 y_1, & y_1 > 0, \\ -x_1 y_1, & y_1 \leq 0; \end{cases}$$

$$\varphi_2(x, y) = \max_{x \in X} x_2 y_2 - x_2 y_2 = \begin{cases} a y_2 - x_2 y_2, & y_2 > 0, \\ -x_2 y_2, & y_2 \leq 0; \end{cases} \tag{3.12}$$

By the accepted approach, pass on from the problem (3.11) to the problem with a vector criterium -risk vector

$$<X, Y, (\varphi_1(x, y), \varphi_2(x, y))> \tag{3.13}$$

where $x = (x_1, x_2) \in X$, $y = (y_1, y_2) \in Y$ and criteria $\varphi_i(x, y)$ are determined in (3.12).

Now we construct r-minimax. For this we fix the solution $x^* = (x_1^*, x_2^*) \in X$. From (3.12) follows that the set $\varphi(x^*, Y) = \{\varphi(x^*, y) / y \in Y\}$ is a rectangle, for which the sides length correspondingly are equal to $\mu_1(x^*) = \max_{y \in Y} \varphi_1(x^*, y)$ and $\mu_2(x^*) = \max_{y \in Y} \varphi_2(x^*, y)$.

Define a set of Pareto (Slater) solutions of the problem $\Gamma(x^*) = <x^*, Y, f(x^*, y)>$ by $Y^P(x^*)$ (correspondingly $Y^S(x^*)$). By [6, p. 31] $f(x^*, Y^P(x^*))$ will be called a set of Pareto maximums for the problem $\Gamma(x^*)$, analogously $f(x^*, Y^S(x^*))$ is a set of Slater maximums.

A set of Pareto (Slater) minima is defined like this.

It is easy to prove the following three Lemmas:

Lemma 3.1.

For the Vector Risk function from (3.12) the following equality

$$\min_{x \in X}\left[\max_{y \in Y} \varphi_i(x,y)\right] = ab(1+b)^{-1}, (i = 1,2) \tag{3.14}$$

is valid and the minimum is achieved only for $x = (a(1+b)^{-1}, a(1+b)^{-1})$.

Lemma 3.2.

Solution $x_p^r = (a(1+b)^{-1}, a(1+b)^{-1})$ is Pareto r-minimax for the problem (3.11).

Remark 3.4. $x_p^r = (a(1+b)^{-1}, a(1+b)^{-1})$ is the unique Pareto r-minimax solution of the problem (3.11).

With the help of the given example we'll compare two approaches to the concept of the guaranteeing solution in the multicriterium problem: the maximin principle [2] and the r-minimax risk which was considered in this paper.

According to [2] solution x^p of the problem $<X, Y, f(x, y)>$ is Pareto maximin if there exists an uncertainty $\hat{y}(x_p) \in Y_p(x^p)$, such that

$$f(x^p, \hat{y}(x^p)) \nleq f(x, y(x)), \forall x \in X, y(x) \in Y_p(x).$$

Here $Y_p(x)$ is a set of Pareto minimum uncertainties for the following problem: $<x, Y, f(x, y)>$, in which the solution x is fixed.

Lemma 3.3.

The solution $x^p = (0, 0)$ is Pareto maximin for the problem (3.11).

Note peculiarties of Pareto r-minimax solution x_p^r(lemma 3.2) and Pareto maximin solution x^p (lemma 3.3) of the problem (3.11). Denote a set of the vector criteria $f(x,y)$ values for the fixed uncertainty y by $f(X, y) = \{f(x,y) / x \in X\}$. Then we'll change the parameter b, which was given when determining the set Y_p, and compare the guaranteeing solutions x_p^r and x^p.

Case I. ($b = 0$). When realizing an arbitrary uncertainty $y^* \in Y$ we get a set $f(X, y^*)$, which is represented in the fig. 3.2 (for $y_1^* > 0, y_2^* > 0$).

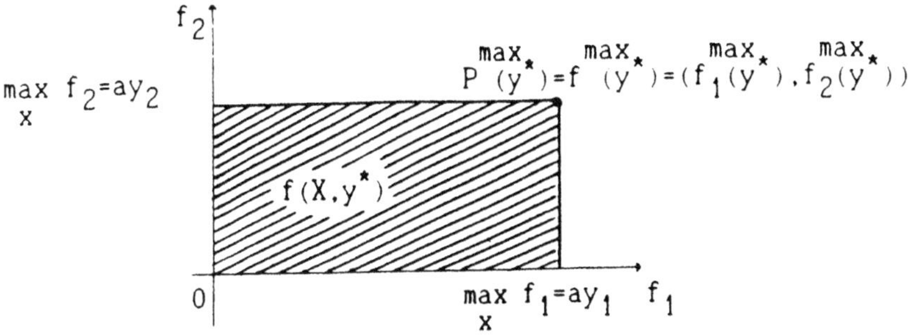

Fig. 3.2

The Pareto maximum point $P^{max}(y^*)$ is achieved for the solution $x = (a, a)$, which doesn't depend upon the uncertainty y* and is Pareto r-minimax solution of the problem (3.11) in the case $b = 0$ (lemma 3.2). Note that the risk function $\varphi(x, y)$ from (3.12) is equal to 0 with $x = x_p^r$ for any $y \in Y$.

Case II. (b is a small positive number, such that $0 < b < 1$).

Denote a set of all possible values of the vector criterium $f(x, y)$ by $f(X, Y) = \{f(x, y) / x \in X, y \in Y\}$ (fig. 3.3)).

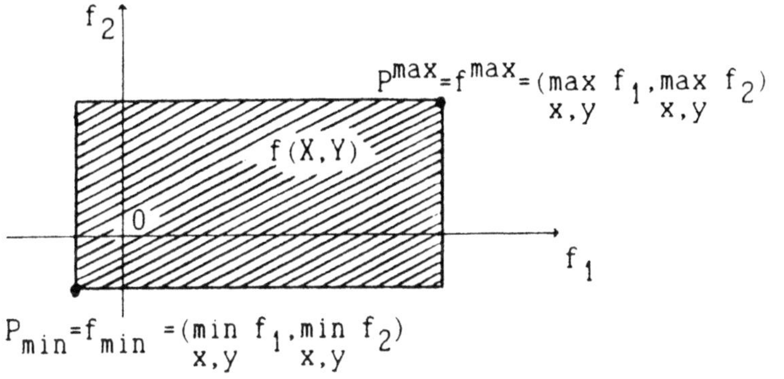

Fig. 3.3

"Careful" Pareto maximin strategy gives the unique solution $x^P = (0, 0)$, for which the value of criterium vector $f(x^P, \hat{y}(x^P)) = (0, 0)$ is a little bit "better" than the Pareto minimum (of the points P_{min}), i.e. the distance (in the Euclidean metric) between the points 0 and P_{min} is equal to $|0 P_{min}| = ab\sqrt{2} << 1$ (fig. 2.3).

In this case Pareto maximin solution x^P is guaranteeing in the sense that for any uncertainty $y \in Y$ all components of the vector $f(x, y)$ can not be simultaneously negative.

But, when realizing the uncertainty $\bar{y} = (\bar{y}_1, \bar{y}_2) \in Y$, such that $\bar{y}_1 > 0$. $\bar{y}_2 > 0$, the solution x^P again becomes the worst, i.e. $f(x^P, \hat{y}(x^P)) << f(x, \bar{y})$ for any $x = (x_1, x_2)$, where $x_1\ x_2 > 0$. Choosing the Pareto r-minimax solution, x_p^r a decision maker gets a negative value at least for one of the components of the vector $f(x,y)$, which nevertheless in the worst situation for a decision maker (with $y = (-b, -b)$) are equal to $-ab/(1+b)$, i.e. they are very close to zero under the inequality condition $0 < ab << 1$.

Case III. ($0 < b < \infty$).

According to the above mentioned, by using the r-minimax risk strategy, the decision maker "loses" in both criteria (in comparison with the maximin) when realizing uncertainties $y^* = (y_1^*, y_2^*) \in \{y \in Y / y_1^* < 0, y_2^* < 0\}$, but "wins" by the same criteria, if $y^* \in Y \cap \mathbf{R}_+^2$. Note, that each component of the vector criterium (in comparison with the maximin) doesn't increase the costant value a. This is where the guaranteeing property of the r-minimax is revealed.

So, in the given example, the decision maker when choosing between the maximin strategy and the r-minimax strategy must take into account the following:

1° The risk value (in the first case the risk is zero, in the second one it is negligibly small and in the third one it is limited);

2° The possibility of realization of unfavourable uncertainties (it is essentially only in the third case);

3° Personal inclination to risk.

References

1. **ZHUKOVSKIY V.I. and SALUKVADZE M.E.,** Multicriteria Control Problems under Conditions of Uncertainty, "Metsniereba", Tbilisi, 1991 (in Russian).
2. **ZHUKOVSKIY V.I. and MOLOSTVOV V.S.,** Multicriterion Optimization of Systems in Conditions of Incomplete Information, MNIIPU, Moscow, 1990 (in Russian).
3. **LUCE R. and RAIFFA H.,** Games and Decisions, Wiley, New York, 1957.
4. **SALUKVADZE M.E.,** Vector-Valued Optimization Problems in Control Theory, Academic Press, New York, 1979.
5. **VOROBIEV N.N.,** Theory of Games, Nauka, Moscow, 1985 (in Russian).
6. **PODINOVSKIY V.V. and NOGIN V.D.,** Pareto-Optimal Solutions of Multicriteria Problems, Nauka, Moscow, 1982 (in Russian).
7. **KRASOVSKIY N.N. and SUBBOTIN A.I.,** Positional Differential Games, Nauka, Moscow, 1974 (in Russian).

Utopian Efficient Strategies in Multicriteria Matrix Games

Fernández, F.R..[1], Mármol, A.M.[2], Monroy, L.[2], Puerto, J.[1]

[1] Departamento de Estadística e Investigación Operativa. Universidad de Sevilla, España.
[2] Departamento de Economía Aplicada I. Universidad de Sevilla, España.

Abstract. In this paper we present a new solution concept for multiple objective problems which generalizes the balance points introduced by Galperin (1992). We apply this solution concept, called utopian efficiency, for solving multiple criteria matrix games. For these games it is shown how to get the whole set of utopian efficient strategies by means of multiobjective linear programs. Finally, we derive a decision criterion which is based on no "a priori" information for choosing a strategy out of the proposed set.

1 Introduction

Real world decision processes involve generally more than one decision criterion. Therefore, multicriteria decision making and multiobjective programming play a very important role in any decision making environment. However, one of the main drawbacks when dealing with multiple criteria is the concept of solution to be used.

The concept of nondominated vectors (in image space) or Pareto-solutions (in variable space) is the most widely accepted solution concept for this kind of problems. The importance of Pareto-solutions comes from the fact that these solutions have the property that for improving one of their criteria some other will get worse values. Thus, Pareto-solutions represent a sort of compromise solutions.

Different methods are known for obtaining the whole set of Pareto-solutions or even some subset with some specific properties. Many of these methods are based on scalarizations. One of those kinds of scalarizations consists of measuring some distance from the ideal point. Steuer and Choo (1983) characterize an optimal criterion vector by means of the weighted Tchebychev distance to an ideal point. Kalizecski (1987) presents a method for obtaining the whole set of nondominated vectors using a modified weighted Tchebychev metric with regards to the ideal point.

Several attempts have been made to introduce different solution concepts for Multiple Objective Problem (MOP). In this line, some refinements as the proper Pareto-solution (Geoffrion, 1968) or the cone Pareto-solutions (Benson, 1979) among others can be considered; recently, Galperin (1992) introduced the concept of balance point as a new concept of solution for MOP although K.H.News (1996) shows the equivalence between balance points and Pareto solution in MOP.

In this paper, we introduce the concept of utopian-efficient solution and state the relationships of this concept with Pareto solutions and with the methodologies that use distances to ideal points for computing nondominated vectors. This analysis shows that balance points and the methods based on distances from an ideal point are particular instances of our methodology based on utopian efficient solutions. In addition, we also apply these ideas to introduce an alternative solution concept for multicriteria matrix games: the concept of utopian efficient strategy (UES), based on the security levels defined by Goshe and Prassad (1989). Finally we present an evaluation function for choosing one out of the whole set of UES.

The paper is organized as follows. Section 2 states the general MOP and the concept of utopian point. The equivalence between balance, utopian and Pareto-optimal points is shown. Section 3 applies the definition of utopian points to introduce a new solution concept for multicriteria matrix games: utopian efficient strategies (UES). In Section 4 we deal with the problem of choosing among the whole set of UES. We introduce a globalizing function which measures the expected deviation with respect to a given set of goals. In addition, we derive a scheme for optimizing this function on the considered set of strategies. The paper ends with some conclusions.

2 Efficiency and Utopia in Multiobjective Programming

We consider the general multiobjective programming problem

$$(MOP) \quad \max \; f(x) \quad \text{s.t.} \; x \in S$$

where $S \subset \mathbb{R}^n$, and $f : \mathbb{R}^n \longrightarrow \mathbb{R}^k$, $f = (f_1, \ldots, f_k)$. Let us denote the maximum value of functions f_i over S by c_i^0 , i = 1,...,k, and let $c^0 = (c_1^0, \ldots, c_k^0)$.

Definition 2.1 $P = (P_1, \ldots, P_k)$ *is an utopian point for the problem* (MOP) *if* $P_i \geq c_i^0$, $\forall i = 1, \ldots, k$.

For each $x \in S$, we define $d_i^P(x) = P_i - f_i(x)$, $\forall i = 1, \ldots, k$. $d^P(x) = (d_1^P(x), \ldots, d_k^P(x))$ is the deviation vector with respect to the utopian point P.

Definition 2.2 *Let P be an utopian point for (MOP). $\bar{x} \in S$ is an utopian-efficient solution if there is no other $x \in S$ such that $d^P(x) \leq d^P(\bar{x})$, and $d^P(x) \neq d^P(\bar{x})$.*

It follows that $\bar{x} \in S$ is an utopian-efficient solution if and only if it is an efficient solution of the problem

$$\begin{array}{ll} \min & d^P(x) \\ \text{s.t.} & x \in S \end{array}$$

The idea of utopian-efficiency is closely related to the goal-efficiency introduced by Steuer. Nevertheless, as we will see, the first one is not associated with preemptive approaches for solving the problem as the second one is.

We denote by U^P the set of utopian-efficient solutions associated to an utopian point P.

Theorem 2.1 *The set of efficient solutions of (MOP), E, coincides with the utopian-efficient solution set, U^P , for any utopian point P, i.e.,*

$$E \equiv U^P \quad \forall P \geq c^0$$

Proof: $\bar{x}$ is an efficient solution for (MOP) iff there exists no $x \in S$, $f(x) \geq f(\bar{x})$, and $f(x) \neq f(\bar{x})$, iff for any utopian point P, there exists no $x \in S$, such that $P - f(x) \leq P - f(\bar{x})$, and $P - f(x) \neq P - f(\bar{x})$, iff there exists no $x \in S$ such that $d^P(x) \leq d^P(\bar{x})$ and $d^P(x) \neq d^P(\bar{x})$, iff $\bar{x}$ is an utopian-efficient point. □

The set U^P is independent of P as it follows easily from the proof of Theorem 2.1. Notice that if P is not a utopian point, with this approach we do not get all the efficient solutions of (MOP). Moreover, we may not obtain any efficient point.

News (1996) shows that the concept of balance point introduced by Galperin (1992) is equivalent to the concept of Pareto-optimality. The following result relates the concept of utopian point and the concept of balance point.

Definition 2.3 *Given $\eta = (\eta_1, \ldots, \eta_k) \in \mathbb{R}^k$, with $\eta \geq 0$, let*

$$S^0_{\eta_i} = \{x \in S, \quad c^0_i - f_i(x) \leq \eta_i\}$$

η is a balance point if $S^0_\eta = \bigcap_{i=1}^k S^0_{\eta_i} \neq \{\phi\}$, and for every $\eta' \in \mathbb{R}^k$, such that $0 \leq \eta' \leq \eta$, $\eta' \neq \eta$, $S^0_{\eta'} = \{\phi\}$

Theorem 2.2 *$\eta \in \mathbb{R}^k$ is a balance point if and only if $\eta = d^P(x)$, where x is an utopian-efficient solution for $P = c^0$.*

Proof: It follows from the equivalence established by News and from Theorem 2.1. □

Remark that both, the concept of utopian-efficient solution and the concept of balance point are related to the concept of compromise solution introduced by Zeleny (1974).

3 Utopian-Efficient Strategies in Multicriteria Matrix Games

In this section we extend the concept of utopian-efficient solutions of MOP to a new solution concept for multicriteria games (MG).

We consider a multicriteria two-person finite game in normal form (matrix game) with a n x m payoff matrix $A = (a_{ij})$, of k-tuples, $a_{ij} = (a_{ij}(1), \ldots, a_{ij}(k))$. This leads us to define individual matrices of dimension n x m as

$$A(s) = (a_{ij}(s)), \quad s = 1, \ldots, k.$$

The players are represented by PI, (the maximizer, who chooses rows), and PII (the minimizer, who chooses columns). The mixed strategies spaces for PI and PII are, respectively,

$$X = \{x \in \mathbb{R}^n, \quad \sum_{i=1}^{n} x_i = 1, \quad x_i \geq 0, \quad i = 1, \ldots, n\}$$

$$Y = \{y \in \mathbb{R}^m, \quad \sum_{j=1}^{m} y_j = 1, \quad y_j \geq 0, \quad j = 1, \ldots, m\}$$

In this paper we consider multicriteria games using an approach based on security levels as defined by Goshe and Prassad (1989), and in the concept of utopian-efficient points defined in Section 2.

Definition 3.1 *The expected payoff of the game, when PI chooses $x \in X$ and PII chooses $y \in Y$ is*

$$v(x, y) = x^t A y = [v_1(x, y), \ldots, v_k(x, y)]$$

where $v_s(x, y) = x^t A(s) y, \quad s = 1, \ldots, k.$

Definition 3.2 *The security level vector for PI is defined ,for each $x \in X$, by*

$$v(x) = (v_1(x), \ldots, v_k(x))$$

where $v_s(x) = \min_{y \in Y} \quad v_s(x, y) = \min_{j} \quad x^t A(s), \quad s = 1, \ldots, k.$

$v_s(x), s = 1, \ldots, k$, is the guaranteed expected payoff level for the k associated scalar games when PI chooses strategy x.

Definition 3.3 *A strategy $x^* \in X$ is a Pareto-optimal security strategy (POSS) for PI if there exists no $x \in X$, such that $v(x^*) \leq v(x)$ and $v(x^*) \neq v(x)$.*

From this definition, it follows that the set of POSS for PI is the set of efficient solutions of the multiobjective problem

$$
\begin{array}{lll}
(MGP): & \max & v_1(x), \ldots, v_k(x) \\
 & \text{s.t.} & x \in X
\end{array}
$$

Fernández and Puerto (1996) proved that the set of Pareto-optimal security strategies (POSS) for player I, coincides with the set of efficient solutions of a Linear Multicriteria Problem.

In this paper we obtain another characterization of POSS, based on the differences between the security levels obtained and some ideal security levels (utopian point).

Now, we establish the concept of utopian payoff vector and utopian-efficient strategy for a multicriteria game.

Let $v_s^0 = \max_{x \in X} \quad v_s(x), \quad s = 1, \ldots, k$, i.e., the values of the k scalar games induced by the multicriteria game.

Definition 3.4 *$P = (P_1, \ldots, P_k)$ is an utopian payoff vector for the multicriteria game if $P_s \geq v_s^0, \quad s = 1, \ldots, k$.*

For each $x \in X$, the vector of deviations with respect to the utopian payoff vector $P = (P_1, \ldots, P_k)$ is $d(x) = (d_1(x), \ldots, d_k(x))$, where $d_s(x) = P_s - v_s(x), \forall s = 1, \ldots, k$.

Definition 3.5 *Let $P = (P_1, \ldots, P_k)$ be an utopian payoff vector for (MG). A strategy $x^* \in X$ is an utopian-efficient strategy for PI if there exists no $x \in X$, such that $d(x) \leq d(x^*)$, and $d(x^*) \neq d(x)$.*

Hence, $x^* \in X$ is an utopian-efficient strategy for PI if and only if it is an efficient solution for problem

$$
\begin{array}{lll}
(UMG): & \min & d_1(x), \ldots, d_k(x) \\
 & \text{s.t.} & x \in X
\end{array}
$$

The following theorem characterizes utopian-efficient strategies for PI as the efficient solutions of a Multiobjective Linear problem that depends only on the matrix of deviations of the payoff and the utopian values.

For each $s = 1, \ldots, k$, we consider the deviation matrix $\bar{A}(s) = (\bar{a}_{ij}(s))$, where $\bar{a}_{ij}(s) = P_s - a_{ij}(s)$, and let $e^t = (1, \ldots, 1) \in \mathbb{R}^k$. The deviations with respect to P can be written as:

$$\begin{aligned} d_s(x) &= P_s - v_s(x) = P_s - \min_j x^t A(s) = \max_j (P_s e^t - x^t A(s)) \\ &= \max_j x^t (eP_s e^t - A(s)) = \max_j x^t \bar{A}(s). \end{aligned}$$

Thus, $d_s(x)$ represents the guaranteed maximum expected deviation with respect to P_s in the s-th scalar game, when PI chooses strategy x.

We consider the linear multiobjective problem

$$\begin{array}{lll} (ULG) & \min & d_1(x), \ldots, d_k(x) \\ & \text{s.t.} & x^t \bar{A}(s) \leq (d_s, \ldots, d_s) \quad s = 1, \ldots, k \\ & & x \in X \end{array}$$

Theorem 3.1 *$x^* \in X$ is an utopian-efficient strategy and $d^* = d(x^*)$ its deviation vector for PI if and only if (x^*, d^*) is an efficient solution for (ULG).*

Proof: The result follows from the equivalence of problems (UMG) and (ULG). □

Next theorem of this section establishes the equivalence between POSS and utopian-efficient strategy.

Theorem 3.2 *$x^* \in X$ is a POSS for PI if and only if $x^* \in X$ is a utopian-efficient strategy for PI.*

Proof: $x^* \in X$ is a POSS for PI iff x^* is an efficient solution for problem (MGP), iff for all utopian payoff vector of (MGP), x^* is utopian efficient iff x^* is efficient for (UMG) iff x^* is an utopian-efficient strategy for PI. □

When obtaining POSS by this approach we need an utopian payoff vector. In a game, in order to get one of them, we do not need to calculate the value of each scalar game, v_s^0 , $s = 1, \ldots, k$. We can obtain an utopian payoff vector easily as the maximum payoff in each scalar game, i.e., $P_s = max_{ij}\{a_{ij}(s)\}$, $s = 1, \ldots, k$. Needless to say that if $P = v^0$ and x^* is an utopian-efficient strategy for PI, then the associated deviation vector $d(x^*)$ is a balance point.

The following result gives a characterization of POSS in terms of the deviations, which is based on the equivalence between efficient solutions of a multiobjective linear problem and the solutions of the associated weighted sum problems.

Theorem 3.3 *$x^* \in X$ is a POSS for PI if and only if $\exists \omega > 0$, such that (x^*, d^*) is an optimal solution of the linear weighted problem*

$$\begin{array}{ll} \min & \sum_{s=1}^{k} \omega_s d_s \\ s.t. & x^t \bar{A}(s) \leq (d_s, \ldots, d_s) \quad s = 1, \ldots, k \\ & x \in X \end{array}$$

This approach is related to the classical approach followed by Cook (1976) and Hannan (1982), which consists of minimizing the total weighted underachievement from a fixed set of goals (Cook (1976), Hannan (1982)).

Notice that due to Theorem 3.3, POSS can be obtained as those strategies that minimize the L_1-weighted distance to the utopian payoff vector. The result relates our approach with the computation of efficient points using the so called Archimedian Methods (Steuer, 1986) when PI has determined a vector of penalty weights associated with the deviations. However, while the Archimedian approach may generate solutions that are not efficient, our approach always generates a POSS for any vector of weights.

4 Evaluation of Utopian-Efficient Strategies

Once we have a set of utopian-efficient strategies and their deviation vectors, we need some criteria to choose among them. Using the probability to get these deviation vectors, we present a procedure in order to decide which strategy to play.

Let $P = (P_1, \ldots, P_k)$ be an utopian payoff vector for the multicriteria game, $x^* \in X$ an utopian-efficient strategy for PI, and d^* the associated deviation vector.

If PI chooses strategy x^* , and PII chooses $y \in Y$, the probability that the deviation from P_s is at most d_s^* in the s-th scalar game is

$$p(x^*, y, d_s^*) = x^{*t} T(s) y$$

where $T(s) = (t_{ij}(s))$, $i = 1, \ldots, n, j = 1, \ldots, m, s = 1, \ldots, k$.

$$t_{ij}(s) = \begin{cases} 1 & \text{if } \bar{a}_{ij}(s) \leq d_s^* \\ 0 & \text{if } \bar{a}_{ij}(s) > d_s^* \end{cases}$$

As $p(x^*, y, d_s^*)$ depends on the strategy that PII plays, we will consider this probability in the worst case, i.e.,

$$p(x^*, d_s^*) = \min_{y \in Y} p(x^*, y, d_s^*) = \min_{y \in Y} x^{*t} T(s) y = \min_j x^{*t} T(s)$$

We will call $p(x^*, d_s^*)$ the probability to obtain at most d_s^* when PI plays x^* .

Hence, given a utopian-efficient strategy for PI, x^* , and d^* the associated deviation vector, we can get a vector of probabilities $(p(x^*, d_1^*), \ldots, p(x^*, d_k^*))$ that will be evaluated in order to choose one of the utopian-efficient strategies.

In this sense, we consider a function that evaluates globally each utopian efficient strategy and the associated deviation vector.

If for each (x, d) we calculate the probability to get each deviation, d_s , $s = 1, \ldots, k$, PI may select among the utopian-efficient strategies the one that gives the smallest expected deviation. We establish the expression for this expected deviation.

If $x \in X$ is an utopian-efficient strategy for PI, and d the associated deviation vector, the expected deviation is

$$\begin{aligned} E[x, d] &= \sum_{s=1}^{k} d_s p(x, d_s) = \sum_{s=1}^{k} d_s (\min_{y \in Y} x^t T(s) y) \\ &= \sum_{s=1}^{k} d_s (\min_j x^t T(s) e^j) = \sum_{s=1}^{k} d_s \min_j (\sum_{i=1}^{k} x_i t_{ij}(s)) \end{aligned}$$

Notice that function $E[x, d]$ is defined on (x, d) where x is a utopian efficient strategy and d its associated deviation vector. Any of this points,(x, d), can be expressed as a convex combination of a finite number of utopian efficient strategies and deviations $(x^1, d^1), \ldots, (x^r, d^r)$ that are extreme efficient solutions of (ULG), then

$$x_\lambda = \lambda_1 x^1 + \ldots + \lambda_r x^r, \quad d_\lambda = \lambda_1 d^1 + \ldots + \lambda_r d^r$$

where $\lambda \in \Lambda = \{\lambda \in^r, \lambda \geq 0, e^t \lambda = 1\}$

The utopian efficient strategy and the associated deviation vector set is the union of convex sets that are convex combinations of extremes. If we consider $E[x, d]$ defined on each of those convex sets, it can be written as a function of λ:

$$E : \lambda \in \Lambda \quad \longrightarrow \mathbb{R}, \quad E[\lambda] = E[x_\lambda, d_\lambda]$$

Matrices $T(s)$ and the expression of $\min_j x^t T(s) e^j$ change in the range of λ, what produces variations on the expression of the function $E[\lambda]$, determining a partition of Λ in polyhedral subsets:

$$\Lambda = \bigcup_j \Lambda^j, \quad \Lambda^i \cap \Lambda^j = \{\phi\}, \quad i \neq j$$

Theorem 4.1 *The function $E[\lambda]$ is bilinear and convex on Λ^j.*

Proof:

$$E[\lambda] = E[x_\lambda, d_\lambda] = \sum_{s=1}^{k} d_s(\lambda) \min_j (\sum_{i=1}^{k} x_i(\lambda) t_{ij}^{\lambda}(s)).$$

$\min_j(\sum_{i=1}^k x_i(\lambda)t_{ij}^{\lambda}(s))$ is a linear expression of λ, because for each j, the expression $(\sum_{i=1}^k x_i(\lambda)t_{ij}^{\lambda}(s))$ is linear. $d_s(\lambda)$ is also linear. It follows that $E[\lambda]$ is a bilinear expression of λ over each subset Λ^j , with nonnegative coefficients. □

The final choice is that utopian-efficient strategy, x_λ with the smallest value of $E[x_\lambda, d_\lambda]$. This means that we look for strategies minimizing the expected value of the achieved deviations.

In general, the sets Λ^j are not closed, then a minimum may not be attained. In this situation, in order to choose the best strategy, we will consider a strategy in the neighborhood of the strategy that gives the infimum value.

We have used the expected value criteria in order to choose a utopian-efficient strategy. Nevertheless, in the same way other decision criteria under uncertainty could be used.

5 Conclusions

In this paper we present an equivalent formulation for the problem of finding the set of Pareto-optimal solutions which includes as particular instances the Archimedian approach of Steuer (1986) and the balance points of Galperin (1992). We apply this approach as a solution concept for solving multiple criteria matrix games. We show that this new concept reduces to the well-known POSS, so that a multiobjective linear program can be used to get the whole set of utopian efficient strategies.

A decision criterion is also proposed to select a specific strategy out of the set of utopian efficient strategies. The advantage of this selection is that it minimizes the expected deviation with respect to some values fixed by one of the players, therefore it tries to satisfy in the best way his/her goals.

6 References

Benson,H.P.(1979): "An Improved Definition of Proper Efficiency for Vector Minimization with respect to Cones", Journal of Mathematical Analysis and Applications, 71, 232-241.

Cook,W.D.(1976): "Zero-Sum Games with Multiple Goals", Naval Research Logistics Quarterly, 23, 615-621.

Fernández,F.R., Puerto, J.(1996): "Vector Linear Programming in Zero-Sum Multicriteria Matrix Games", Journal of Optimization Theory and Applications, 89, 115-127.

Galperin, E.A.(1992): "Nonscalarized Multiobjective Global Optimization", Journal of Optimization Theory and Applications, 75, 69-85.

Geoffrion, A.M.(1968): "Proper Efficiency and the Theory of Vector Maximization", Journal of Mathematical Analysis and Applications, 22, 618-630.

Ghose, D.(1991): "A Necessary and Sufficient Condition for Pareto-Optimal Security Strategies in Multicriteria Matrix Games". Journal of Optimization Theory and Applications, 68, 463-481.

Ghose, D., Prassad, U.R.(1989): "Solutions Concepts in Two-person Multicriteria Games", Journal of Optimization Theory and Applications, 63, 167-189.

Hannan,E.L.(1982): "Reformulating Zero-Sum Games with Multiple Goals", Naval Research Logistics Quarterly, 29, 113-118.

News, K.H.(1996): "A Note on the Equivalence of Balance Points and Pareto Solutions in Multiple Objective Programming", accepted in Journal of Optimization Theory and Applications.

Kaliszewski,I.(1987): "A Modified Weighted Tchebycheff Metric for Multiple Objective Programming". Computer Operations Research, 19, 315-323.

Steuer,R.E.(1983): Operating Manual for the ADBASE Multiple Objective Linear Programming Package. College of Business Administration, University of Georgia, Athens, Georgia.

Steuer, R.E.(1986): Multiple Criteria Optimization: Theory, Computation and Application, Willey, New York.

Steuer,R.E.,Choo,E.U.(1986): "An Interactive Weighted Tchebycheff Procedure for Multiple Objective Programming", Mathematical Programming, 26, 326-344.

Zeleny, M.(1974): "A Concept of Compromise Solutions and the Method of the Displaced Ideal", International Journal of Computers and Operations Research, 1, 479-496.

About the Existence of Nash-Slater Equilibrium for a Non-Cooperative Game under Uncertainty

Larbani Moussa

Institut des sciences exactes, universite de Tizi-Ozou,
Tizi-Ozou. Algeria. FAX: 213-3-20-23-61.

Abstract. This paper consists of two parts .In the first one we introduce the notion of the N-S equilibrium for a non-cooperative game under uncertainty.The second one is devoted to the problem of the existence of this equilibrium.

Keywords. Non-cooperative game ,uncertainty,equilibrium. Ky Fan inequality.

Introduction. Real problems usually contain undetermined parameters which are generally due to lack of information. In economy , the undetermined parameters may be the rates of currencies or costs of raw materials , behavior of competitors. They may have natural origins for example climatic like rainfall and temperature.In industry, uncertainty may be the result of errors of measurement caused by the apparatus or any perturbations.

When the influence of undetermined parameters is not important ,their negligence in mathematical resolution leads to satisfactory approximate solution. However, when this influence is important their negligence may give erroneous solutions . In this case decision makers apply to experts who generally only give a domain where these parameters can vary. Consequently, in order to help them to choose the optimal decision , it is necessary to use mathematical methods of the theory of decision under uncertainty.

In this paper we are concerned with problems whose mathematical model are within the context of the non-cooperative games under uncertainty theory initiated by professor V.I. Zhukovskii in [5].The author introduced the notion of Nash-Slater equilibrium (N-S equilibrium) for a non-cooperative game under uncertainty. This notion is fairly wide, it contains the notion of Nash equilibrium for a non-cooperative game [2] and the notion of weak Pareto (or Slater) optimal solution of a multi-objective problem [3,4] as particular cases. The existence of the N-S equilibrium is proved by two approaches. In this paper, using the Ky Fan inequality, we prove the existence of N-S equilibrium with more general conditions than in [5]and , furthermore , our approach provides relatively easier method for its practical determination.

1. PROBLEM STATEMENT

Let us consider the non-cooperative game under uncertainty

$$\langle I, X, Y, f(x,y) \rangle \tag{1}$$

where

I={1,2,...,n} is the set of players , n ≥ 1 ; $X = \prod_{i=1}^{n} X_i$ is the set of situations of the game, X_i is the set of strategies of the i-th player, $X_i \subset R^{n_i}$, $n_i \geq 1$, $i = \overline{1,n}$; $Y \subset R^p$, p ≥ 1; Y is the set of undetermined parameters ; $f = (f_1, f_2, \ldots, f_n)$, $f_i : X * Y \to R$, is the objective function of the i-th player; $x = (x_1, x_2, \ldots, x_n) \in X$, x is a situation of the game; x_i is the strategy of the i-th player $y = (y_1, y_2, \ldots, y_p) \in Y$, y is the vector of undetermined parameters.The aim of each player ,by choosing his strategy , is to maximise his objective function with taking into account his complete ignorance of the uncertainty $y \in Y$.

We suppose that the following conditions are met :

a) Y, X_i, $i = \overline{1,n}$ are convex and compact sets;

b) $f_i : X * Y \to R$, $i = \overline{1,n}$ are continuous on $X * Y$.

We shall use the notation

$$\forall x = (x_1, x_2, \ldots, x_n) \in X, \quad \forall i \in I, \quad \forall z_i \in X_i$$

$$(x // z_i) = (x_1, x_2, \ldots, x_{i-1}, z_i, x_{i+1}, \ldots, x_n).$$

Now let us proceed to the definition of the equilibrium for the game (1).

DEfINITION 1 . A pair $(x^0, y^0) \in X * Y$ is referred to as an N-S equilibrium for the game(1) if following conditions are met:

1) $\forall i \in I$, $f_i\left(x^0 // x_i, y^0\right) \leq f_i\left(x^0, y^0\right)$, $\forall x_i \in X_i$;

2) $\forall y \in Y$, the system of inequalities

$$f_i\left(x^0, y\right) < f_i\left(x^0, y^0\right) \quad , \quad i \in I$$

is non simultaneous.

REMARK 1 . The first condition of the definition 1 means that x^0 is a Nash equilibrium [2] for the non-cooperative game

$$\left\langle I, X, f\left(x, y^0\right)\right\rangle$$

with fixed uncertainty $y = y^0$.

The second one means that y^0 is a Slater optimal (i.e weak Pareto optimal) solution for the multi-objective problem

$$\left\langle Y, f\left(x^0, y\right)\right\rangle$$

with fixed situation $x = x^0$.

REMARK 2 . The notion of N-S equilibrium is fairly wide and includes as particular cases the concept of Nash equilibrium for a non-cooperative game and Pareto (or weak Pareto) optimal solution of a multi-objective problem. Indeed,
1) if $Y = \varnothing$ (or y is fixed) in the game (1) we obtain a non-cooperative game ;
2) if $X = \varnothing$ (or x is fixed) in the game (1), we obtain a multi-objective problem ;
3) if $I = \{1\}$ (i.e there is one player) in the game (1), we obtain the antagonistic game

$$\langle X, Y, f(x, y)\rangle \ .$$

2. EXISTENCE OF N-S EQUILIBRIUM

In [5] the existence of N-S equilibrium is proved using two approaches. In the first one the author constructs an antagonistic game which saddle points give N-S equilibrium for the game (1). In the second one he introduced a n+1-th player representing the objective function of the undetermined parameters. In this paper the existence of N-S equilibrium is proved with more general conditions than in [5], by using the Ky Fan inequality [1].Furthermore, we obtain a relatively easier method for
practical research of this equilibrium.

For every $j \in I$, let us consider the function

$$\Phi_j(.,.): Z * T \to R$$

defined by

$$\Phi_j(z,t) = f_j\left(x // x'_j, y\right) - f_j(x, y') + \sum_{\substack{i=1 \\ i \neq j}}^{n} f_i\left(x // x'_i, y\right) - f_i(x, y) \ .$$

where

$$Z = T = X * Y \quad , \quad z = (x, y) \in Z \quad , \quad t = (x', y') \in T.$$

Note that

$$\Phi_j(t, t) = 0 \quad , \quad \forall t \in X * Y \tag{2}$$

The following lemma establishes the relation between N-S equilibrium of the game (1) and functions

$$\Phi_j(.,.) \quad , \quad j \in I.$$

LEMMA 1 . If there exist $i_0 \in I$ and $\bar{z} = (\bar{x}, \bar{y}) \in X * Y$ such that

$$\Phi_{i0}(\bar{z}, t) \leq 0 \quad , \quad \forall t \in X * Y . \tag{3}$$

Then $\bar{z} = (\bar{x}, \bar{y})$ is an N-S equilibrium of the game (1).

Proof. Let $l \in I$ substuting

$$t = (\bar{x} // x_l, \bar{y}) \quad \text{in (3)}$$

according to the definition of $\Phi_{i0}(...)$, we obtain

$$\Phi_{i0}(\bar{z}, t) = f_l(\bar{x} // x_l, \bar{y}) - f_l(\bar{x}, \bar{y}) \leq 0, \quad \forall x_l \in X_l$$

(the remaining terms of the sum $\Phi_{i0}(\bar{z}, t)$ are equal to zero).

Hence

$$f_l(\bar{x} // x_l, \bar{y}) \leq f_l(\bar{x}, \bar{y}) \quad , \quad \forall x_l \in X_l . \tag{4}$$

Since l is arbitrarily chosen in I , by (4) we conclude that $\bar{z} = (\bar{x}, \bar{y})$ satisfies the first condition of the definition 1.

Substituting

$$t = (\bar{x}, y) \quad \text{in (3)},$$

we obtain

$$\Phi_{i0}(\bar{z}, t) = f_{i0}(\bar{x}, \bar{y}) - f_{i0}(\bar{x}, y) \leq 0 \quad , \quad \forall y \in Y .$$

Hence

$$f_{i0}(\bar{x},\bar{y}) \le f_{i0}(\bar{x},y) \;, \qquad \forall y \in Y .$$

The last inequality proves that $\bar{z} = (\bar{x},\bar{y})$ verifies the second condition of definition1 , which completes the proof of lemma 1.

According to lemma 1, let us now prove the theorem of existence of N-S equilibrium for the game (1).

THEOREM 1. Suppose that the following conditions are satisfied :

1) the function $t_i \to f_i(x//t_i, y)$ is concave on X_i, $\forall y \in Y$, $\forall x \in X$, $\forall i \in I$;

2) $\exists i_0 \in I$ such that the funtion $y \to f_{i0}(x,y)$ is convex on Y, $\forall x \in X$.

Then the game (1) has at least one N-S equilibrium.

Proof. According to lemma 1, one has to prove that there exist $\bar{z} \in Z$ such that the inequality (3) holds.
Obviously, by conditions of theorem 1 and assumptions a),b) we deduce that the function

$$\Phi_{i_0}(z,t) = f_{i_0}(x//x'_{i_0}, y) - f_{i_0}(x,y') + \sum_{\substack{i=1 \\ i \neq i_0}}^{n} f_i(x//x'_i, y) - f_i(x,y)$$

verifies the following properties:

1) the function $z \to \Phi_{i0}(z,t)$ is continuous $\forall t \in X * Y$;

2) the function $t \to \Phi_{i0}(z,t)$ is concave $\forall z \in X * Y$.

Thus the function $(z,t) \to \Phi_{i0}(z,t)$ satisfies the conditions of the Ky Fan inequality [1]. Hence there exist $\bar{z} \in X * Y$ such that

$$\Phi_{i0}(\bar{z},t) \;\le\; \Phi_{i0}(t,t). \qquad \forall t \in T .$$

Using (2) ,we obtain

$$\Phi_{i0}(\bar{z},t) \;\le 0 , \qquad \forall t \in T .$$

By the lemma1,we deduce that $\bar{z} = (\bar{x}, \bar{y})$ is an N-S equilibrium for the game (1).

Let us illustrate this theorem by an example.

EXAMPLE. Suppose that in the game (1)

$$n = 3,\ X_i = [\alpha_i, \beta_i] \subset R,\ \alpha_i < \beta_i,\ i = \overline{1,3}$$

$$Y = [\lambda_1, \theta_1] * [\lambda_2, \theta_2],\ Y \subset R^2,\quad \lambda_i < \theta_i,\ i = \overline{1,2}\ ,$$

$f_i(x, y) = (a_i, x) + (b_i, y)$ where $(.,.)$ is the scalar product of vectors;

$$a_i = (a_{i1}, a_{i2}, a_{i3}) \in R^3,\ b_i = (b_{i1}, b_{i2}) \in R^2,\ i = \overline{1,3}\ .$$

with conditions

$$a_{11} \ge 0,\ a_{22} \le 0,\ a_{33} \ge 0, \qquad b_{11} \ge 0,\ b_{12} \le 0\ . \tag{5}$$

The conditions of theorem 1 are satisfied for any $i_0 \in I$ .We choose for example $i_0 = 1$. Then

$$\Phi_1(z, t) = f_1(x // x'_1, y) - f_1(x, y') + f_2(x // x'_2, y) - f_2(x, y) +$$

$$+ f_3(x // x'_3, y) - f_3(x, y) =$$

$$= a_{11}(x'_1 - x_1) + b_{11}(y_1 - y'_1) + b_{12}(y_2 - y'_2) + a_{22}(x'_2 - x_2) + a_{33}(x'_3 - x_3)$$

the theorem 1 guarantees the existence of $\bar{z} = (\bar{x}, \bar{y})$ such that

$$\Phi_1(\bar{z}, t) \le 0, \qquad \forall t \in X * Y\ .$$

According to assumptions (5) if $\bar{z} = (\bar{x}_1, \bar{x}_2, \bar{x}_3, \bar{y}_1, \bar{y}_2)$ with

$$\bar{x}1 = \sup X_1 = \beta_1\ , \quad \bar{x}2 = \inf X_2 = \alpha_2\ , \quad \bar{x}3 = \sup X_3 = \beta_3\ ,$$

$$\bar{y}_1 = \inf[\lambda_1, \theta_1] = \lambda_1\ , \ \bar{y}_2 = \sup[\lambda_2, \theta_2] = \theta_2\ .$$

then

$$\Phi_1(\bar{z},t) \le 0 , \qquad \forall t \in X * Y .$$

By the lemma 1 ,we deduce that

$$\bar{z} = ((\beta_1,\alpha_2,\beta_3),(\lambda_1,\theta_2))$$

is an N-S equilibrium for the game (1).

REMARK 3 .

1) If there is $K \subset I$ such that functions $y \to f_i(x,y)$, for $i \in K$ are convex $\forall x \in X$, the result of theorem 1 remains true if we use the function

$$\Phi_K(z,t) = \sum_{i \in K} f_i(x // x'_i , y) - f_i(x,y') + \sum_{i \in I-K} f_i(x // x'_i , y) - f_i(x,y) .$$

2) If function $y \to f_i(x,y)$ is convex on Y , $\forall i \in I$, $\forall x \in X$ the result of theorem 1 remains also true , if we use the function

$$\Phi(z,t) = \sum_{i=1}^{n} f_i(x // x'_i , y) - f_i(x,y')$$

furthermore, we obtain an N-S equilibrium $\bar{z} = (\bar{x},\bar{y})$ where $\bar{y}$ is a Pareto optimal solution of the multi-objective problem

$$\langle Y, f(\bar{x},y)\rangle .$$

REMARK 4 . Theorem 1 suggests a practical method to determine N-S equilibrium for the game (1). Indeed, if conditions of theorem 1 are satisfied, it is sufficient to determine
$(\bar{z},\bar{t}) \in Z * T$ such that

$$\Phi_{i_0}(\bar{z},\bar{t}) = \min_{z \in Z} \max_{t \in T} \Phi_{i0}(z,t) .$$

According to theorem 1 and lemma 1 , we conclude that $\bar{z} = (\bar{x},\bar{y})$ is an N-S equilibrium for the game (1). Thus, the problem of determining N-S equilibrium of game (1) is transformed to the well-known problem of minimax. In the case where the conditions of theorem 1 are not satisfied, we can use the lemma 1, but the existence of N-S equilibrium is not guaranteed.

REFERENCES

[1]- Aubin J.P. L' analyse non lineaire et ses motivations économiques . Masson, Paris , 1984.
[2]- Nash J.F. Non-cooperative games -Ann. Math.1951, 54 , N° 5 pp. 286-295.
[3]- Slater M. Lagrange multipliers revisited. A contribution to non linear programming " cowls commission discussion paper,mahtematics, 403, 1950.
[4]- Steuer R. Multiple criteria optimization :theory , computation and application , Wiley, New York, 1986.
[5]- Zhukovskii V.I. , Tchikry A.A. Linear-quadratic differential games. Naoukova Doumka,Kiev, 1994.

Part 3

Implementation and Practice of Multiple Objective Programming

An Algorithmic Package for the Resolution of Dynamic Multiobjective Problems

R. Caballero, T. Gómez, M. González, L. Rey and F. Ruiz.*

Department of Applied Economics (Mathematics). Campus El Ejido s/n. University of Málaga. 29071-Málaga. Spain.

Abstract. In this paper we develop an algorithmic package to solve dynamic multiobjective problems in the convex case, as well as comments and computational results analysis for the quadratic problem, which we consider as the most natural extension of the linear problems in economic modeling. This package includes generating techniques (weighting and constraint methods) and Lexicographic Goal Programming.

Keywords. Multiple Objective Programming, Dynamic Optimization, Goal Programming.

1 Introduction

Many of the features that take place in general decision making processes in the economic world have somewhat dynamic components, for the decisions that are made in the present moment are influenced by those taken in the past, and toady's decisions will in turn affect the future ones. This is why we think it is worth studying decision making processes which occur along a definite period of time, and to provide the decision maker with computational tools that enable such decision making process, allowing him to obtain efficient solutions, as well as satisfying solutions for given target values.

Although some aspects of Dynamic Multiobjective Programming were originally treated in papers like Zadeh (1963), a further development of these ideas cannot be found until Wierzbicki (1991), Salukvadze (1979) and Li and Haimes (1990). On the other hand, the Dynamic Goal Programming problem was not taken in deep consideration until the works by Levary (1984), Trzaskalik (1996) and Caballero et al. (1996-b), where the concept of dynamic target values

---

* This paper was written while this author was visiting the University of Georgia, USA. His research is supported by the grant number PR95-287, of the Spanish Ministry of Education and Science, Dirección General de Investigación Científica y Técnica (DGICYT).

can be pointed out among the generalizations of the elements of the static problems.

2 Statement of the problem and basic definitions

The general problem that we face takes the form of a dynamic multiobjective program. If we consider it from a continuous point of view, then, given a time interval $[t_0, t_f] \subset R$, we have:

$$\begin{cases} min & \mathbf{J}(\mathbf{u}) = (J_1(\mathbf{u}), J_2(\mathbf{u}), \ldots, J_p(\mathbf{u})) \\ s.t. & \mathbf{x}'(t) = \mathbf{h}(\mathbf{x}(t), \mathbf{u}(t), t), \quad t \in [t_0, t_f] \\ & \mathbf{x}(t_0) = \mathbf{x}_0 \\ & \left. \begin{array}{l} g_i(\mathbf{x}(t), \mathbf{u}(t), t) = 0, \quad i = 1, \ldots, s_I \\ g_i(\mathbf{x}(t), \mathbf{u}(t), t) \le 0, \quad i = s_I + 1, \ldots, s \\ L_i \le u_i(t) \le U_i \quad i = 1, \ldots m \end{array} \right\} t \in [t_0, t_f] \end{cases}$$

where

$$J_i(\mathbf{u}) = \int_{t_0}^{t_f} f_i(\mathbf{x}(t), \mathbf{u}(t), t)dt + F_i(\mathbf{x}(t_f)), \; i = 1, \ldots, p$$

are the objective functionals, $\mathbf{u}$: $[t_0, t_f] \longrightarrow R^m$, are the control variables, which we will assume to be piecewise continuous, and $\mathbf{x}$: $[t_0, t_f] \longrightarrow R^n$, the state variables, which will be supposed to be continuously differentiable in the whole planning interval.

From the discrete point of view, given a positive integer N, we will consider the problem:

$$\begin{cases} min & \mathbf{J}(\mathbf{u}) = (J_1(\mathbf{u}), J_2(\mathbf{u}), \ldots, J_p(\mathbf{u})) \\ s.t. & \mathbf{x}(k+1) = \mathbf{h}(\mathbf{x}(k), \mathbf{u}(k), k), \quad k = 0, \ldots, N-1 \\ & \mathbf{x}(0) = \mathbf{x}_0 \\ & \left. \begin{array}{l} g_i(\mathbf{x}(k), \mathbf{u}(k), k) = 0, \quad i = 1, \ldots, s_I \\ g_i(\mathbf{x}(k), \mathbf{u}(k), k) \le 0, \quad i = s_I + 1, \ldots, s \\ L_i \le u_i(k) \le U_i \quad i = 1, \ldots m \end{array} \right\} k = 0, \ldots, N-1 \end{cases}$$

where

$$J_i(\mathbf{u}) = \sum_{k=0}^{N-1} f_i(\mathbf{x}(k), \mathbf{u}(k), k) + F_i(\mathbf{x}(N)), \; i = 1, \ldots, p.$$

In what follows in the present section, the dynamic problem, in any of its two forms, will be shortly denoted by:

$$(MODP) \quad \begin{cases} min & \mathbf{J}(\mathbf{u}) \\ s.t. & \mathbf{u} \in U \end{cases}$$

Under this formulation, the classical definitions of efficient, weakly efficient and properly efficient solutions can be used, and the basic results stating the relations between these solutions and the optimal solutions of several scalarized problems (basically weighting and constraint problems) also hold.

With the aim of guaranteeing the convexity of the problems, and so as to make possible the utilization of subroutines to solve convex single objective problems, it will be assumed that, in what follows, the following additional hypothesis hold:

- The dynamic system of the problem is linear, that is, it takes the form:
  $$\mathbf{x}'(t) = A(t)\mathbf{x}(t) + B(t)\mathbf{u}(t),$$
  in the continuous case, or
  $$\mathbf{x}(k+1) = A(k)\mathbf{x}(k) + B(k)\mathbf{u}(k),$$
  in the discrete case.
- All the equality constraints are linear.
- All the inequality constraints are convex in $\mathbf{x}$ and $\mathbf{u}$.
- All the functions f_i are convex in $\mathbf{x}$ and $\mathbf{u}$, and autonomous, that is, they do not explicitly depend on t.

In the following section, the computational package developed to find efficient solutions for problem (MODP) is described, with an especial attention to the discrete quadratic problem, for which computational results are shown. This particular problem takes the following form:

$$\begin{array}{ll} \min & \mathbf{J}(\mathbf{u}) = (J_1(\mathbf{u}), J_2(\mathbf{u}), \ldots, J_p(\mathbf{u})) \\ \text{s.t} & \mathbf{x}(k+1) = A(k)\mathbf{x}(k) + B(k)\mathbf{u}(k),\ \ k = 0, \ldots, N-1 \\ & \mathbf{x}(0) = \mathbf{x}_0 \\ & C(k)\mathbf{x}(k) + D(k)\mathbf{u}(k) \leq \mathbf{b}(k),\ \ k = 0, \ldots, N-1 \\ & \mathbf{L} \leq \mathbf{u}(k) \leq \mathbf{U},\ \ k = 0, \ldots, N-1, \end{array}$$

where, for each k, $A(k)$, $B(k)$, $C(k)$ y $D(k)$ are real matrices of dimensions $n\times n$, $n\times m$, $s\times n$ and $s\times m$, respectively, and $\mathbf{b}(k) \in R^s$, and

$$J_i(\mathbf{u}) = \sum_{i=0}^{N-1}\left\{\frac{1}{2}\left[\mathbf{x}(k)' Q_i(k)\mathbf{x}(k) + \mathbf{u}(k)' R_i(k)\mathbf{u}(k)\right] + \mathbf{q}_i(k)'\mathbf{x}(k) + \mathbf{r}_i(k)'\mathbf{u}(k)\right\} +$$
$$+\frac{1}{2}\mathbf{x}(N)' S_i\mathbf{x}(N) + s_i'\mathbf{x}(N),$$

where Q_i, R_i, S_i are positive semidefinite real matrices, of dimensions $n\times n$, $m\times m$ and $n\times n$, respectively, and $\mathbf{q}_i$, $\mathbf{r}_i$, $\mathbf{s}_i$ are vectors in R^n, R^m y R^n respectively.

From the algorithmical point of view, this formulation has the additional advantage of letting us turn it into a static quadratic multiobjective problem. Ruiz et al. (1997), prove that the above stated dynamic quadratic problem, is equivalent to the static problem:

$$\begin{cases} min & \mathbf{J}(\tilde{\mathbf{u}}) = (J_1(\tilde{\mathbf{u}}), \ldots, J_p(\tilde{\mathbf{u}})) \\ s.t. & \tilde{\mathbf{A}}\tilde{\mathbf{u}} \leq \tilde{\mathbf{b}} \end{cases},$$

where $\tilde{\mathbf{u}} = (\mathbf{u}(0), \mathbf{u}(1), \ldots, \mathbf{u}(N-1))'$. The explicit forms of the objective functionals and the constraints of this problem can also be found in the above mentioned paper.

In the general convex case, this reasoning can be carried out in a theoretical form, that is, the problem is equivalent to a static one, but there does not exist a closed expression of the functions that form this static problem, and that is why the algorithmic resolution of the general convex problem is slightly different, as it will be seen in the following section.

3 Algorithm for the generation of efficient controls

In this section, the different steps of the computational algorithm developed to obtain efficient controls for problem (MODP) are schematically displayed, distinguishing, when necessary, among the different possible cases, and showing the results obtained for the quadratic discrete problem.

3.1 Data introduction

The problem data introduction substantially varies depending on whether the problem is linear-quadratic or a general convex one. In the first case, all the data a read through a data file, to which the main program has automatic access in running time, and this way the user can change the problem to be solved just by building another data file. In the general convex case, a mixed scheme must be used, which combines the above mentioned data file with a FORTRAN subroutine that must be compiled and linked to the main program, previous to its execution. Therefore, in this case, the program itself is affected by a change in the problem to be solved, which makes its utilization more complicated. In general, the data introduction procedure is the same for discrete and continuos problems.

3.2 Construction of the scalar problems

Next, the program builds the scalar problems that must be solved to obtain efficient solutions for the original multiobjective problem. To this end, the user must choose the method he wants to use, that is, weighting or constraint. Previously, the algorithm has calculated the payoff matrix of the problem, determining as well the ideal and antiideal values for each objective functional, which will be denoted by $J_i(\mathbf{u}^{(i)})$ y $J_i(\mathbf{u}^{-(i)})$ respectively, for each $i = 1, \ldots, p$. For the weighting method, making use of the values obtained in the payoff matrix, the algorithm builds the normalizing factors for the weights:

$$n_i = \frac{1}{J_i(\mathbf{u}^{-(i)}) - J_i(\mathbf{u}^{(i)})}, \quad i = 1, \ldots, p$$

Next, the user is offered two options to generate weighting problems for the original multiobjective problem:

- Manual method. The user must introduce the vector of weights that he wishes to use, and the programs builds and solves the corresponding weighting problem.
- Automatic method. Given the nature of the weighting method, considered as a generating technique, it is very probable that the user just wishes to obtain a meaningful number of efficient controls which let him gather information about the structure of the efficient set. For this reason, it can be more useful that the program itself generates a definite number of vectors of weights, solving the corresponding problems. If the user chooses this options, he will be asked to give the number q of subintervals, in which he wishes to divide the weighting interval [0, 1], for each functional. Then, the program generates and normalizes the corresponding weights.

A similar scheme is carried out for the constraint method.

3.3 Resolution of the problems

Once the weighting or constraint problems have been built, the algorithm proceeds to the resolution of such scalar problems. To this end, we must distinguish between the discrete and the continuous case. In the former one, the subroutines used to solve the scalar problems where taken from the NAG subroutine library (see NAG, 1991), and they are basically quadratic sequential schemes. In the latter case the problem that has to be solved after the corresponding scalarizing process is also a single objective one, that is, a traditional optimal control problem. To solve it, in both the quadratic and the general convex case, numerical techniques for the resolution of optimal control problems can be used, which use parametrizations of the control trajectory trough piecewise constant functions that turn the problem into a traditional mathematical programming one.

3.4 Remarks and results analysis

Taking into account the experience of the authors in the resolution of static problems of big dimensions, it seems advisable to us to use the weighting method when the user wishes exclusively to generate efficient controls for he original problem, and to leave the constraint method only for the case when the user wants to obtain specific information relative to a definite bound vector, or to use it as an efficiency test. This is why the results displayed in this section are exclusively referred to the weighting method, and more precisely, to the discrete quadratic case with constant coefficients. Prior to the presentation of the computational results, some remarks must be made concerning to the weighting method:

- In any of the cases under consideration, the subroutine that has been used to solve the scalar problem asks the user to give an initial estimate of the final optimal solution. Although the subroutine will eventually reach the optimal solution of the problem, regardless of this initial estimate, it is clear that the time that it will take it to do so can depend on how accurate the initial estimate is. At least, it is very desirable that this initial estimate is a feasible control for the problem, for this will prevent the algorithm from having to use the feasibility phase, which is usually more time consuming than the optimality phase itself. For this reason, the algorithm uses as an initial estimation in each weighting problem, the solution of the previous problem, and this way, on the one hand the feasibility of the initial estimate is guaranteed, and, on the other hand, if the vectors of weights are close to each other, we indeed have an accurate estimation of the optimal solution.
- It is possible that for some problems very big differences can be observed between the results corresponding to two close weighting vectors. In these cases, the user may wish to observe in more detail this particular zone, and to this end the algorithm provides him with the possibility to "magnify" it. To do so, the user must introduce the interval $[a_i, b_i]$ where he wishes the weights of each functional to vary, and the program then proceeds to a loop similar to the general case one.

The algorithms has been tested on a series of linear quadratic problems with constant coefficients, randomly generated using a schemed which is based on the procedure proposed by Rosen and Suzuki (1965), and using a VAX computer with an Alfa coprocesor, with the aid of the N.A.G. subroutine library. The results can be regarded as acceptable, taking into account the global C.P.U. times taken in the resolution and the dimensions of the problems.

4 Goal Programming

With the aim of adapting the Goal Programming philosophy to the dynamic problem, we have admitted the possibility that the Decision Maker wishes to set target values not only on the final value of the objective functionals after the whole time period, but he would rather impose them on the behavior of such functionals along the whole dynamic process (following the scheme developed in Caballero et al. 1996-b). In other words, the target values will also be dynamic, and they will thus take the form of a function:

$$\alpha_i : [t_0, t_f] \longrightarrow R.$$

This way, the goals of our model, assuming that the objective functions where to be minimized, take the form:

$$J_i(\mathbf{u},t) = \int_{t_0}^{t} f_i(\mathbf{x}(\tau),\mathbf{u}(\tau),\tau)d\tau + F_i(\mathbf{x}(t)) \leq \alpha_i(t) \quad t \in \left[t_0, t_f\right], \quad i = 1,\ldots,p$$

in the continuous case, or

$$J_i(\mathbf{u},k)=\sum_{\kappa=0}^{k} f_i(\mathbf{x}(\kappa),\mathbf{u}(\kappa),\kappa)+F_i(\mathbf{x}(k))\leq \alpha_i(k) \quad k=0,\ldots,n-1, \quad i=1,\ldots,p$$

in the discrete case.

With the aim of maintaining the model convexity, as developed in Caballero et al. (1996-a) for the static case, only positive deviation variables will be considered. But, in our model, this variables will also be functions of time, so that, in each instant t, $\pi_i(t)$ measures the deviation of the accumulated value of the objective functional up to this moment, from the target value $\alpha_i(t)$. This way, the dynamic goals take the form:

$$J_i(\mathbf{u},t)-\pi_i(t)\leq \alpha_i(t) \quad t\in[t_0,t_f], \quad i=1,\ldots,p$$

$$J_i(\mathbf{u},k)-\pi_i(k)\leq \alpha_i(k) \quad k=0,\ldots,n-1, \quad i=1,\ldots,p$$

To keep on with our dynamic goals scheme, it is also necessary to adapt the definition of the achievement function, which, logically, will have now to consider the behavior of the deviation variables along the whole planning period, so that the temporal sum of such deviations is minimized. For a given functional J_i, this effect is measured through the utilization of the new functional

$$\int_{t_0}^{t_f} R_i(\pi_i(t))dt \quad \text{or} \quad \sum_{k=0}^{N-1} R_i(\pi_i(k)),$$

where R_i is any function, strictly increasing with respect to π_i, and such that attains its minimum in 0. The Goal Programming scheme developed is the lexicographic one, because we understand that, from the algorithmical point of view, it includes as a particular case the weighting scheme. So, it will be assumed that the Decision Maker groups his goals in r priority levels, $r \leq p$, and he has the possibility to assign weights to these functionals that share the same priority level. Following the mentioned scheme, the algorithm solves first a feasibility problem

$$\min \quad R_0(\pi) \qquad \text{s.t.} \quad \mathbf{u}\in U,$$

and then proceeds to solve the problems corresponding to the different priority levels, and where the problem corresponding to level q is:

$$\begin{aligned} \min \quad & R_q(\pi) \\ \text{s.t.} \quad & \mathbf{u}\in U \\ & J_i(\mathbf{u}, t) - \pi_i(t) \leq \alpha_i(t), \quad i \in N_q \\ & J_i(\mathbf{u}, t) \leq \alpha_i(t), \quad i \in N_1, \ldots, N_{q-1} \end{aligned}$$

4.1 Election of the achievement functional

In this step, the user can choose among several options for the achievement functionals that are to be used by the algorithm to solve the problems corresponding to the different priority levels, by giving a certain value to a parameter β, where the functionals are :

$$\sum_{i\in N_q}\mu_i\sum_{\kappa=0}^{N-1}\left(\pi_i(\kappa)\right)^\beta, \quad \text{or} \quad \sum_{i\in N_q}\mu_i\int_{t_0}^{t_f}\left(\pi_i(\tau)\right)^\beta d\tau,$$

In the discrete case, the authors experience in the resolution of nonlinear problems with big dimensions (Caballero et al., 1995), makes us suggest to minimize a constant functional for level 0, and $\beta = 2$ for the subsequent priority levels, provided that the quadratic sequential programming scheme is used to solve the scalar nonlinear problems.

4.2 Resolution of the intermediate problems

The resolution of the scalar intermediate problems, in both the feasibility level 0 and in the subsequent priority levels, is carried out following the same schemes that were described in section 3 for the algorithms to obtain efficient control, for both the conversion of the dynamic problem into a static one in the discrete case, and for the subroutines used to solve the scalar problems, for all the reasoning followed in section 3 applies for any dynamic scalar problem.

4.3 Priority level loop

After solving the feasibility problem corresponding to level 0, the algorithm enters a loop for the priority levels, whose aim is to sequentially solve the problems corresponding to each level. With respect to this loop, the following remarks must be made:

- With the aim of providing the subroutine with a feasible initial estimate for the problem corresponding to level q, the algorithm build the point $(\mathbf{u}^0, \boldsymbol{\pi}^0)$, such that:

  $$\mathbf{u}^0 = \mathbf{u}^{(i-1)*}$$

  $$\pi_i^0(t) = \begin{cases} \int_{t_0}^{t} f_i(\mathbf{x}(\tau),\mathbf{u}(\tau),\tau)d\tau + F_i(\mathbf{x}(t)) - \alpha_i(t) > 0 \\ 0 \quad \text{in other case.} \end{cases} \quad i \in N_q$$

  for the continuous case, and

  $$\pi_i^0(k) = \begin{cases} \sum_{\kappa=0}^{k} f_i(\mathbf{x}(\kappa),\mathbf{u}(\kappa),\kappa) + F_i(\mathbf{x}(k+1)) - \alpha_i(k) > 0 \\ 0 \quad \text{in other case.} \end{cases} \quad i \in N_q$$

  for the discrete case.
- After solving the problem of the current level, the algorithm tests whether its goals have been achieved. If so, it proceeds to the next level. If not, a redefinition of the target values is carried out:

$$\widetilde{\alpha}_i(t) = \alpha_i(t) + \pi_i{}^*(t), \quad t \in [t_0, t_f]$$

or

$$\widetilde{\alpha}_i(k) = \alpha_i(k) + \pi_i{}^*(k), \quad k \in \{0, 1, \ldots, N-1\}$$

so that the procedure can be carried out for the following levels, although the user is informed that the goals corresponding to the current level have not been achieved.

References

Caballero, R., González, M., Rey, L., Ruiz, F. (1995). *"An Algorithmic Package for the Resolution and Analysis of Convex Multiple Objective Problems".* Presented in "The Twelfth International Conference on MCDM". To Appear in the Proceeding volume of the Conference.

Caballero, R., Rey, L., Ruiz, F. (1996-a). "Determination of Satisfying and Efficient Solutions in Convex Multi-Objetive Programming". Optimization 1996 June/July. pp. 1-13.

Caballero, R., Gomez, T., González, M., Rey, L., Ruiz, F. (1996-b). "Goal Programming with Dinamic Goals". Sent To Journal of Multicriteria Analysis.

Levary, R.R. (1984). "Dynamic programming models with goal objectives", International Journal System Sciences, 15, 3, 309-314.

Li, D. and Haimes, Y.Y. (1990). "New approach for nonseparable dynamic programming problems", Journal of Optimization Theory and Applications, 64, 2, 311-330.

N.A.G. (Numerical Algorithms Groups Limited) (1991). The N.A.G. Fortran Library Introductory Guide, Mark 15.

Rosen, J.B. Y Suzuki, S. (1965). "Construction of Nonlinear Programming Test Problems". Comunications of ACM, 113.

Ruiz, F., Caballero, R., Gómez, T., González, M. and Rey, L. (1997). "The Static Approach to Quadratic Dynamic Goal Programming". Presented to the 13th International Conference MCDM, held in January, 1997 in Cape Town (Soth Africa).

Salukvadze, M. (1979). Vector-Valued Optimization Problems in Control Theory. Academic Press. Nueva York.

Trzaskalik, T. (1996). "Dynamic goal programming models", Presented to the Second International Conference in Multi-objective Programming and Goal Programming, Málaga (Spain). To appear in the proceedings volume.

Wierzbicki, A.P. (1991). "Dynamic Aspects of Multi-objective Optimization", en Lewandowski, A. and Volkovich, V. (eds). Multiobjective Problems of Matematical Programming. Berlin. Springer-Verlag.

Zadeh, L.A. (1963). "Optimality and Non-Scalar-Valued Performance Criteria", IEEE Transactions on Automatic Control, AC-8, 1.

Closest Solutions in Ideal-Point Methods

E.Carrizosa[1], E. Conde[1], A. Pascual[1] and D. Romero-Morales[2]

[1]Facultad de Matemáticas, Universidad de Sevilla, Sevilla, Spain
[2]E.U. Empresariales de Jerez, Universidad de Cádiz, Cádiz, Spain

Abstract. In ideal-point methods one seeks the feasible solution minimizing in the value space the distance to some ideal point given in advance.

A crucial step is the choice of the distance measure, since such choice strongly influences the output of the method. In this paper we propose a description of the set of closest-points to the ideal when the distance measure in use is allowed to vary in a certain family of metrics chosen according to some rationality principles.

Connections with the concept of compromise set of Yu and Zeleny are briefly discussed.

Keywords: Ideal points; Compromise set; Norms.

1 Introduction

Given a nonempty polyhedron X in $\mathbb{R}^d$ and an $n \times d$-matrix C, consider the linear vector optimization problem $(LVOP)$,

$$\text{vector-optimize}_{x \in X}\, Cx \qquad (LVOP)$$

In ideal-point methods, e.g. [3], it is customary to assume the existence of

1. A vector of weights $\omega = (\omega_1, \ldots, \omega_n)$ to reflect the relative importance of the different criteria.

2. An *ideal point* $z^* = (z_1^*, \ldots, z_n^*)$ of targets associated with the different criteria.

3. A norm γ in $\mathbb{R}^n$ used to measure the degree of deviations of the alternatives with respect to the ideal point,

and then one solves the scalar problem

$$\min_{x \in X} \gamma(\omega_1(c_1'x - z_1^*), \ldots, \omega_n(c_n'x - z_n^*)) \qquad (1)$$

It is evident that the optimal solution of (1) strongly depends on the three abovementioned ingredients. However, whilst the determination of weights and the ideal point can be driven with some rationale – weights reflect the relative importance of different criteria and the ideal point has a clear managerial meaning– the choice of the distance measure seems to be much more controversial.

In this paper we will consider the weights and ideal point as fixed, and devote our attention to studying the set of optimal solutions to problems of the form (1) when the distance measure is allowed to vary in a certain set $\mathcal{G}$ of norms.

This strategy was already used twenty years back by Yu, [8] and Zeleny, [9], defining the *compromise set* as the set of optimal solutions of problems of the form (1), i.e., the set of points closest to the ideal, when the norm varies in the class of l_p norms ($1 \leq p \leq \infty$),

$$\|(z_1, \ldots, z_n)\|_p = \begin{cases} \left(\sum_{i=1}^n |z_i|^p\right)^{1/p}, & \text{if } 1 \leq p < \infty \\ \max_{1 \leq i \leq n} |z_i|, & \text{if } p = \infty \end{cases}$$

Our aim is also to construct a set of closest points from a class $\mathcal{G}$ of norms in $\mathbb{R}^n$ (containing in particular the l_p norms), chosen according to rationality principles (monotonicity with respect to deviations and symmetry in the arguments). However (and this is the main difference with the class of l_p norms), finding such closest points amounts to solving linear programs, instead of the nonlinear nonsmooth problems found with l_p norms.

2 AS-Norms

In this section we review some basic properties of norms. For a detailed introduction to norms and their properties, we refer the reader to [6].

Definition 2.1 *A norm γ in $\mathbb{R}^n$ is said to be* absolute *iff*

$$(|u_i| = |v_i| \quad \forall i) \Rightarrow \gamma(u) = \gamma(v)$$

Remark 2.1 Recall, e.g. [1, 4], that γ is absolute iff it is *monotonic*, i.e.,

$$(u, v \in \mathbb{R}^n, |u_i| \leq |v_i| \quad \forall i) \Rightarrow \gamma(u) \leq \gamma(v)$$

□

Let Perm denote the family of permutations of the set $\{1, 2, \ldots, n\}$.

Definition 2.2 *A norm γ in $\mathbb{R}^n$ is said to be* symmetric *iff*

$$\gamma(u_1, \ldots, u_n) = \gamma(u_{\sigma(1)}, \ldots, u_{\sigma(n)}) \quad \text{for all } \sigma \in \text{Perm} \tag{2}$$

Definition 2.3 *We will say that a norm γ in $\mathbb{R}^n$ is an AS-norm iff γ is an absolute and symmetric norm.*

Hence, assuming that overall deviation with respect to the vector of targets is measured by an AS-norm just means that such deviation is componentwise nondecreasing in the criteria deviations and also that all such deviations are considered as equally important, a reduction which, (recall that the weight ω is assumed to be fixed!) implies no loss of generality once the objective functions are properly rescaled.

Typical examples of AS-norms are l_p norms $(1 \leq p \leq \infty)$. A much less familiar class of AS norms will play a central role in our development. Define Λ as the set of n-tuples in the unit simplex whose coordinates are sorted in nondecreasing order,

$$\Lambda = \{(\lambda_1, \ldots, \lambda_n) : \sum_{i=1}^{n} \lambda_i = 1, \; 0 \leq \lambda_1 \leq \ldots \leq \lambda_n\}$$

Moreover, for a vector $z = (z_1, \ldots, z_n) \in \mathbb{R}^n$, let $\delta(z) \in \mathbb{R}^n$ denote the vector of deviations of z with respect to the origin once such deviations are sorted in nondecreasing order, namely, $\delta(z)$ is the permutation of $(|z_1|, \ldots, |z_n|)$ such that

$$\delta_1(z) \leq \delta_2(z) \leq \ldots \leq \delta_n(z)$$

As a direct consequence of Hardy's principle, e.g. [5], one obtains the following

Proposition 2.1 *Let $z \in \mathbb{R}^n$.*

1. *For any $k \in \{1, 2, \ldots, n\}$, one has*

$$\sum_{i=k}^{n} \delta_i(z) = \max\{\sum_{i \in I} |z_i| : |I| = n - k + 1\}$$

2. *For any $\lambda \in \Lambda$, one has*

$$\sum_{i=1}^{n} \lambda_i \delta_i(z) = \max\{\sum_{i=1}^{n} \lambda_{\sigma(i)} |z_i| : \sigma \in \mathrm{Perm}\}$$

3. *The function $\|\cdot\|_\lambda$,*

$$\|\cdot\|_\lambda = \sum_{i=1}^{n} \lambda_i \delta_i \tag{3}$$

is a (piecewise linear) AS norm.

3 Compromise sets

Since the weights and the targets in (LVOP) are assumed to be given, we can impose (after rescaling the data) without loss of generality that all the weights are equal (to one), thus (1) becomes

$$\min_{x \in X} \gamma(Cx - z^*) \qquad (P_\gamma)$$

When the problem (P_γ) is solved for different norms γ, one obtains a family of optimal solutions, which constitutes what we call a *compromise set*:

Definition 3.1 *Let $\mathcal{G}$ be a nonempty family of norms in $\mathbb{R}^n$. $\bar{x} \in X$ is said to be a* compromise solution *for $(LVOP)$ with respect to the class of norms $\mathcal{G}$ if $\bar{x}$ is an optimal solution of a problem of the form (P_γ) for some γ in $\mathcal{G}$,*

$$\bar{x} \in \arg\min_{x \in X} \gamma(Cx - z^*).$$

*The set of compromise solutions for $(LVOP)$ with respect to the family $\mathcal{G}$ will be called the $\mathcal{G}$-*compromise set *for $(LVOP)$, and denoted $X^*(\mathcal{G})$.*

If one takes the class $\mathcal{L}_p$ of l_p norms,

$$\mathcal{L}_p := \{\|\cdot\|_p : p \in [1, +\infty]\},$$

one obtains $X^*(\mathcal{L}_p)$, the so-called compromise set of Yu and Zeleny, [8, 9]. From a technical viewpoint, determining such compromise set amounts to solving an infinite series of nonlinear nonsmooth problems. Hence, in practice one cannot obtain exactly $X^*(\mathcal{L}_p)$, and must replace it by a *surrogate* obtained by solving problem (1) for a finite number of parameter values $p_1, \ldots, p_t$.

Instead, we propose to take as class of norms the set of AS-norms. Contrary to the class of l_p norms, the construction of the compromise set for AS norms can be performed by using standard multiobjective LP techniques. Moreover, such set also enjoys rationality principles, since, as mentioned in Section 2, taking AS norms corresponds to accepting that all criteria are equally important and also that the distance to the ideal point is nondecreasing in the deviations with respect to the different targets.

In what follows let $X^*(\mathcal{AS})$ denote the compromise set with for $(LVOP)$ with respect to the family of AS norms.

First we give a characterization of $X^*(\mathcal{AS})$; using the same scheme of proof than in Lemma 4.2 of [2] and the monotonicity of $X^*(\cdot)$, one easily derives the following

Theorem 3.1 *Let $X^*(\Lambda)$ denote the compromise set for the class of norms $\{\|\cdot\|_\lambda : \lambda \in \Lambda\}$, as defined in (3). Then,*

$$X^*(\Lambda) = X^*(\mathcal{AS}) \supset X^*(\mathcal{L}_p)$$

Hence, in order to find the compromise set $X^*(\mathcal{AS})$ one just needs to solve the problem (Q_λ),

$$\min_{x \in X} \sum_{i=1}^{n} \lambda_i \delta_i (Cx - z^*) \qquad (Q_\lambda)$$

for λ varying in Λ. This leads to a different characterization of the compromise set:

Theorem 3.2 *Given $\bar{x} \in X$, the following statements are equivalent*

1. $\bar{x} \in X^*(\mathcal{AS})$.
2. *$\bar{x}$ is a weakly efficient solution for the multiple-objective problem*

$$\text{vector-min}_{x \in X} \left(\sum_{i=1}^{n} \delta_i (Cx - z^*), \sum_{i=2}^{n} \delta_i (Cx - z^*), \ldots, \delta_n (Cx - z^*) \right) \tag{4}$$

Proof. By Theorem 3.1 $\bar{x} \in X^*(\mathcal{AS})$ iff there exists a nonzero nonnegative ω such that $\bar{x}$ is an optimal solution of

$$\min_{x \in X} \sum_{i=1}^{n} \left(\omega_i \sum_{j=i}^{n} \delta_j (Cx - z^*) \right) \tag{5}$$

On the other hand, (4) is, by Proposition 2.1, a convex multiple-objective problem, the set of weakly efficient points of which must then coincide with the set of optimal solutions to problems of the form (5), [7]. □

Theorem 3.2 provides an easy way of testing whether a certain x^* is in $X^*(\mathcal{AS})$:

Corollary 3.1 *Given $x^* \in X$, the following statements are equivalent*

1. $x^* \in X^*(\mathcal{AS})$
2. *For any $\eta \in \{-1, 1\}^n$ such that*

$$\eta_i (Cx^* - z^*)_i \geq 0 \qquad \forall i = 1, 2, \ldots, n$$

and any $\sigma \in$ Perm such that

$$|c_{\sigma(1)} x^* - z^*_{\sigma(1)}| \leq \cdots \leq |c_{\sigma(n)} x^* - z^*_{\sigma(n)}|,$$

then x^* is a weakly efficient solution to the multiple-objective linear program

$$\begin{array}{ll} \text{vector-min} & \left(\sum_{i=1}^{n} \left(\eta_{\sigma(i)} (c_{\sigma(i)} x^* - z^*_{\sigma(i)}) \right), \ldots, \sum_{i=n}^{n} \left(\eta_{\sigma(i)} (c_{\sigma(i)} x^* - z^*_{\sigma(i)}) \right) \right) \\ \text{s.t.} & \eta_i (c_i x^* - z_i) \geq 0 \quad \forall i \\ & \eta_{\sigma(1)} (c_{\sigma(1)} x^* - z^*_{\sigma(1)}) \leq \cdots \leq \eta_{\sigma(n)} (c_{\sigma(n)} x^* - z^*_{\sigma(n)}) \\ & x \in X \end{array}$$

Remark 3.1 For the two-objective case, Theorem 3.2 leads to the characterization of compromise solutions as weakly efficient solutions for the bi-objective problem

$$\text{vector-min}_{x\in X}\left(\|Cx - z^*\|_1, \|Cx - z^*\|_\infty\right), \tag{6}$$

a problem which has been proposed as an approximation for the compromise set of l_p norms.

However, when the number of criteria is greater than two, there may exist elements of $X^*(\mathcal{L}_p)$ (and $X^*(\mathcal{AS})$!) not captured by (6). This is shown in the following example.

Example 3.1 Consider an one-dimensional approximation problem, in which the points $0, 2.5, 5, 6, 10$ must be aggregated into a single scalar x^*. We can pose this problem in our context, by considering the one-dimensional problem ($d = 1$) with 5 objectives ($n = 5$), $C = (1,1,1,1,1)$, and targets $0, 2.5, 5, 6, 10$.

In order to find *an approximation* to the compromise set $X^*(\mathcal{L}_p)$, we have solved problem (P_γ) for $\gamma = \|\cdot\|_p$, with p in the grid $1, 1.1, \ldots, 9.9, 10$, yielding (an approximation to) the interval $[4.70, 5]$.

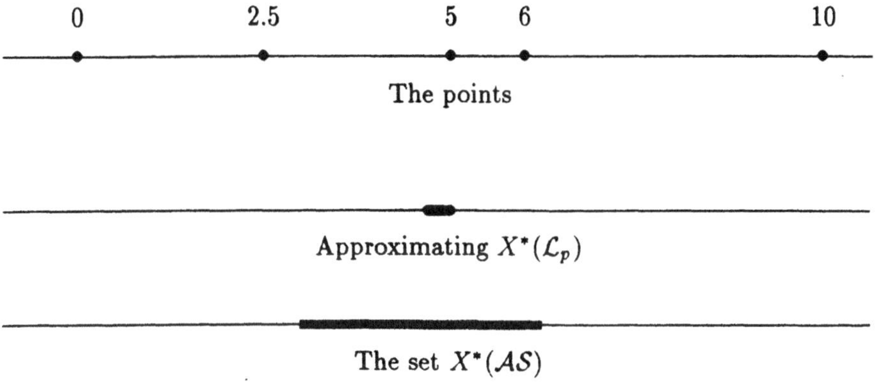

On the other hand, point $x^* = 5$ is the unique optimal solution to (P_γ) for both the $\|\cdot\|_1$ and $\|\cdot\|_\infty$. Hence, the output provided by the biobjective problem (6) consists of the singleton $\{5\}$.

Finally, by using Corollary 3.1, it is easily checked that the compromise set $X^*(\mathcal{AS})$ is the interval $[3, 6.25]$ (in fact, the set of minimizers of the *two* highest deviations), which strictly contains $X^*(\mathcal{L}_p)$.

Since (4) is a polyhedral multiple-objective problem, it can be transformed into a linear one. Indeed, it is straightforward to check that $\bar{x}$ is a weakly efficient solution to (4) iff there exists $\bar{t} \in \mathbb{R}^n$ such that the pair $(\bar{x}, \bar{t})$ is a weakly efficient solution of

$$\begin{array}{rl} \text{vector-min} & (t_1, \ldots, t_n) \\ \text{s.t.} & t_k \geq \sum_{j \geq k} \delta_j (Cx - z^*) \quad \text{for each } k \\ & x \in X \end{array}$$

By Proposition 2.1 such problem can be transformed into an equivalent linear problem, such as

$$\begin{array}{rrll} \text{vector-min} & (t_1, \ldots, t_n) & & (7) \\ \text{s.t.} & t_k \geq \sum_{i \in I} s_i & \text{for each } I,\ |I| = n - k + 1,\ k = 1, \ldots, n & \\ & s_i \geq (Cx - z^*)_i & \text{for each } i & (8) \\ & s_i \geq -(Cx - z^*)_i & \text{for each } i & (9) \\ & x \in X & & (10) \end{array}$$

Thus if $X^*(\mathcal{AS})$ is seen as the projection on the x-space of the set of weakly efficient points of (7) one obtains the following

Proposition 3.1 *$X^*(\mathcal{AS})$ is closed, connected and piecewise polyhedral.*

4 Acknowledgements

The research of E. Carrizosa and E. Conde has been partially supported by DGICYT grant PB93-0927. This support is gratefully acknowledged.

References

[1] Bauer, F.L., Stoer, J., y Witzgall, C. "Absolute and monotonic norms." *Numerische Mathematik* **3** (1961) 257-264.

[2] Carrizosa, E., Conde, E., Romero-Morales, D. "Maxcovering Decisions. Theory and Applications". Prepublicación de la Facultad de Matemáticas. Universidad de Sevilla. Presented at the XXth Meeting of the SEIO, Seville, 1995.

[3] Chankong, V., Haimes, Y.Y. *Multiobjective Decision Making: Theory and Methodology.* North-Holland, 1983.

[4] Johnson, C.R., Nylen, P. "Monotonicity properties of norms." *Linear Algebra and its Applications* **148** (1991) 43-58.

[5] Marshall, A.W., Olkin, I. *Inequalities: Theory of majorization and its applications.* Academic Press, 1979.

[6] Rockafellar, R.T. *Convex Analysis.* Princeton University Press, 1970.

[7] LOWE, T., THISSE, J.F., WARD, J.E., WENDELL, R.E. "On efficient solutions to multiple objective mathematical programs", *Management Science* **30** (1984) 1346-1349.

[8] YU, P.L. "A class of solutions for group decision problems", *Management Science* **19** (1973) 936-946.

[9] ZELENY "Compromise programming", in *Multiple Criteria Decision Making* (J.L. Cochrane and M. Zeleny eds.), University of South Carolina, Columbia, 373-391, 1973.

Methods of Multicriteria Decision Support Based on Reference Sets

Andrzej M. J. Skulimowski
Dept. of Operations Research, Institute of Automatic Control,
University of Mining & Metalurgy, Cracow, Poland

Abstract. In this paper we present the foundations of the multiple reference points and reference set method of solving multicriteria decision problems and discuss its properties. To select a compromise solution to the multicriteria optimization problem, the decision-maker is requested to provide the additional information consisting of several families of reference points representing the values of criteria being of a special importance for the problem solution. The supplementary information concerning the preferences in the criteria space is aggregated in such a manner, so that the problem is reduced to the bicriteria trade-off between the proximity measures to the sets of desired and avoidable values of criteria. The solution process is interactive, the decision-maker may define at each step new reference points, trade-offs, and search directions.

1 Introduction

In this paper, we will study the solutions to the vector optimization problems

$$(F : U \to E) \to \min(\theta). \tag{1}$$

U and E are called the decision space and the criteria space, respectively, $F = (F_1, F_2, \ldots, F_N)$ is the vector objective function, and θ is a closed and convex cone which introduces the partial order in E. In the most common case where $\theta := I\!R^N_+$, the corresponding partial order is the natural (coordinatewise) partial order in $I\!R^N$. The decision space U may be a list of discrete alternatives, a subset of a Euclidean space $I\!R^k$, or a subset of a function space. The criteria space E is assumed to be a partially ordered linear metric space, and the minimum in (1) relates to this partial order. The sets of nondominated decisions, and nondominated values of criteria with respect to θ will be denoted by $P(U, \theta)$, and $FP(U, \theta)$, respectively.

Solution choice methods applied to the Pareto-optimal subset of U use additional knowledge about the decision-maker's preferences not included in the formulation of the problem (1).

Here, we present a method of deriving preference structures from a set of points in the criteria space, called *reference points*, and their application to select a compromise solution to (1). A reference point is an element of the criteria space representing the values of criteria being of a special importance to the decision-maker. When regarded as a potential solution to the original decision problem, it may be evaluated as a "desired", "acceptable", or "wrong" optimization result. Using suitable mathematical methods one can find the element $u \in U$ so that its values of criteria, $F(u)$, most closely approximate the desired results, or are most distant to a "wrong" one.

The methods based on reference points constitute one of the most important classes of multicriteria decision-making procedures. The idea of reference points representing the desired (or ideal) values of criteria has been extensively studied in the past, cf. e.g. Wierzbicki (1986), Yu and Leitmann (1974), Weistroffer (1984), and many others. The drawback of the classical approach based on a single reference point consists mainly in the fact that the decision-maker is allowed to define one such point at a time only. This leads to an inconsistency if two different criteria values are equally desired, but a separate approximation of each one of them results in different admissible solutions. Moreover, the classical single reference point methods neglect usually the existence of the criteria-space constraints.

We assume that, reference points can be determined by the decision-maker, or are results of experts' judgments. The evaluations of experts are assumed independent from each other and from the constraints occurring in the multicriteria problem formulation, and have the same credibility to the decision-maker. The method bases on aggregation of reference points into sets of desired and avoidable values, and on considering some of them as the criteria-space constraints. Thus we obtain a model of preferences consisting of a family of reference sets in the criteria space, and distance functions modelling the decision-maker's wish to reach one of the sets and to avoid another. To select a compromise solution, we find the set D containing the elements of E nondominated with respect to both distance functions. Then we confine the search for a compromise solution to the intersection of D and of the set of nondominated points to the problem (1).

2 The Decision–Making Problem Statement

In this paper we will study the problem of selecting a single solution to (1) based on the additional information provided by experts or the decision-maker. Namely, the reference points will serve to define the function $v(x)$ called the *utility* or *value function.* We will show that v thus obtained is ***strongly monotonically increasing***, i.e.

$$\forall x, y \in E \ (x \leq_\theta y,\ x \neq y \Rightarrow v(x) < v(y)),$$

consequently, the minimal value of v can be achieved only on the nondominated set $FP(U, \theta)$ and determines the best-compromise solution to (1). Taking into account that the information on hand is often insufficient to evaluate the global estimate of v, nor is such estimate necessary to select a single compromise alternative, one is usually satisfied with a local estimate which minimum on $FP(U, \theta)$ coincides with the minimum of the hypothetical global estimate. Thus, the multicriteria decision making problem for (1) consists in finding or estimating v, and solving the minimization problem

$$(v : F(U) \to \mathbb{R}) \to \min, \tag{2}$$

while from the strong monotonicity property it follows that

$$\arg\min\{v(x) : x \in F(U)\} \subset FP(U).$$

The main features of the here presented multiple reference point approach to solve (1)-(2) can be summarized as follows:

- After formulating the problem (1), the decision–maker makes statements concerning the selected elements of the criteria space (reference points). Thus, to select a compromise solution one uses exclusively the information B actually available, in contradistinction to a large class of MCDM methods which require an evaluation of subsequently generated nondominated solutions which need not be of special interest to the decision–maker.
- We assume that to each reference point one can associate a utility value which, however, need not be given explicitly. There is no need for a linearity or additivity assumptions which are replaced by the distance utility principle (cf. Sec. 4).
- The original vector optimization problem (1) will be replaced by the *limited optimization problem* with the criteria space constraints Q,

  $$(F : U \cap F^{-1}(Q) \to E) \to \min, \; F_{opt} \in Q, \tag{3}$$

  where $Q \subset E$ is the set of values of F, where the minimization of F makes sense, and F_{opt} is the compromise value of F.
- Each iteration of the decision–making process consists of two stages: at the first stage all additional information available is used to confine the set of nondominated points. At the second one a compromise solution is generated and presented to the decision–maker.
- The observation that the constraints on the substitution coefficients between the criteria can be expressed equivalently as the partial order introduced by a convex cone different from $\mathbb{R}^N_+$, lets us consider the trade–off constraints and the reference points in one preference model jointly.

3 Utility Interpretation of Reference Points

According to the previous remarks, distinguished points or sets of points in the criteria space will serve as a source of additional information while modelling the decision maker's preferences. All these points are called *reference points*. Each reference point is characterized by two different types of information:

- its meaning to the decision–maker, which is set up a priori by experts involved in the decision–aid, usually without taking into account the constraints in the underlying vector optimization problem;

and

- the relation to the set of attainable criteria values in the vector optimization problem (1).

The latter relation may be taken into account after gaining some information about the location of the attainable set $F(U)$ in the criteria space and may result in re–classifying the first class of relations.

Both kinds of information generate different classifications of reference points. First, we will consider this one which is based on external information conveyed to the decision maker by experts. In the sequel we will denote the classes of reference points consecutively by A_0, A_1, $A_2, \ldots$.

A_0 – **Bounds of optimality** – the reference points which determine the lower bound of the region Q where the optimization of criteria makes sense (cf. the problem with the criteria space constraints (3)).

The existence of these reference points illustrate a limited scope of assumption that every point dominating a target point is still desired by the decision–maker.

Estimated utility: To the elements of A_0 we will assign the same utility *deterministic* value as to the target points (see below), i.e. $v(A_0) = a_1 > 0$.

A_1 – **Target points** – the elements of E which model the ideal solutions desired by the decision–maker.

The coordinates of target points express the desired values of criteria without paying attention to constraints and properties of the objective functions. Whenever possible, the target points should be achieved or surpassed by the solutions to (3). If it is impossible, a compromise solution should be as close as possible to this set.

Other notation: aspiration levels, ideal points.

Estimated utility: To each element of A we assign the same utility a_1.

A_2 – **Status–Quo Solutions** – attainable values of criteria which must be surpassed during the decision–making process.

If multiple status–quo solutions are defined, it is sufficient to find a non-dominated solution dominating at least one of them. The interpretation of status–quo solutions can be (at least) twofold; either as

- the values corresponding to the alternatives chosen in one of previous decision problems which are still available or as
- the experience-based judgments concerning the minimal satisfying levels of criteria ("bounds of satisfaction").

Other notation: reservation levels (Wierzbicki, 1986), required values (Weistroffer; 1984), solutions available at the pre-decision stage.

Estimated utility: To the status-quo solutions we will associate a utility value a_2, with $a_2 < a_1$.

A_3 – **Anti-ideal reference points** – elements of the criteria space which express the wrong choice of solution.

These points should be avoided by the decision-maker, by choosing a solution most distant to A_3.
Other notation: failure levels, avoidance points.
Estimated utility: To each element of A_3 we will assign a utility value a_3, with $a_3 < a_2 < a_1$.

In some situations, e.g. when the information is provided by experts, a finer classification may be available. For instance, the required and satisfying values may be distinguished. If the reference points are defined as the interval subsets of E, one may similarly consider the upper and lower edges of the estimation intervals. In general, one can split a class A_i into the subclasses $A_{i,1}, \ldots, A_{i,k(i)}$ assigning to each subclass a utility value $a_{i,j}$, for $0 \leq i \leq 3$, $1 \leq j \leq k(i)$, with

$$a_{i,j+1} \leq a_{i,j} \leq a_{i,j-1}$$

if $a_{i,j-1}$ or $a_{i,j+1}$ exist for $1 \leq j \leq k(i) - 1$.

An example of reference points for certain multicriteria optimization problem, and satisfying the consistency assumptions is shown in Fig. 1.

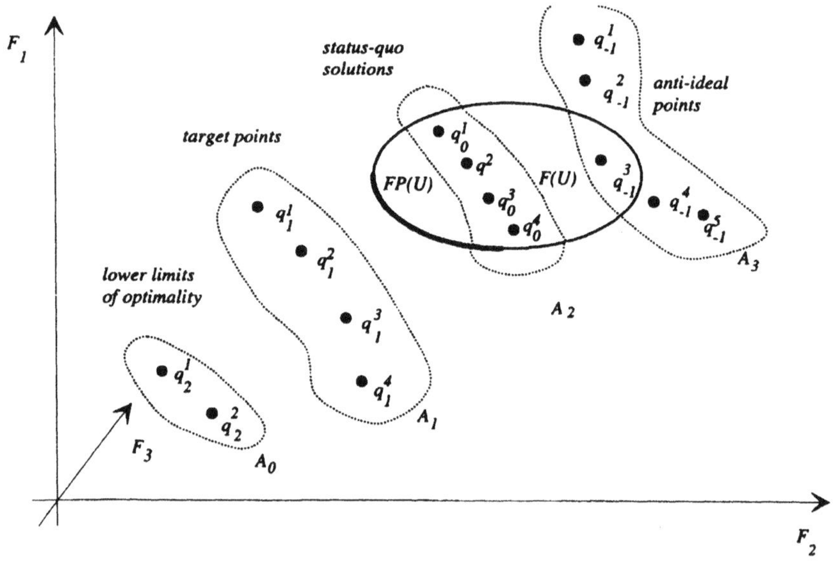

Fig. 1. *An example of well-defined reference points.*

Now, let us present an example of a real–life industrial design problem illustrating the application of the above ideas.

Example 3.1. An industrial company is planning to construct a new model of an elevator. Each considerable alternative is described by the set of non-compensatory criteria, including the reliability coefficient, the amount of capital necessary to start the production, the minimal time of evaluation of the project, the lifting capacity, the maximal range of the arm, the cost of production per unit, and others.

To make a compromise decision the following information is available:

- the parameters of the elevators already produced by this company (A_2),
- the parameters of the products which are expected to sell best at the market, the parameters of best elevators produced by other companies, or both (A_1),
- the bounds for the optimal values of parameters which are justified by the expected scope of the use of elevators (A_0),
- the parameters of elevators which turned out a market failure for the companies producing them (A_3).

To consider all that information jointly, the choice of a decision should be made according to the rules described in the next section. ■

4 The Metric Utility Principle

Having defined the goals, and/or avoidance region in the criteria space we have to formalize the meaning of "approaching the target" or "avoiding the dangerous decisions".

From the definition of the sets $A_0, \ldots, A_3$ it may be derived the following verbal description of the decision–maker's interpretation of the classes of reference points:

- the set of target points A_1 should be approached,
- the set of anti–ideal points A_3 should be avoided,
- the set of status–quo solutions A_2 should be exceeded,
- the solutions better than the bounds of optimality A_0 should be avoided.

The utility estimation process will consist of three steps:

- the approximation of the level sets of v,
- the determination of the estimation domain E where the utility function estimate $\tilde{v}$ is defined,
- the interpolation of $\tilde{v}$ in the regions bounded by the level sets.

As the first step to estimate the utility function one should approximate those level sets of $\tilde{v}$ which correspond to the values associated to the sets A_i. The level sets so found divide the criteria space E into the regions

$$R_i := \{x \in E : a_i \leq \tilde{v}(x) < a_{i-1}\}, \text{ for } i = 1, \ldots, K \tag{4}$$

with $a_0 := -\infty$, $a_K := +\infty$. From the assumed continuity of v it follows that the regions R_j are disjoint, i. e. $R_i \cap R_j = \emptyset$ for $i \neq j$, $i, j = 1, \ldots, k$, therefore it exists the unique

$$j := \min\{i : R_i \cap FP(U, \theta) \neq \emptyset\}. \tag{5}$$

The utility values in R_j are highest possible for the elements of $F(U)$, so it is reasonable to restrict the further search for a compromise solution to the set $R_j \cap FP(U, \theta)$.

Hence one can conclude that to complete the model of the decision-maker's preferences one has to find two distance functions, the first one modelling the degree of avoidance of A_j, the second one modelling the achievement of A_{j+1} within the region R_j. The measures of proximity from the sets A_j take a form of seminorm-like functions g_{-j} (for distance maximization), and g_{+j} (for distance minimization), defined as follows:

$$g_{-j}(x) := d(x, Q_-) := \inf\{\| x - r \| : r \in Q_-\}, \tag{6}$$

$$g_{+j}(x) := d(x, Q_+) := \inf\{\| x - s \| : r \in Q_+\}, \tag{7}$$

where $Q_- := A_j - \theta$, and $Q_+ := A_j + \theta$ for j determined by (9), and the norm in E is strongly monotonically increasing on θ. Thus, the compromise solution should belong to the intersection of two sets of nondominated points: first one to the initial problem (1), the second one to the bicriteria problem

$$[(-g_{-j}, g_{+j}) : R_j \to I\!R^2] \to \min(I\!R^2_+). \tag{8}$$

To evaluate such intersection it is necessary to find the attainable elements which minimize g_{+j} or maximize g_{-j}.

To choose the compromise solution from the set $FP(U, \theta)$ we will use the bicriteria trade-off method based on the following theorem:

Theorem 4.1. *For each $\lambda > 0$ the utility estimation function in the region R_j,*

$$G_j(x) := g_{+j}(x) - \lambda g_{-j}(x)$$

is strongly monotonically increasing. Consequently, the set $P(G_j, F(U, \theta), I\!R^2_+)$ is contained in $FP(U, \theta)$ and

$$\arg\min\{G_j(F(u) : u \in U\} \subset P(F, U, \theta),$$

i.e. $G_j \circ F$ can be used as a scalarizing function for the problem (1).

Proof. Based on the optimality conditions given in Skulimowski (1988), it is sufficient to observe that the functions $g_{+j}(x)$ and $-g_{-j}(x)$ are both monotonically increasing in R_j. Then so is any their positive linear combination. ■

For the basic decision-making problem with four classes of reference points $A_0, \ldots, A_3$, the solution thus obtained fulfills the following compromise rule:

it is as distant as possible from the set A_3, exceeds one of elements of A_2, it is as close as possible to the convex hull of the set $A_1 + \theta$, and does not exceed any element of A_0. Varying the positive trade–off parameter λ and the scaling coefficients g_{+j} and g_{-j} (cf. (6)-(7)), lets us interactively modify the resulting compromise solution $x_c \in U$ and $F_{opt} = F(x_c)$ within a consistent decision process.

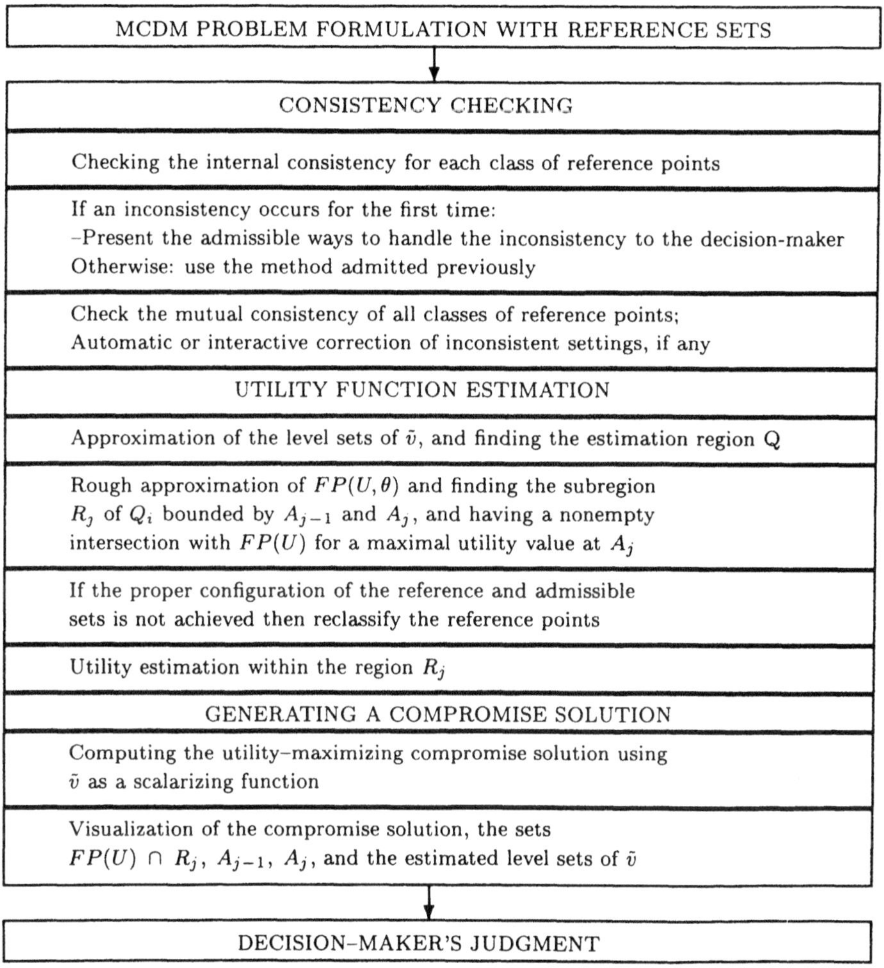

Fig. 2. *A scheme of establishing the preference model and generating the compromise solutions in a decision support algorithm based on reference sets.*

5 An Interactive Decision–Making Algorithm

The numerical methods applied to generate the nondominated solutions within the decision–making procedure presented in Secs. 3–4 use Pareto–optimality test and an a posteriori correction of results, so that the solution proposed to the decision–maker is actually nondominated. An application of the above described issues in a decision–choice procedure, is present in Fig. 2.

As the input information for the procedure, the user is requested to define the constraints, criteria and as the additional preference information for the continuous optimal control problems with the terminal–time values of objectives: the reference values of criteria representing the most desired optimization results (ideal points), those satisfactory, the limits of optimality, and the values identified with a failure which should therefore be avoided;

The solution process is interactive, the results of each step are illustrated graphically, the decision–maker may use a pointing device to pick up the relevant values, or define numerically new reference points, trade–offs, and search directions. Additionally, for multi-step solutions with the time horizon not strictly defined, we propose a learning scheme which allows for examining the currently defined reference points versus the a posteriori judgments related to the solutions obtained in the past with similar reference objectives. The method has been implemented in the MS WINDOWS^{TM} environment under the trade name MREF.

References

[1] Górecki H., Skulimowski A.M.J. (1988). Safety Principle in Multiobjective Decision Support in the Decision Space Defined by Availability of Resources. *Arch. Automatyki i Telem.*, **32**, 339-353.

[2] Jacquet-Lagrèze E., Meziani R., Słowiński R. (1987). MOLP with an interactive assessment of a piecewise linear utility function. *EJOR*, **31**, 350-357.

[3] Michałowski W., Szapiro T. (1992). A Bi-Reference Procedure for Interactive Multiple Criteria Programming. *Operations Research*, **40**, No.2, 247-258.

[4] Skulimowski A.M.J. (1988). A Sufficient Condition for θ-Optimality of Distance Scalarization Procedures. *Found. Contr. Engrg.*, **13**, No.1, 27- 41.

[5] Skulimowski A.M.J. (1990). Classification and Properties of Dominating Points in Vector Optimization. *Methods of Operations Research*, **58**, 99- 112.

[6] Skulimowski A.M.J. (1996). Decision Support Systems Based on Reference Sets, UMM Publishers: Monographs, No. 40, Kraków,p. 165.

[7] Weistroffer H.R. (1984). A combined over- and under- achievement programming approach to multiple objectives decision-making. *LSS*, **7**, 47-58.

[8] Wierzbicki A.P. (1986). On the Completeness and Constructiveness of Parametric Characterizations to Vector Optimization Problems. *OR Spektrum*, **8**, 73-87.

[9] Yu P.-L., Leitmann G. (1974a). Compromise Solutions, Domination Structures and Salukvadze's Solution. *JOTA*, **13**, No.3, 362-378.

A Tabu Search Procedure to Solve MultiObjective Combinatorial Optimization Problems

Xavier GANDIBLEUX, Nazik MEZDAOUI and Arnaud FRÉVILLE

Op.Res./Dcs.Aid Unit, LAMIH & LIMAV, Université de Valenciennes
Le Mont Houy - BP311, F-59304 Valenciennes Cedex, FRANCE
Tel : +33 (0)3 27 14 14 72, Fax : +33 (0)3 27 14 12 94
email : {gandibleux,mezdaoui,freville}@univ-valenciennes.fr

Abstract. Several studies have considered metaheuristics, especially simulated annealing, for solving combinatorial optimization problems involving several objectives. Yet, few works have been devoted to tabu search approaches. In this paper, we present a heuristic based upon tabu search principles to generate a good approximation of the set of the Pareto-optimal (efficient) solutions.

Keywords. MultiObjective Combinatorial Optimization, Tabu Search, efficient solutions

1 Introduction

1.1 MultiObjective Combinatorial Optimization problems

In this paper, we consider MultiObjective Combinatorial Optimization (MOCO) Problems stated as following :

$$\textbf{(MOCO)} \quad \left[\begin{array}{ll} \max & \{f_1(x), f_2(x), .., f_p(x)\} \\ \text{s.t.} & x \in D \end{array} \right.$$

where D is a discrete (and finite) subset of IR^n defining the *decision space* (or the feasible domain), and $f_1(x), f_2(x), .., f_p(x)$ are p real valued functions well-defined on D ($p \geq 2$).

MOCO is a class of Multiple Criteria Decision Making problems (MCDM) for which we can assume the situation where all the decision-maker preferences are known at the beginning of the decision-making process (a priori case; [17],[4]). MOCO problems take place in the context of Pareto preference. No ambiguity exists in preferences, and greater performances for each component $f_j(x)$ are more preferred. Let $z = f(x) = (f_1(x), f_2(x), .., f_p(x))$ be the objective vector associated to the decision variable x, and $Z = \{ z \mid z = (f_1(x), f_2(x), ..., f_p(x))$ for any $x \in D\} \subseteq IR^p$ be the *objective space*. Then $x^1 \in D$ is Pareto preferred to $x^2 \in D$ iff component-wise

$z_j^1 \geq z_j^2$, $j = 1,...,p$, and $z^1 \neq z^2$. The main goal for this kind of problems is to determine the set E(P) of nondominated (or efficient) solutions; $x' \in E(P)$ iff it does not exist any $x \in D$ such that x is Pareto preferred to x' (a survey on MOCO problems and their complexity properties is given in Ulungu and Teghem [15]).

The goal in the single objective case (p=1), where an optimal solution is investigated, is quite different. Exact methods exist in both cases to achieve these tasks. Yet, as in the single objective case, when the MOCO problem is NP-hard, a reasonable alternative is to try to derive a heuristic approach for solving large-scale instances, that yields a good tradeoff between the quality of the approximation $\hat{E(P)}$ of the efficient solutions set E(P), and the time and memory requirements.

1.2 Metaheuristics and MOCO problems

Dedicated heuristics can been derived for solving special cases of MOCO problems. Often based on a Knowledge-Based System or Rule-based System, they are specialised to the studied problem and not easily generalizable (Gandibleux et al. [3]).

Metaheuristics are approximate optimization techniques which can be applied to a great variety of problems. In the single objective case, they encountered widespread successes in attacking many difficult combinatorial optimization problems ([10],[11]). Up to now, the use of metaheuristics for solving MOCO problems has been explored not much, and as far as we know, only four papers have been recently proposed.

Dahl et al. [2] used the link between efficient solution and convex combinaison of the objectives. Under appropriate conditions, it is well-known that an efficient solution is optimal in (P_λ) for some $\lambda = (\lambda_1,...,\lambda_p) \geq 0$, $\lambda \neq 0$, where (P_λ) is the following scalar maximum problem:

$$(P_\lambda) \quad \left[\begin{array}{ll} \max & \sum_{j=1}^{p} \lambda_j f_j(x) \\ \text{s.t.} & x \in D \end{array} \right.$$

Tabu search is invoked to find near-optimal solutions for (P_λ). Hertz et al. developed in [9] an other approach also based on tabu search, consisting in the solving of a sequence of single-objective problems. Each objective f_j is considered in turn and associated to a penalty term, taking into account imposing threshold values on those which have already been examined.

Two others papers are developed a simulated annealing framework to approximate the efficient solutions set ([1],[16]). The main idea consists in using a weighted norm component in the evaluation of the probabilty under which a solution of less quality can be accepted [13]. In [16] several lists of potential efficient solutions are generated according to different sets of weights and merged to provide the final approximate set. On the other hand the weights are adjusted dynamically in [1].

2 A MultiObjective Tabu Search Procedure - MOTS

Tabu Search (TS) ideas were introduced by Hansen [8] and Glover [6] who has derived its modern form [7]. In the single objective context, TS method shares with the SA algorithm the ability to guide the local search descent method to avoid bad local optima. At each iteration, TS moves to the 'best' admissible neighbor using a deterministic criterion. This criterion may lead to cycling, i.e. returning to solutions already visited. Attributes of accepted solutions are stored in a *tabu list* and declared tabu for a number of iterations to avoid cycling. A neighboring solution is considered forbidden and deemed not admissible if it has attributes on the tabu list. Storing attributes rather than the complete solutions may cause some non-tabu solutions to be wrongly prevented and *aspiration criteria* are normally used to correct such errors. Others mechanisms using flexible memory are employed to improve the results. *Intensification* strategy concentrates the search in promising regions, whereas *diversification* strategy tries to drive the search into new regions.

We now present a procedure based on tabu search principles for solving MOCO problems, and able to determine the efficient set E(P), or at least a good inner approximation $E(\hat{P})$. Our MOTS procedure is based on a skeleton of tabu search, including the classic tools of the method (see [8]). Yet, regarding to the single objective case, all of them need to be stated in a suitable manner according to the quite different goal of the MOCO problems.

2.1 Search strategy

We assume that a suitable *move* $x \rightarrow x'$ has been defined according to the structure of the feasible domain D. The move defines the neighborhood V(x) of any current solution x and a first tabu list *TL_move* is managed to avoid cycling.

The task is then to determine the 'best' solution $\hat{x}$ in V(x), which will become the successor of x in the iterative TS process. For that, we use the concept of *compromise solution* for determining $\hat{x}$. Typically, since each objective f_j, $j = 1,..,p$, is characterized by 'more is better', one consider the utopian point $\tilde{z} = (\tilde{z}_1,...,\tilde{z}_p)$ over V(x), where $\tilde{z}_j = \sup\{f_j(x') \mid x' \in V(x)\}$. Now, given $z = f(x) \in Z$, the regret of using z instead of obtaining the utopian point $\tilde{z}$ may be approximated by the distance $S(z,q,\omega)$ between z and $\tilde{z}$, according to some specified norm l_q-norm and a weight vector $\omega = (\omega_1,\omega_2,..,\omega_p)$, $\omega \geq 0$ [14] :

$$S(z,q,\omega) = \|z - \tilde{z}\|_{q,\omega} = [\sum_{j=1}^{p} | \omega_j^q (z_j - \tilde{z}_j)^q |]^{1/q}$$

The weight vector ω may be assigned to signal the different degrees of importance of the objectives. Finally, the compromise solution with respect to the *lq*-norm and the weight vector ω, is $\hat{x} \in V(x)$ which minimises $S(f(x),q,\omega)$ over V(x) and such that the move $x \rightarrow \hat{x}$ is not tabu. This compromise solution is chosen as the new current solution in the iterative TS process.

As $V(x)$ is generally a finite enumerable subset of D, the compromise solution can be found in an obvious way. However, the time complexity depends on the size of the neighborhood $V(x)$.

2.2 Efficient solutions update

The main goal of MOTS procedure is to generate a good approximation $E(\hat{P})$ of the efficient set $E(P)$. The set $E(\hat{P})$ is built step by step. At each iteration of the TS, we consider the M better solutions x^m, $m = 1,...,M$ ($x^1 = \hat{x}$ and $1 \leq M \leq$ size of $V(x)$) according to the scalarazing function $S(f(x),q,\omega)$. Then for each of them, we check if the solution x^m is nondominated over the set $E(\hat{P})$. If it is the case, x^m is added to $E(\hat{P})$ and all solutions of $E(\hat{P})$ which are Pareto dominated by x^m are eliminated from $E(\hat{P})$. Since $M \geq 1$, the update of $E(\hat{P})$ is always processed with $\hat{x}$ at each iteration. But if $\hat{x}$ seems to be a new promising efficient solution, then M is increased. The promising characterisation of $\hat{x}$ is evaluated by regarding the number of solutions of $E(\hat{P})$ which were Pareto dominated by $\hat{x}$. This step can be considered as an *intensification* strategy when M is large enough.

An *aspiration criterion* may be employed. Any solution $x^\# \in V(x)$, considered forbidden according to its tabu status, and such that $S(f(x^\#),q,\omega) \leq S(f(\hat{x}),q,\omega) - \delta$ (a threshold value) can also be compared as above with the set $E(\hat{P})$.

At last, and to answer to the problematic choice [12], the MOTS procedure proposes at the end of the search a solution $\hat{x} \in E(\hat{P})$ by using S(.). This solution realizes a suitable compromise on each objective.

2.3 Weight update

The compromise solution is not scale independent and the weight vector ω has the effect of changing the scale of each objective. The concept of scalarazing function $S(z,q,\omega)$ to determine a compromise solution leads to explore efficient solutions only in the direction of the utopian point associated to the whole decision set D. Changing the weight vector ω may help to browse the border of the set $f(E(P))$, image in the objective space of the efficient solutions. This can be viewed as a *diversification* strategy controled at the space objective level.

To perform this idea, we use the pseudo-criterion preference model introduced in [12] to update the weights at each iteration. Let $t_{min,j}$ and $t_{max,j}$ two threshold parameters for each objectives such that $0 < t_{min,j} < t_{max,j}$, which measure the level of improvement $z_j(\hat{x}) - z_j(x)$ according to the objective j. Four cases can be distinguished :

D_j dominated	$z_j(\hat{x}) - z_j(x) < 0$	the performance on objective j is not improved
I_j indifferent	$0 \leq z_j(\hat{x}) - z_j(x) < t_{min,j}$	the improvement on performance of objective j is too weak to be interesting

Q_j weak improvement	$t_{min,j} \leq z_j(\hat{x}) - z_j(x) < t_{max,j}$	the improvement on performance of objective j is significant
P_j strong improvement	$z_j(\hat{x}) - z_j(x) \geq t_{max,j}$	the improvement on performance of objective j is strong

Then, when the objective j is significantly [case Q] (respectively strongly [case P]) improved, the weight ω_j decreases by the quantity $\Delta_{t_{min,j}}$ (resp. $\Delta_{t_{max,j}}$). On the contrary, the weights related to the objectives of cases D and I increase.

To reinforce the diversification strategy, the concept of tabu list with respect to the objectives is introduced (*TL_objective*). In the case Q (resp. case P), the objective j is declared tabu for a number of τ_j^{min} (resp. τ_j^{max}) iterations. Then its weight cannot be increased as long as its tabu status is not equal to 0.

The procedure to update the weight vector and the management of the tabu list *TL_objective* are summarized in the following procedure:

```
update_weight : procedure

  J+ ← ∅;  J- ← ∅ ;  Δ ← 0

  for j = 1 to p do

     case <objective j is improved> of                    ! case Q and P !
        Q do   Δ ← Δ+ Δtmin,j ; ωj ← ωj - Δtmin,j ; J+ ← J+ ∪ {j}
               if j ∉ TL_objective then
                   TL_objective ← TL_objective ∪ {j}
               endIf
        P do   Δ ← Δ+ Δtmax,j ; ωj ← ωj - Δtmax,j ; J+ ← J+ ∪ {j}
               if j ∉ TL_objective then
                   TL_objective ← TL_objective ∪ {j}
               endIf
     endCase
  endFor

  J- ← ( j | j ∉ J+, j ∉ TL_objective )

  for j ∈ J- do ωj ← ωj + Δ / max{1,card(J-)} endFor

endProc
```

2.4 Skeleton of MOTS procedure

The following algorithm describes the main steps of our MOTS procedure for solving MOCO problems.

```
MOTS : procedure
  ! initialization !
    <Select an initial solution x0 in D>
    E(P̂) ← {x0} ; TL_move ← ∅ ; TL_objective ← ∅
    <Choose ω>                    ! an initial weight vector          !
    <Choose maxIter>              ! a maximum number of iterations !
    <Choose stepIter>             ! a fixed number of iterations     !
    k ← 0
  ! search !
    while k ≤ maxIter do
      repeat
          <Calculate V(xk) and its associated utopian point>
          <Select the compromise solution x̂ using the scalarizing function S:
            x̂ = argmin { S(f(x),q,ω) | x ∈ V(xk) } and xk → x̂ ∉ TL_move >
          <Update TL_move with the move xk → x̂>
          <Update E(P̂) with x̂ and
            evaluate Mk according to results of the update>
          ! intensification phase !
          if Mk > 1 then
             <update E(P̂) with the best Mk non tabu solutions in V(xk)
               according to the scalarizing function S>
          endif
          ! aspiration criterion !
          for x# such that  xk → x# ∈ TL_move and
                            S(f(x#),q,ω) ≤ S(f(x̂),q,ω) - δ   do
             Update E(P̂) with x#
          endfor
          k ← k + 1 ; xk ← x̂
      until <stepIter times are done>
      ! diversification phase !
        update_weight                   ! update ω and TL_objective !
    endWhile
    <Propose a solution x̂ ∈ E(P̂) using the scalarizing function S>
endProc
```

3 The MultiObjective Permutation problem (MOP)

We considered the MultiObjective Permutation problem (MOP) adapted from Reeves [11]. Each person a_i of a set of N people is characterized according to p dimensions y_j, j=1...p (for example, weight, age, ...). For each objective, the order in which these people are arranged determines the value of the resulting sequence. Then, we considered the problem of determining a sequence of N people that satisfies p objectives. This problem can be stated as the following MOCO problem where:

$$f_j(x) = \sum_{i=1}^{N-1} \left| y_j(a_{\sigma(i)}) - y_j(a_{\sigma(i+1)}) \right| \qquad j = 1,...,p$$

and $D = \{ x \mid x = (a_{\sigma(1)}, a_{\sigma(2)}, ..., a_{\sigma(N)})$ for any permutation σ of $\{1,2,...,N\} \}$

For the MOP problem, a move $x \to x'$ corresponds to a swap ($a_i \leftrightarrow a_j$) between two persons in the sequence and its attributes are $\{i,j\} \in$ TL_move. The size of a neighborhood V(x) is N(N-1)/2=(N-1)×(N-2)...2×1. By example, the neighborhood V(x) of x = (a,b,c,d) is defined by the six potential moves $\{a \leftrightarrow b, a \leftrightarrow c, a \leftrightarrow d, b \leftrightarrow c, b \leftrightarrow d, c \leftrightarrow d\}$.

4 Numerical experiments

We apply our MOTS procedure to a small-size instance of MOP problem (N = 7, p = 2). The task chased in this limited numerical experiment is to validate the robustness of our approach, i.e. its ability to determine a suitable approximation of the efficient set. For this small-size instance, the efficient set E(P) is determined by using an explicit enumeration procedure. We obtain card(E(P))=7 (table 3).

The data of the instance are given in table 1.

Table 1. MOP Data

	a_1	a_2	a_3	a_4	a_5	a_6	a_7
y_1	10	20	110	30	6	19	2
y_2	5	1	20	45	100	75	10

Table 2. MOTS parameters

MaxIter ← 15 ; TL_move ← 5

	$t_{min,j}$	$t_{max,j}$	$\Delta_{tmin,j}$	$\Delta_{tmax,j}$	$\tau_j^{t_{min}}$	$\tau_j^{t_{max}}$
f_1	6	12	5	15	2	3
f_2	6	12	5	15	2	3

Table 3. Potential Efficient set and efficient solution at the end of TS viewed in objective space

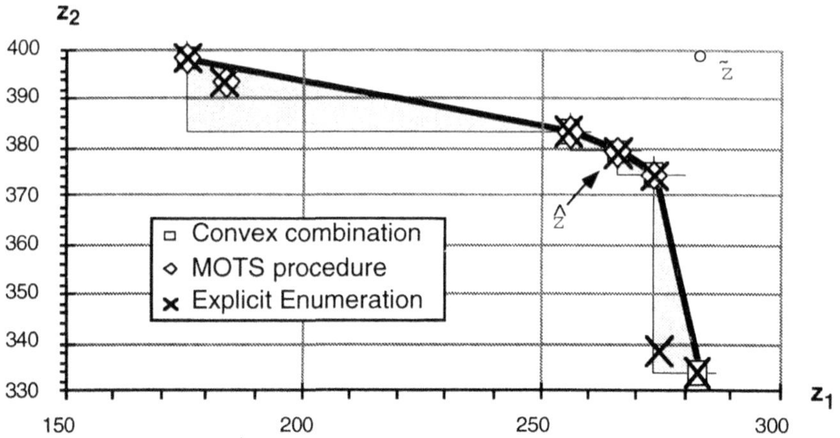

$E(\hat{P})$ = 5 efficient solutions and $\hat{x}$ (denoted by $\hat{z}$ in Z_X) :

$\hat{x}$							f_1	f_2
a_2	a_5	a_3	a_6	a_1	a_4	a_7	266	379

The comparison of solutions obtained (table 4) is given taking into account :

- the initial solution x_0. As all solution is feasible, x_0 is the trivial solution defined by the data.
- the supported efficient border is given by using a convex combinaison of the objectives
- the final solution $\hat{x}$ is the result obtained by our MOTS procedure. We remark that the solution $\hat{x}$ is a supported efficient solution obtained by the convex combinaison of the objectives.

Table 4. Comparison of solutions obtained

sol.	initial (x_0)	final ($\hat{x}$)	impr.(%)	optimal value
f_1	234	266	13	283 (6,7,4,1,3,5,2)
f_2	193	379	96	398 (7,5,1,6,2,4,3)
	(1,2,3,4,5,6,7)	(2,5,3,6,1,4,7)		

5 On-going research

First, we plan to test our procedure on large numerical instances and on NP-hard problem, in particular on a scheduling problem of n tasks with one machine. In addition, several aspects must be studied and thoroughly evaluated, for example, the algorithm of the weights, the MOTS procedure complexity, the calibration of control parameters (on the movement -swap- and on the objectives), and the tools used (the norm for example). Second, we customize the MOTS procedure on the MultiObjective Knapsack Problem. For this problem, the MOTS algorithm is a tiny different because the referenced point z is easily available.

6 References

[1] Czyzak P. and Jaszkiewicz A.(1996) "Pareto Simulated Annealing - a metaheuristic technique for multiple objective combinatorial optimization", Working paper, Poznan University of Technology, Poland.

[2] Dahl G., Jörnsten K. and Lokketangen A. (1995) "A Tabu Search Approach to the Channel Minimization Problem", ICOTA'95, 9 pages.

[3] Gandibleux X., Libert G., Cartignies E., et Millot P. (1994) "SMART: Étude de la faisabilité d'un solveur de problèmes de mobilisation de réserve tertiaire d'électricité.", Revue des Systèmes de Décision, vol. 3, n°1, pp.45-67.

[4] Gandibleux X. (1995a) "Système d'aide à la décision pour la conduite de processus perturbés: une approche hybride fondée sur l'intelligence artificielle, la programmation linéaire et l'aide multicritère à la décision. Application à la mobilisation de réserve tertiaire d'Électricité De France", Thèse de Doctorat, Janvier 1995, Université de Valenciennes.

[5] Gandibleux X., Fréville A., Hanafi S. (1995b) "Modelization of the MRT Problem on the Short-Term Horizon: An Integer Programming Approach", International Conference on Industrial Engineering and Production Management Proceedings., vol.1, pp.423-433, April 4-7, 1995, Marrakech, Morocco.

[6] Glover F. (1986) "Future paths for integer programming and links to artificial intelligence", Computers & Operationss Research., 5, pp.533-549.

[7] Glover F. (1995) "Tabu search fundamentals and uses", Technical report, April 1995, 85 pages, University of Colorado, Boulder, Colorado, USA.

[8] Hansen P. (1986) "The steepest ascent mildest descent heuristic for combinatorial programming", Congress on Numerical Methods in Combinatorial Optimization, Capri, Italy.

[9] Hertz A., Jaumard B., Ribeiro C.C., Formosinho Filho W.P. (1994) "A MultiCriteria Tabu Search approach to cell formation problems in group technology with multiple objectives", Recherche Opérationnelle/Operations Research, vol.28, n°3, 1994, pp.303-328.

[10] Pirlot M. (1992) "General local search heuristics in combinatorial optimization : a tutorial", JORBEL, vol.32 (1,2), pp.7-67.

[11] Reeves C. (1995) Modern Heuristic Techniques for Combinatorial Problems, McGrawHill, London, 320 pages.

[12] Roy B. et Bouyssou D. (1993) Aide multicritère à la décision: méthodes et cas, collection gestion, Économica, Paris.

[13] Serafini P. (1992) "Simulated Annealing for Multiple Objective Optimization Problems", 10th MCDM International Conference Proceedings, Tapei 19-24 July 1992, vol. 1, 1992, pp.87-96.
[14] Steuer R. (1986) Multiple Criteria Optimization : theory, computation and application, Wiley, New York.
[15] Ulungu E. and Teghem J. (1994) "Multi-objective Combinatorial Optimization Problems: A Survey", Journal of Multi-Criteria Decision Analysis, vol.3, pp.83-104.
[16] Ulungu E. L.B., Teghem J. and Fortemps P. (1995) "Heuristics for Multi-objective Combinatorial Optimization Problems by simulated annealing", MCDM : Theory and applications (Gu J., Wei, G.C.Q & Wang Sh. Eds), SCI-TECH Information Services, pp.228-238.
[17] Vanderpooten D. (1990) "L'approche interactive dans l'aide multicritère à la décision", Thèse de doctorat, Université de Paris IX - Dauphine, France.

Linking Production Theory and Multiobjective Fractional Programming as a Support Tool for Animal Diet Formulation

Pablo Lara[1]

[1]University of Córdoba, ETSIAM, Apdo. 3048, CORDOBA, SPAIN

Abstract. The purpose of this paper is to show how Multiobjective Fractional Programming can improve the process of diet formulation. The aim of least cost feed formulation is to provide a technically efficient feed, i.e., to obtain the best animal performance (growth, milk production) at the minimun cost. But with different combination of nutrient supplies, mainly energy and protein, we can get the same performance. So, several least cost diets with different levels of energy and protein can be computed to choose the minimun least-cost formulation. Our thesis is that we can compute with Multiobjective Fractional Programming the set of least-cost diets, making the process more efficient. We show an example with a coefficient matrix of a real industry and outline the benefits for improving the whole formulation process.

1 Introduction

Today there is a great variety of livestock production methods. From the animal feeding point of view they range from the most traditional based on foraging to intensive methods that supply livestock with a controlled quantity of food of a desired quality. Although other operations involved are important, the economic efficiency of the system rely mainly on an adequate provision of food. Conceptually, the provision of food is made as a diet or as a ration. Following the convention of Glen (1986), a diet is defined as the proportions of constituent feedstuffs and a ration is defined by the quantities of constituent feedstuffs. Feed-mix is a general term encompassing both ration and diet. Physically, the food can be supplied as a set of feeds on as-is basis or as a unique compound-feed produced in a factory or even self-produced by the farmer. The diet is formulated to meet a specified set of nutrients requirements expressed as proportions in the food and the ration is formulated to meet a set of nutrient requirements specified as quantities. Due to the characteristics of production, process of diets and compound feeds. In the second section, a theoretical model of diet formulation with a production (growth) function basis is presented. In the third section, the desired properties of the growth function are outlined. Fourth section is devoted to develop a multiobjective fractional programming framework suitable for treating the diet formulation problem. Finally, the framework is applied to a real example taken from the literature.

2 A theoretical model for diet formulation

In the process of livestock production, as in other production processes, we are interested in maximising the benefit per unit of product (meat, milk, fiber). In the case of meat production, the quantity of product obtained in a period of time is measured by the liveweight gain or growth (**g**) of the animal. Valuing the meat by its market price (**p**) and taking account of a cost function (c), we can write the criterion as:

$$Max. \frac{p*g - c}{g} = Max. \; p - \frac{c}{g} \qquad (1.1)$$

As **p** is given to the producer, the criterion is the same as:

$$Max. \; - c/g \approx Max. \; g/c \qquad (1.2)$$

what is tantamount to maximise the productivity of the money invested in making the animal grows. The growth of the animal depends on the quantities of the relevant set of nutrients supplied to the animal through a set of available feeds. Although particular nutrients are qualitatively relevant for particular animal species, the growth of all of the species is considered mainly as a function of the energy (**e**) and protein (**p**) supplied. So, our criterion is:

$$Max. \frac{g}{c} = Max. \frac{g(e,p)}{c} \qquad (1.3)$$

The major component of the cost of growing the animal is the expense in feeding it. Given a set of n feeds let $\mathbf{x} = (x_1,x_2,...,x_n)$ be the vector of feeds quantities. Let c_i be the cost of the i-th feed. The cost of feeding the animal is a function:

$$c = c(x) = \sum_{i=1}^{n} c_i * x_i \qquad (1.4)$$

In the same way, the energy and the protein ingested by the animal is a function of the enregy (e_i) and protein (p_i) contents in the n feeds:

$$e = e(x) = \sum_{i=1}^{n} e_i * x_i \qquad (1.51)$$

$$p = p(x) = \sum_{i=1}^{n} p_i * x_i \qquad (1.52)$$

Our criterion function is expressed as:

$$Max.\ F = \frac{g[e(x),p(x)]}{c(x)} \qquad (1.6)$$

i.e., our interest is to choose the quantities of available feeds x_i that maximise the ratio function F.

The application of this theoretical model to livestock diet formulation problems has been restricted mainly because there is no a mechanistic growth function **g** performing for all the livestock species, but only empirical approximations to specific species in particular growth stages. To avoid this problem, research organizations in animal nutrition have established tables of nutrient recommendations in which a given level of performance g^* is associated with a unique value for each nutrient. Based on this tables, the algorithmic approach to the livestock diet formulation problem has consisted in:

1º) To choose a desired (and biologically possible) level of productivity g^*.

2º) To select from the tables of nutrient recommendations the values of energy, protein and other nutrient intakes needed to reach g*.

3º) Let the value for the i-th nutrient be b_j, and let the i-th nutrient content in the j-th feedstuff be a_{ij}. To establish the set of restraints:

$$\sum_{i=1}^{n} a_{ij} * x_{ij} \geq b_i$$

4º) To minimise the cost function c(x) over the feasible region defined by the previous set of restraints, i.e.:

$$Min.\ c(x)$$

$$A * x \geq b$$

which is a classical linear programming framework. What we intend with this paper is to propose an operative framework for livestock diet formulation nearest to this theoretical model than the linear programming framework. The desired properties for the approach are to be useful in a decision support environment and to avoid the difficulties associated with the estimation of the growth function g.

3 The properties of the growth function

The models of nutrient requirements based on tables suppose that energy and protein as inputs of livestock growth are economic complements, instead of economic substitutes. However, the empirical evidence support the idea of substitution between energy and protein within a range of economic and technical interest. This issue is illustrated, for example, in the work of Miller et al.(1.986). Several diets including different levels of energy and protein were formulated using least cost for producing broilers of 4 lb. of weight. The kilograms of feed consumed by the birds were recorded (feed conversion) and so the cost per kg. of bird were calculated. A given size of bird can be obtained with several technically efficient diets, but only one will be economically efficient, i.e., it will provide the minimum cost per kg. of bird. The substitution between energy and protein is biologically possible because the capacity of the animals to regulate the intake of feed based on the energy density of the diet. The more the energy density the less the intake.

Several authors have made use of a production function in a mathematical programming framework (Dent (1964), Miller et al., 1.986). As standard production functions, g is a concave function in the region of relevant protein and energy variation, a condition that holds for the experimental functions reported in the literature. From a practical point of view, we have to use a framework for the economic analysis that assure: 1º) the optimum of F to be between the solution set, 2º) a set of economically suitable solutions allowing us interchanging between energy and protein according to their prices.

It is shown that a Bicriterion Linear Fractional Model (BLFM) meet the two previous requirements. The first point is demonstrated as a natural extend of well known results. The second point is accomplished through the use of energy and protein prices as explicit criteria.

4 The multiobjective fractional programming framework

Due to the concavity of g, F pertains to the class of concave-convex fractional programs, in which the numerator is concave, the denominator is convex and the set of restrictions is also convex. Schaible (1976) has shown that every concave fractional program is transformable into a concave program using the variable transformation suggested by Charnes & Cooper (1.962):

$$y = \frac{1}{c(x)} * x, \qquad t = \frac{1}{c(x)}$$

Problem F is transform to:

$$F_t: \quad Máx.\ g[\ e(y),\ p(y)\ ]$$

$$A * y - b * t \leq 0$$

$$c(y) = 1$$

$$y,t \geq 0$$

which is a concave program. If (y',t') is an optimal solution of F_t, x'=y'/t' is an optimal solution of F.

Consider now the problem:

$$Max.\quad h[f_1(x),f_2(x)]$$

$$x \in X$$

with h a real-valued increasing (monotone nondecreasing in each argument) ordinal utility indicator function defined on the pair of achievable values for f_1 and f_2, which are real-valued concave criterion (payoff) functions of the n-vector x of decision variables that are constrained to lie in a convex subset X of E^n.

LEMMA 1. (From Geoffrion, 1.967). At least one point at which h[f(x)] achieves its maximum over X is efficient with respect to the vector $f(x)=[f_1(x),f_2(x)]$.

As g, e and p have the properties of g, f_1, and f_2:

COROLLARY 1. In problem F_t, *at least one point at which g(y,t) achieves its maximum is efficient with respect to the vector* $f(y)=[e(y),p(y)]$.

(Note that e and p are linear functions but not affine, so t is an argument of g but not of f).

Consider now the parametric problem:

$$F_\mu: \quad Max.\ \mu * e(y) + (1-\mu) * p(y)$$

$$A * y - b * t \leq 0$$

$$c(y) = 1$$

$$y,\ t \geq 0 \qquad 0 \leq \mu \leq 1$$

LEMMA 2. (From Geoffrion, 1.967). If (y,t') is efficient for f(y), then there exist a scalar μ in the unit interval such that (y,t') is an optimal solution of F_μ.

THEOREM 1. (From Geoffrion, 1967). An optimal solution of F_t, is found among the optimal solutions of F_μ, for some μ in the unit interval.

COROLLARY 2. An optimal solution for F is found among the optimal solutions of $F\mu$, for some μ in the unit interval.

But F_μ is a parametric problem associated with the bicriteria linear program:

$$BF_t: \qquad Eff.\ [\ e(y)\ ,p(y)\ ]$$

$$A * y - b * t \leq 0$$

$$c(y) = 1$$

$$y,\ t \geq 0$$

which is the "Charnes-Cooper transformed" problem of:

$$BF: \qquad Eff.\ [\ \frac{e(x)}{c(x)}\ ,\ \frac{p(x)}{c(x)}]$$

$$A * x \leq b$$

$$x \geq 0$$

The peculiarities of bicriteria linear fractional programs were studied by Choo & Atkins (1.982, 1.983). Although in general, the efficient sets of multicriteria linear fractional programs do not have the nice linear properties as in the case of multicriteria linear programs, when there are only two criteria the efficient sets are finite unions of linearly constrained sets. They also described and algorithm for evaluating the efficient frontier based on the variable transformation of Charnes & Cooper (1.962) and parametric linear programming techniques.

The connections between multiobjective fractional programming (MFP) and related multiobjective linear programs (MLP) have been suggested by Nykowski & Zolkiewski (1985). Dutta, Rao & Tiwari (1.993) showed that a restricted class of multiobjective fractional programming problems in the sense that the denominators of the ratios are the same can be completely solve resorting to a related a MLP arising after a "Charnes-Cooper variable transformation" due to a one-one mapping of one feasible set onto the other that also maps one nondominated set onto the other. They proved:

THEOREM 2. The nondominated solutions obtained for problem BF_t will also be the nondominated solutions for BF.

COROLLARY 3. If x' if efficient in problem BF, there exist a μ in the unit interval such that x' is optimal in F_{μ}.

Let x^{*} be the optimal solution of problem F. Let X_{BF} be the set of efficient solution of problem BF. With the conditions stated along the text, the following theorem is self-evident now:

THEOREM 3. $x^{*} \in X_{BF}$

If we have an explicit representation of function g, the optimum could be computed with a concave programming algorithm solving F_t. But if we do not have an explicit represention of g, or even with it, we search for the efficent solutions of BF and we know that x^{*} is in it and we have an explicit representation of the trade-offs between energy and protein.

5 The diet problem

The example diet problem used to illustrate the approach is a slightly modified version of the problem presented by Pesti et al. (1.986). Fifteen ingredients are used for formulating a diet for broilers. The level of variation for the energy and the protein hasbeen restricted to a biologically suitable range. One of the usual recommendation for this type of animal is to set the level of energy in the diet to 13.37 Kilojules/gr. and the level of protein to 200 gr./kg. The restrictions needed to maintain both energy and protein between ± 20% of these recommended values are also included. (The coefficient matrix can be obtained upon request from the author).

The table 1 presents the efficient extreme points obtained after the transformation to the x space, together with the least cost solution of setting the levels of energy and protein to 13.37 and 200 respectively.

Table 1. Efficient extreme diets and least cost diet for energy and protein fixed.

	E1	E2	E3	E4	Least cost solution
Energy (KJ./kg. diet)	13.505	13.73	13.84	13.99	13.37
Protein (gr./kg. in diet)	239.87	240.08	203.17	182.47	200
Cost (cents/kg. diet)	18.58	18.72	17.98	17.54	17.79
Energy/cost (KJ./cents)	0.72	0.73	0.77	0.79	0.75
Protein/cost (gr./cents)	12.91	12.83	11.30	10.40	11.24
Marginal rate of technical substitution$_{e,p}$ (gr. protein/KJ.)	17.93	17.57	14.67	13.16	14.98

The problem has been solved in a Toshiba T2130 CS portable computer with the MLP package (Computing & Systems Consultant BV, 1987). The results show that the efficient set is characterized by four extreme efficient points. The whole efficient set represents the relevant portion of *any* production function biologically suitable to represent the growth of broilers in response to the intake of energy and protein for the feasible set stated.

Pesti et al. (1986), using a quadratic growth response, reported an optimum diet which is in the range of the efficient set. Their quadratic growth response had been obtained empirically. They compared the savings per bird with the resulting diet and the least cost diet resulting from the same matrix, the typical diet used in the industry in the USA in that year. This was 1.54 cents/bird, an important figure because it results in 800,800 $ for a typical broiler complex slaughtering one million birds/week.

Possibly, a portion of the efficient set computed with our approach would also results in important savings per bird, but empirical work need to be done to test the performance of the diets. However, the results can serve as a guide for nutrition experts that could make use of the information about trade-offs between energy and protein together with their knowledge in a support computer environment to advance efficient diets able to render substantial savings for the industry.

Acknowledgement

The author wishes to thank Professor Stancu-Minasian his helpful comments for the improvement of the manuscript.

References

Charnes, A. and W.W. Cooper (1962) "Programming with Linear Fractional Functionals", *Naval Research Logistics Quarterly* 9:181-186.

Computing & Systems Consultants BV (1987) *Multiobjective Linear Programming (Reference Manual)*. Eindhoven.

Choo, E.U. and D.R. Atkins (1982) "Bicriteria Linear Fractional Programming" *Journal of Optimization Theory and Applications* 36:203-220.

Choo, E.U. and D.R. Atkins (1983) "Connectedness in Multiple Criterion Linear Fractional Programming", *Management Science*, 29:250-255.

Dent, J.B., (1964) "Optimal rations for livestock with special refernce to bacon pigs" *Journal of Agricultural Economics* 16:68-87.

Dutta, D. Rao, J.R. and R.N. Tiwari (1993) "A restricted class of Multiobjective Linear Fractional Programming problems", *European Journal of Operational Research* 68:352-355.

Glen J.J. (1987) "Mathematical models in farm planning: a survey" *Operations Research*, 35:641-660.

Miller, B.R., Arraes, R.A. and G.M. Pesti, (1986) "Formulation of broiler finishing rations by Quadratic Programming" *Southern Journal of Agricultural Economics* 18:141-150.

Nykowski, I. and Z. Zolkiewski, "A Compromise Procedure for the Multiple Objective Linear Fractional Programming Problem" *European Journal of Operational Research* (1985) 19:91-97.

Pesti, G.M., Arraes, R.A. and B.R. Miller (1986) "Use of the Quadratic Growth Response to dietary protein and energy concentrations in Least-Cost Feed Formulation" *Poultry Science* 65:1040-1051

Schaible, S., (1.976) "Fractional Programming, I, Duality" *Management Science* 22:858-867.

Multiobjective Energy-Environmental-Economy Model for ISRAEL *(ENMIS)*

David Soloveitchik

Economics & Planning Department, Ministry of National Infraestructures, Jerusalem 91130, ISRAEL.

Abstract. This paper reports on an energy optimization model (ENMIS) which has been developed[1] for planning the energy supply sector in Israel. We have applied the multiple objective linear programming techniques to analyze an energy-environmental-economy model for policy evaluation. The simplex lattice design (SL) for simplex of weights coefficients is used here to obtain the set of nondominated solutions.

The model is based on integration of input-output analysis, linear programming and optimization models in the electricity and refinery sectors.

Keywords. Multiobjective optimization, Energy economy, Environment.

1 Introduction

The **ENMIS model** is a multiple objective linear programming model consisting of the following major modules (Fig. 1).

Energy sector- Energy products are represented in the model in the terms of various types of crude oils, coals and refinery products.

Non-energy sectors- 43 sectors of the economy. Final demand was desegregated into private consumption, government expenditure, private investment, export and changes in stocks. Some additional constraints (competitive imports, foreign trade, labor force, etc) were added to describe the total gross output pollutants (Input-Output optimization model of the economy).

Energy-environmental module- Four pollutants (SO_2, NOx, SPM, CO_2) and accordingly pollution emission coefficients are considered.

Optimization model (ENMIS) has been developed for planning the energy supply sector in Israel. The model is based on integration of input-output analysis, linear programming and optimization models in the electricity and refinery sectors.

[1] The author would like to thank Dr. N. Levin, Dr. I. Danin, Dr. Y. Bargur, Dr. N. Ben-Aderest, Mrs M. Grinman and Mr. A. Ichiel for working with them on some parts of ENMIS.

The objective functions of ENMIS can be exogenously chosen by the user of the model. The role of objectives and constraints may be changed during the solution process. Three objectives were selected to illustrate our multiobjective format of this model: GDP growth, total energy requirements and amounts of SO_2 pollutants.

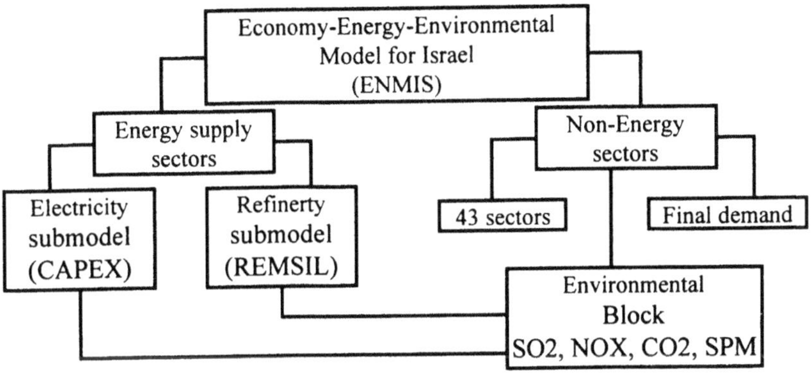

Fig 1. Economy-Energy-Environmental Model (ENMIS)

The electricity sector of the economy is modeled in ENMIS by the optimization computer package CAPEX (CAPacity Expansion)[2].

The CAPEX model finds the optimal electric generation expansion plan, the levels and timing of investments and the installation of the new units. This package is capable of taking into account up to ten objectives. Minimum power generation cost and minimun emission cost are examined in detail.

REMSIL (Refinery Modeling System for Israel), an optimization-based modeling system, will be used to assist in planning for a Israeli refinery sector. REMSIL is a single-period nonlinear optimization model which uses a variety of nonlinear equations to calculate the properties of blends from component properties. In the nearest future the model can be used for long-term planning as well. This model includes several objective functions that the user can select.

ENMIS is being used to investigate the interrelationships between energy, environment and economic welfare in Israel. The paper is organized as follows. Section 1 provides the input-output model description. Section 2 reviews the electricity sector submodel. Section 3 sets out the applications to energy planning. Section 4 concludes the paper.

---

[2] The computer package CAPEX was developed by Prof. J. Zahavi and Dr. N. Levin (Tel Aviv University. Tel Aviv. Israel).

2 Input - Output Model

The Input-Output model can be expressed as the matrix equation

$$\mathbf{X = AX + Y} \qquad \text{or} \qquad \mathbf{X = (I - A)^{-1}\, Y} \tag{1}$$

where

- **Y** is the n-vector of goods to satisfy final demand with
-

$$\mathbf{Y = PC + GC + EXP + CF;}$$

- **PC** is private consumption;
- **GC** is government consumption;
- **EXP** is export;
- **CF** is capital formation;
- **X** is the *n*-vector of goods required for total output of the economy;
- **A** is $n \times n$ matrix of technological coefficients for the economy;
- $\mathbf{(I - A)^{-1}}$ is known as "Leontief Inverse" matrix.

Vector of total use of various fuels (TF) is:

$$\mathbf{TF = C\,X} \tag{2}$$

where C is fuel-use coefficients matrix per unit total output by each sector.
Total emissions output from fuel use (TEM) is given by:

$$\mathbf{TEM = EM\ \ TF} \tag{3}$$

where **EM** is the matrix of coefficients relating pollution emissions per unit of fuel.

This model can be used to estimate the effect of meeting the Israeli Ministry of Environment controls on pollution.

Total energy intensities which denote the total amount of each energy type required for each energy sector can be calculated through the use of the "Leontief Inverse" matrix. Impact of fuel prices changes, energy demand level and impact of the new technologies and air polluting emissions reduction from the changes of electricity intensities can also be examined.

The multiobjective input-output model can be formulated as follows:

$$\textbf{Max (Min) GDP, LABOR, TAX, EMISSIONS} \tag{4}$$

s.t.

$$\sum_j a_{ij}x_j + (PC_i^{\bullet}x_{pc} + GC_i^{\bullet}x_{gc} + EXP_i^{\bullet}x_{exp} + CF_i^{\bullet}x_{cf}) - x_i = 0 \tag{5}$$

where

$$x_{pc}, x_{gc}, x_{exp}, x_{cf}$$

are changes in private and government consumptions, export and capital formation

i, j ∈ [sectors, import, tax, labor].

$$\sum_j c^{\bullet}_{fj} x_j \leq TF_f \tag{6}$$

f ∈ [fuels, electricity].

$$\sum_j em^{\bullet}_{lj} x_j \leq TEM_l \tag{7}$$

l **∈ [SO_2, CO_2, NOx, SPM],**
TEM are the government imposed environmental standard.

The LP model can be applied to evaluate the impact of energy shortages, changes in energy policy and possibilities, changing in the structure of electricity and refinery industries, energy taxes, etc. The impact on the prices of all the sector's outputs of making individual sectors pay all or part of the cost of reduction pollution can be estimated as well.

Interindustry data were obtained from the Israel Central Bureau of Statistics, energy and environmental data were obtained from the Ministry of National Infrastructures and Ministry of Environment.

Some processes of converting coal to gas for use in electricity generation were examined. We have experimented with changes in the Israel electricity sector resulting from the introduction of natural gas generation in place of coal.

We also explored the cost-effectiveness of different strategies for achieving predetermined sets of environmental quality targets in the form of emission standards.

3 Electricity Sector Submodel

We combined the CAPEX model and I - O analysis in the study of the control of air pollutants in Israel. It allows us to estimate the marginal costs per unit of pollution control.

The computer package CAPEX is capable of taking into account up to ten objectives. Two kinds of objectives: minimum power generation cost and minimum emission cost are examined in detail. Generation cost includes the impact of capital investment, fuel and O&M costs and the cost of unsupplied energy. The deterministic version of the model can be formulated as follows [4]:

$$Min\ TC = \sum_{i=1}^{n} c_i^{\bullet} x_i + \sum_{i=1}^{n} b_i^{\bullet} \int_{D_i}^{D_{i+1}} g(x)dx$$

s.t.

$$x_i > 0 \qquad i = 1,\ldots,n$$

$$\sum_{i=1}^{n} x_i > L_{max}$$

where

- **TC** - the total cost function,
- **n** - number of new technologies in the target year,
- **b_i** - total fuel cost plus variable operating and maintenance cost for unit *I* (\$/MWY),
- **c_i** - total fix cost for unit i (fraction of peak demand),
- **D_i** - loading point of unit (fraction of peak demand),
- **L_{max}** - peak load,
- ***g(x)*** - inverted load duration curve (LDC).

In the linear case the inverted LDC is given by:

$$g(x) = \begin{cases} \dfrac{1-x}{1-L} & L < x < 1 \\ 1 & x < L \end{cases}$$

The actual problem that the model has to deal with is a multiobjective functions model, i.e:

-minimum power generation costs, (\$);

-minimum amount of some kind of emissions in tons per kilowatt-hour produced in any regions of the country.

The new objective function in this case can be defined by:

Min TC=GENC + EC

where

- **GENC** - generation cost;
- **EC** - emission cost.

Emission cost is given by:

$$\mathbf{EC} = \sum_{i=1}^{n} E_i^{\bullet} \sum_{j=1}^{k} \alpha_j EMIS(i,j)$$

where

E_i - energy producted from unit *i*,

α_j - weights of emissions, $\alpha_j \geq 0$.

EMIS(i, j) - quantity of j-emissions from unit i.

The ENMIS(i, j) is measured, for example, as the amount of sulfur dioxide emissions in tons per kilowatthour produced under different conditions. Both generation cost and SO_2 emissions are assumed to be linear functions of energy generated. The simplex lattice design (SL) [2] for simplex of weights is used to obtained the set of nondominated solutions.

Some additional constraints concerning the power stations operations and demand level of electricity are taken into account.

The problem is to choose between different sets of policies to encourage more effective ways of pollution reduction under different sets of fuel prices and other restrictions.

The results obtained are based on actual data of the Israeli power generation system and calculations were made for the period 1996-2020.

The information required for the optimization model includes:

- capital cost ($/KW);
- fixed and variable O&M cost;
- maximum capacity of power station (MW) and life time (years);
- equivalent forced outage rates (%);
- scheduled maintenance (days per year);
- heat rates (KCAL/KWH).

The following information from the output of model CAPEX is obtained:

- the total quantity of each emission for the power generation system as a whole;
- the system total cost (which includes capital, O&M cost, fuel cost and cost of unsupplied energy);
- the average production cost per unit of output (c/kwh).

4 An Application to Energy Planning

The ENMIS model was employed to construct seven alternative scenarios for the development of the Israeli Electricity System up to the year 2020:

- the base case scenario reflects the situation when the all power station in Israel operate with FGD technology from the year 2000 and they includes three firing technologies:
  - coal fired;
  - oil fired (bunker - oil - residual fuel oil);
  - gas turbines.

- the natural gas scenario examines the impact on the Israel economy of converting oil fired power stations and gas turbines to the use of natural gas;
- some environmental limit scenarios show how upper limits of the levels of SO_2 and NOx emissions, for example: might affect the economy.

The base year for the scenarios is 1996 but an input - output table exists for 1990 only. So it was decided to use the CAPEX model for projection purposes.

The scenarios have been developed in several stages. Firstly, the CAPEX model was solved for 2010 on the basis of the 1996 as a typical minimum least-cost capacity expansion model and the emissions for SO_2, NOx, SPM and CO_2 were calculated. These levels are considered as a maximum level for each pollutant.

In the next stage the cost of reducing emission levels of SO_2 and NOx are calculated with the help of the CAPEX model as a goal programming or multiple objective functions model.

Upper limits on release of SO_2 and NOx emissions were imposed by the Ministry of Environment - a number of quantitative goals has been set for the year 2010, for example. It was difficult to use the CAPEX model - dynamic programming optimization model - for these purposes because of the unfeasibility of solutions subject to restrictions imposed for specific years. At this stage it was decided to consider imposed restrictions as an objective functions in the model.

There are some ways to achieve emissions reduction in the electric utility industry, as follows:

- to substitute fuels that include less polluting;
- to change the electricity generating technology and emission control system in order to meet prescribed environmental objectives;
- demand-side management (DSM) strategies, etc.

In order to comply with the imposed maximum emission levels some options were considered: either to switch to cleaner fuels, or to install abatement technologies. All these options lower the demand for energy because of a rise in the energy prices and thus an increasing demand for capital. The trade-off curve (for the year 2010) is presented in Fig. 2.

The trade-off curves represent the calculated results for each year considered. This trade-off curve shows the "lowest" (44,000 tons of SO_2) and "highest" levels of SO_2 (141,000 tons), possible range of emissions. Reduction of the level of emissions from the "highest" to the "average" one will raise the average cost. The shape of the curve demonstrates that the changes in cost associated with SO_2 reduction is not proportional. As we can see from Fig. 2, the average cost of electricity is rising sharply after the level of 71,000 tons of SO_2 emissions

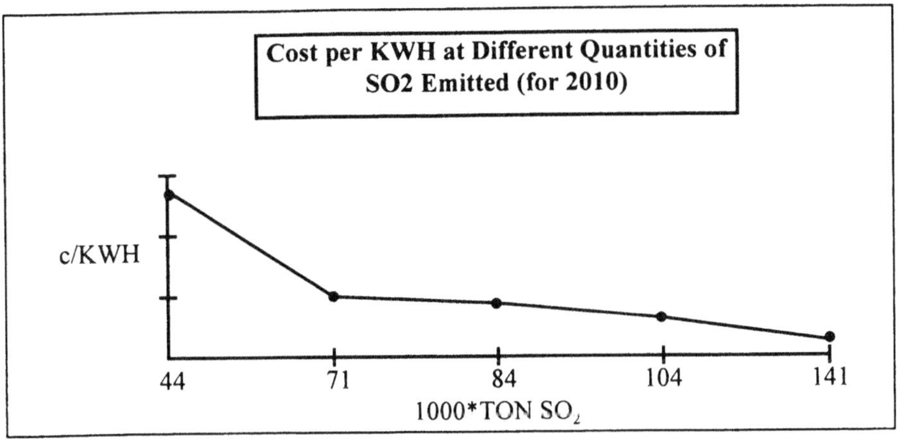

Fig 2. Cost per KWH at Different Quantities of SO_2 (for 2010)

The point of the graph associated with the "highest" level of emissions represents the least cost solution with absolutely no restrictions and complete disregard to the environmental consequences. The extreme point on the left associated with 44,000 tons represents the opposite, i. e., maximum regard to environmental quality and complete disregard to economic cost of the production of electricity.

All the intermediate points represent different policy combinations at different costs. The graph allows us therefore to calculate the "economic cost" of SO_2 emission abatement and allows us to put the price tag on the social value of pollution control.

The model presented can be considered as a tool to provide information to the decision makers on alternatives for reduction of air pollution on a long-term basis in energy in Israel.

5 Conclusions

ENMIS can be considered as a tool for evaluating the impact of energy-environmental issues of decisions made in the energy field.

The ENMIS model should not be regarded as the final version. The model will be extended to investigate the consequences of a policy decision in the energy sector and in the environment on the rest of the economy in Israel.

References

1. **Brovender and D. Soloveitchik.** (1993). Using the CAPEX Model for Environmental Policy (an Israeli case study), Energy & Environment, v. 4, n. 2, pp. 123-131.
2. **Soloveitchik.** (1986). A new approach to decision making in Multiple - Objective Linear Programming problems, 7th Int. Conf. on Collective Phenomena, Annals of the New York Academy of Sciences.
3. **Hawdon and P. Pearson.** (1995). Input - output simulations of energy, environment, economy interactions in the UK, Energy Economics, v. 17, n. 1, pp. 73-86.
4. **Levin, A. Tishler and J. Zahavi.** (1985). Capacity Expansion of Power Generation Systems with Uncertainty in the Prices of Primary Energy Resource, Management Science, v. 31, n. 2, pp. 175-186.
5. **Ralph E. Steuer.** (1986). Multiple Criteria Optimization: Theory, Computation, and Application. John Wiley & Sons.

BayRes: A System for Stochastic Multiobjective Reservoir Operations

D. Ríos Insua[1], C. Bielza[1], J. Martín[1] and K. Salewicz[2]

[1]Decision Analysis Group, Madrid Technical University, 28660 Madrid, Spain.
[2]IMES, Strobl Gruppe, Austria.

Abstract. Reservoir operation problems are very complex due to factors including multiple sources of uncertainty, multiple objectives, and time changing preferences. We describe a methodology for multiobjective stochastic reservoir operations. Its implementation is not easy, hence the need to develop BayRes, a system for single reservoir management decision support, which we introduce in the paper.

Keywords. Reservoir operations, Decision support systems, Multiattribute utility, Reference trajectory, Dynamic models

1 Reservoir operations

In a time of global change, water resources management remains as one of the most challenging problems. Recall the dependence of the Middle East peace process on the satisfactory allocation of the Jordan river resources, or the massive protests of Spanish farmers demanding proper water supply. These suggest that current methods for water resources management are not adequate to the needs resulting from changes in the climatic conditions hence requiring new developments. Several features make these problems extremely involved: hierarchical structure of various planning processes; sequential nature of the problem; appearance of multiple objectives, stemming from various uses of water resources; effects of time on preferences, with priorities and targets changing from one period to another; multiple sources of uncertainty (inflow processes, water demand, evaporation processes,...), many of them nonlinear and nonstationary; and computational complications arising from the stochastic dynamic programming nature of the problem.

In Ríos Insua and Salewicz (1995) and Ríos Insua et al (1997), we have introduced a methodology for solving multiobjective stochastic management problems for reservoir systems. The performance of our methodology has been extremely good in the case studies there considered. However, its application has been far from simple. To facilitate its use and implementation we are developing BayRes, a system for single reservoir operations, as a project for an electric utility, which we describe below, after introducing the methodology.

2 Methodology

We aim at solving the reservoir operation problem within the Bayesian decision analytic framework, see French (1986). That would require: defining the set of alternatives, in our case release policies u; providing a forecasting model for inflows i to the reservoir; modelling the multiple consequences $c(u,i)$, associated with policy u and inflow i; modelling the decision maker's preferences by means of a utility function F, typically multiobjective and multiperiod, with objective functions depending on several consequences; and computing maximum expected utility controls.

Let h designate the predictive density of inflows, with inflow history D. Let subindices designate periods; s, storages; $G(\cdot)$ a function evaluating the final storage and r the planning horizon. Then, at time t, the reservoir management planning problem consists of finding controls $(u_{t+1},...,u_{t+r})$, maximising the expected utility over the planning period

$$\int [\sum_{j=1}^{r} F(c(u_{t+j}, i_{t+j})) + G(s_{t+r+1})] h(i_{t+1},...,i_{t+r}|D_{t+1}) di_{t+1}...di_{t+r} \quad (1)$$

taking into account the dynamics of the reservoir system, and the constraints over controls and reservoir storage. The problem becomes unmanageable for a realistic planning horizon, e.g. 36 months for some large reservoirs, since we have to solve a long horizon stochastic dynamic program; the evaluation of each control requires the solution of a high dimensional integral; and uncertainty about the inflow process rapidly propagates through time.

Instead, we adopt a strategy based on a 'reference trajectory'. Assume we have found a good 'reference' storage level for each period. Then, at each period, we would like to maximise the expected value of a utility function F taking into account the consequences of interest and the deviation from that reference state, i.e.,

$$\int \left[F(c(u_{t+1}, i_{t+1})) + \delta(s_{t+2}, s^*_{t+2})\right] h(i_{t+1}|D_{t+1}) di_{t+1},$$

where $\delta(s_{t+2}, s^*_{t+2})$ represents the deviation of the final state s_{t+2} from the reference state s^*_{t+2}. Intuitively, if the reference states are defined to account for the dynamic aspects of the problem, we would not loose too much with this approach.

This proposal is, in fact, based on a traditional and well established method of reservoir operation using the concept of *rule curves*, which represent optimal trajectories of the reservoir over a long time horizon, see Loucks and Sigvaldasson (1982). There is also a relation with various reference trajectory methods in multiobjective optimization, see e.g. Romero (1991), and some heuristics for sequential analysis, see e.g. Berger (1985).

A key feature in our methodology is the obtention of a reference trajectory. We use a deterministic version of (1), where inflows are considered known and fixed at their predictive expected values $\bar{\imath}$. Dynamics and constraints over controls and storages remain the same. We set the planning period at r, the typical planning horizon of the reservoir, and adopt an initial volume s_0. We maximise the objective function which includes the main terms of the objective function in (1), together with a term describing deviation from the initial state s_0:

$$\sum_{j=1}^{r} F(c(u_{t+j}, \bar{\imath}_{t+j})) - \rho\,(s_{t+r+1} - s_0)^2.$$

The optimal solution will provide the trajectory of reference states.

3 A guided tour through BayRes

The above approach has been implemented with success to analyse the operation of Lake Kariba (Ríos Insua and Salewicz, 1995) and of the main reservoirs in Zambezi river (Ríos Insua et al, 1997). Its implementation has been involved. As a consequence, we are developing BayRes, a decision support system for single reservoir multiobjective operations. Its aim is to facilitate the implementation of all the activities involved in the methodology described above. In this first version we assume monthly planning, though concepts extend easily to other planning activities.

A key feature of BayRes is the incorporation of a principle of *management by exception*: a set of models is routinely used for processing information, making inferences and predictions, and making decisions, unless exceptional circumstances arise. Examples would include a sudden rainfall or the detection of a dry period. In those cases, the system is open to external intervention, by inclusion of additional information leading to a modification of the forecasting model, the reference trajectory and/or the utility function. We increase interactiveness of our system in such a way. We also embed the idea of iterating through phases of problem structuring, uncertainty modelling, preference modelling, expected utility optimisation and sensitivity analysis, until the model is requisite for decision making.

BayRes is a user-friendly system based on menus and a powerful equation editor. The interface is in VISUAL BASIC, whereas the heaviest computational parts are implemented in C. Many help menus support and facilitate the modelling process. The rest of this section describes a simple session with BayRes. Unless one recovers a previously analysed reservoir, the user is meant to follow a path consisting of introduction of basic data, construction of the forecasting model, construction of the preference model, optimization and sensitivity analysis.

Reservoir description. We start by introducing the basic data describing the reservoir. First, we input the number and types of releases. In our discussion, we shall assume that there are two types of releases, u_1 and u_2, through turbines and spillgates, respectively. We aim at providing optimal releases at each period of time. The user needs to specify also the maximum releases for various purposes ($m_1 = 3820$ mln.m^3 and $m_2 = 10000$ mln.m^3 in our case) and the maximum reservoir storage (say $M = 70980$ mln.m^3). We assume a priority on these releases, in our example: a) If there is not enough water to release u_1, all available water is released for energy production. Otherwise, u_1 is released for that purpose; b) If there is still water available, some water may be additionally released to control the storage level. If there is not enough water to release the volume u_2 above, all available water is released. Otherwise, u_2 is released. Remaining water after both releases exceeding the maximum storage would be spilled.

At time t, the basic consequences stemming from the policies would be the final storage of the reservoir s_{t+1}, the actual releases and the total outflow u_t. Storages in consecutive months are related by a continuity equation relating them with inflow i_t, outflow and evaporation e_t:

$$s_{t+1} = s_t + i_t - u_t - e_t. \tag{2}$$

In our example, with two types of releases u_{1t} and u_{2t}, we would have

$$u_t = u_{1t} + u_{2t}. \tag{3}$$

For evaporation, we use the model

$$e_t = k_t \left(a \times \frac{s_t + s_{t+1}}{2} + b \right), \tag{4}$$

where k_t represents the evaporation intensity during period t. Substitution of (4) in (2) leads to a new version of the continuity equation:

$$s_{t+1} = d_{1t} s_t + d_{2t}(i_t - u_t) + d_{3t}, \tag{5}$$

where d_{1t}, d_{2t}, d_{3t} are periodic constants. The user needs to estimate these evaporation constants through hydrologic analysis external to BayRes, and provide them in a file.

The user needs to communicate also the objectives, the corresponding attributes, their range, how they relate to basic consequences and whether preferences are monotonic decreasing or increasing in the attribute. For our discussion, we shall assume that there are two objectives: the first one is *energy produced*, measured in GWh/month, with maximum production of 950 Gwh/month, which relates, e.g., to the basic consequences according to the relation

$$E_t = \eta u_{1t} \left(\alpha \frac{s_t + s_{t+1}}{2} + \beta \right)$$

with η designating a productivity coefficient of the system and the right hand side describing a model for the head of the reservoir, with appropriate coefficients α and β; the second one is *water spilled*, measured in mln.m^3/month, with maximum value 10000 mln.m^3/month, and, trivially, relates to the basic consequences according to $u_{2t} = u_{2t}$. The first (second) objective is introduced for economic (ecological) reasons, and we want to maximise (minimise) it. Note that we shall be considering an additional objective based on the deviation from the reference trajectory.

Finally, the user needs to communicate a file in which inflows to the reservoir, and other explanatory information, are available to build the forecasting model.

Forecasting inflows. BayRes allows the user to build a model to forecast inflows to the reservoir. The methodology implemented is that of Dynamic Linear Models (DLMs), see West and Harrison (1997) for a thorough introduction. Berger and Ríos Insua (1997) describe their impact in hydrology. In our application it is specially important that model updating and forecasting, one or more steps ahead, is done very fastly. BayRes supports all activities involved in determining the model.

As *exploratory analysis* is concerned, it provides tools like plots, various sample moments, histograms, time series plots, and standard data transformations like `log`.

Once with some intuitions, the user may formulate an initial model. Based on the *superposition principle* (West and Harrison, 1997), we implement a model building strategy allowing the user to incorporate: 1) a *trend term* modelling a constant or a piecewise linear trend in the inflow process; 2) a *seasonal effects term* to model seasonal behaviour of the series, typical in hydrology. Harmonic importance may be evaluated as an aid to decide which harmonics should be included; 3) a *regression term*, based on covariates chosen or transformed from the available variables. Examples could include snowfall, or number of months since last rainfall; 4) a *low order autorregresive term*, to improve short term forecasting. Figure 1a) specifies the model chosen for the case we discuss. The system also facilitates the assessment of prior distributions, including as well an option for using a reference distribution. Observational variances are described by a discount principle, with a discount factor for each component, see West and Harrison (1997).

Finally, BayRes allows the evaluation of the forecasting model, based on standard methods like autocorrelation functions, predictive probability bands, and graphical displays and summaries of residuals. A file stores various models and summaries of their evaluation so that, in the end, we may choose the most adequate model.

Preference model. The next step in using BayRes is the construction of the preference model. Assuming k objectives $c_1, ...c_k$ of primary interest (in our case $k = 2$, energy produced and water spilled) we use an additive multiattribute utility function

$$\lambda_1 F_1(c_1(u,i)) + \cdots + \lambda_k F_k(c_k(u,i)) - \rho\,(s - s^*)^2$$

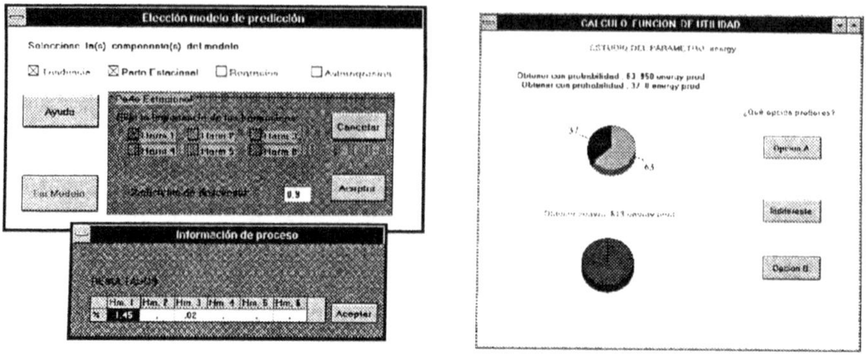

Fig. 1. a) Specification of forecasting model. b) Assessment of utility values.

where F_i is the i-th component utility, λ_i is the i-th weight, s is the final state and s^* is the reference state. This choice serves as a first order approximation. BayRes provides sensitivity analysis tools to check its appropriateness.

The F_i functions are built with the help of the *probability equivalent* (PE) method, see Farquhar (1984). Through an interactive dialog, we obtain the probability equivalents of several levels of each objective. For example, should we be interested in assessing the utility of 650 Gwh/month of energy produced, we would compare a lottery providing us that level for sure with lotteries giving the best value (950 Gwh/month) with a certain probability and the worst value (0 Gwh/month) with the complementary probability, until indifference is achieved. We then adjust by least squares a parametric curve to the elicited values: after visualizing the points, the user may choose an inflection point for the curve and adjust the component utility function F_i in two exponential pieces, to ensure adequate risk properties. Figure 1b) describes a screen for the PE method. In our example, the utility functions assessed were $F_1(E_t) = .0038(-1 + rmexp(.0066E_t))$ if $E_t \leq 765$ and $F_1(E_t) = 1.003475 - .40347\exp(-.02377(E_t - 765))$ otherwise and $F_2(u_{2t}) = -.07171 + 1.08365\exp(-.0001u_{2t})$. Once BayRes has elicited the component utility functions F_i, it assesses the weights λ_i, again with the PE method, see Keeney and Raiffa (1993). In our example, $\lambda_1 = .75$, $\lambda_2 = .25$. BayRes provides an initial estimate of ρ to scale appropriately penalties with other component utilities. In our case, it fixed it at 10^{-10}.

BayRes then aids in finding out a reference trajectory. In our case, it would solve by discrete deterministic dynamic programming the problem

$$\max \quad \sum_{t=1}^{r}(\lambda_1 F_1(c_1(u_t,\bar{\imath}_t)) + \lambda_2 F_2(c_2(u_t,\bar{\imath}_t)) - \rho\,(s_{r+1} - s_0)^2$$

$$\begin{aligned} s.t.\ & 0 \le s_t \le M \\ & 0 \le u_{1t} \le m_1 \\ & 0 \le u_{2t} \le m_2 \end{aligned}$$

with initial storage s_0 (in our case, .5M) and (3) and (5). At each stage, BayRes solves the nonlinear optimisation problems based on multistart stategy. It then displays the obtained discrete set of optimal values, and allows the user to specify a parametric value function, which is adjusted by least squares as an approximation to the optimal value function, for the next optimisation stage. Once stage 0 is reached, BayRes finds out the optimal trajectory starting from s_0 by table look-up. The features of this trajectory are presented in a table with initial storage, releases, consequences and final storage, from stage 1 to r. The user may accept this trajectory as a reference if he feels it corresponds to a satisfactory reservoir performance. If not, he may modify the problem above, introducing additional constraints.

Optimisation. At every time step, the expected utility function has to be maximized with respect to the releases from the reservoir subject to some constraints: releases have to be nonnegative and smaller than the maximum allowed. BayRes internally takes into account both continuity and maximum storage conditions.

In our case, with two controls and two objectives, if the predictive expected utility is designated $\Psi(u_1, u_2) = \int [\lambda_1 F_1(c_1(u_1, u_2, i)) + \lambda_2 F_2(c_2(u_1, u_2, i)) - \rho\,(s - s^*)^2] h(i|D)\, di$, the optimisation problem at each stage is

$$\begin{aligned} \max\ & \Psi(u_1, u_2) \\ s.t.\ & 0 \le u_1 \le m_1 \\ & 0 \le u_2 \le m_2 \end{aligned}$$

Typically, we shall not have an explicit expression for $\Psi(u_1, u_2)$; BayRes uses the Monte Carlo approximation

$$\hat{\Psi}_R(u_1, u_2) = \frac{1}{R} \sum_{j=1}^{R} \left[\lambda_1 F_1(c_1(u_1, u_2, i^j)) + \lambda_2 F_2(c_2(u_1, u_2, i^j)) - \rho\,(s^j - s^*)^2\right]$$

with $(i^1, ..., i^R)$ a sample from the predictive distribution. Then, at each stage BayRes solves the problem

$$\max\ \hat{\Psi}_R(u_1, u_2)$$

under the above constraints, with a guided multistart version of the Nelder Mead algorithm, see Rinnooy Kan et al. (1989), requiring only function evaluations. It is important that this method is robust since the objective functions may change considerably from one stage to another. The multistart strategy is required to mitigate convergence to local optima and is guided, since starting points are chosen from regions where we expect good controls to be found.

Normally, BayRes would be used to study long-term performance of reservoirs. Typically, we shall include a warming up period for the forecasting algorithm, followed by a period in which both forecasting and optimisation take place at each month. The performance of the reservoir is presented through plots of releases, storage and objectives, see Figure 2. These plots may be summarised with standard data analytic tools, similar to those used in the forecasting module.

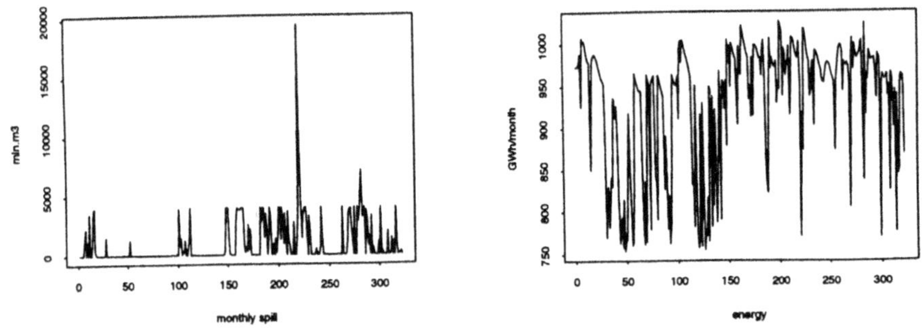

Fig. 2. Optimal trajectories for our example.

Sensitivity analysis Using the initial preference model as a reference, BayRes allows the user to analyse the impact of key parameters in the performance of the reservoir, e.g. the weights ρ and λ_i, comparing with that of the original preference model, suggesting perhaps changes.

BayRes also facilitates other robustness checking via testing of alternative scenarios, e.g. starting from very low or high storages, or simulating the reservoir under a much lower inflow regime. The performance of the reservoir and its adaptation under these extreme conditions may be explored then with the corresponding performance plots and summaries.

4 Conclusions

BayRes has been used to solve the cases described in Ríos Insua et al (1995, 1997). Additional cases studies of Spanish reservoirs are being considered for final tests and refinement of the interface.

It has received acceptance from users, who appreciate the capability of the system to make use of all available information, and allow for interaction in various ways. For example, the forecasting model could require interventions if sudden rainfall suggests an increase in the inflow level. This is also reflected in the use of predictive expected inflows in the definition of the reference trajectory, with obvious differences from wet to dry years, and rules introduced to handle cases in which reservoir storages are very high. Other types of interventions are permitted, making BayRes truly interactive.

Acknowledgments Project supported by grants from Iberdrola, CICYT and NATO. We are grateful to J. Fuertes, M. Farache, E. Pérez and Y. Algaba for their computational work.

References

Berger, J. (1985) *Statistical Decision Theory and Bayesian Analysis,* Springer, New York.

Berger, J., Ríos Insua, D. (1997) Recent developments in Bayesian Inference, with applications in Hydrology, *UNESCO*, Paris.

Farquhar, P. H. (1984) Utility assessment methods, *Management Science, 30,* 1283-1300.

French, S. (1986) *Decision Theory,* Ellis Horwood, Chichester.

Keeney, R. and Raiffa, H. (1993) *Decisions with Multiple Objectives: Preferences and Values Tradeoffs*, Wiley, New York.

Loucks, D. and Sigalvadsson, C. (1982) Multiple reservoir operations in North America. In Kaczmarek and Kinder (eds.), *The Operation of Multiple Reservoir Systems*, IIASA CP82-S02, Laxemburg, Austria.

Rinnooy Kan, A., Todd, J., Nemhauser, G. (1989) *Optimisation*, North Holland, Amsterdam.

Ríos Insua, D. and Salewicz, A. (1995) The operation of Lake Kariba. *J. of Multicriteria Decision Analysis*, 4, 203-222.

Ríos Insua, Salewicz, K., D., Müller, P., and Bielza, C. (1997) Bayesian methods in reservoir operations. To appear in French and Smith (eds.). *Case Studies in Bayesian Analysis*, Arnold, London.

Romero, C. (1991) *Handbook of Critical Issues in Goal Programming*, Pergamon, Oxford.

West, M., Harrison, J. (1997) *Bayesian Forecasting and Dynamic Models,* Springer, New York.

Timber Harvesting Scheduling Problems: A Compromise Programming Approach

Luis Díaz-Balteiro and Carlos Romero[1]

[1] Departamento de Economía y Gestión, E.T.S. Ingenieros de Montes, Avenida Complutense s/n, 28040 Madrid, Spain

Abstract. A timber harvest scheduling model with several criteria was first formulated by considering the net present value as the objetive function and the other criteria as rigid constraints. However this type of formulation generated an empty feasible set. Therefore, the problem was then reformulated as a compromise programming model. The preferential weights incorporated into the models were derived through the application of the Analitical Hierarchy Process (AHP) method after interviewing a group of forestry experts. All the theoretical developments were applied to the forest "Dehesa de la Garganta" in Segovia Mountains ("Sistema Central"), with an area of 2111 hectares covered with Scots pine (*Pinus sylvestris L.)*.

Key words: forest management, compromise programming, analytic hierarchy process

1 Introduction

Since the pioneer work of Field (1973), where a woodland property is planned with the help of a lexicographic goal programming model, an extensive applied literature addressing several forest management and planning problems with a multi-criteria perspective has arisen. This proliferation of multi-criteria problems in forestry is easy to understand given the complexity as well as the multiplicity of purposes and interests involved in this type of decisional context. In fact, every decision taken in this field affects several criteria of very different nature such as: economic, environmental, social, etc.

Among these applications the most widely used multi-criteria approaches in forestry planning have been goal programming (e.g., Arp and Lavigne 1982, Kao and Brodie 1979, ...), fuzzy approaches (e.g., Bare and Mendoza 1994, Mendoza et al. 1993, ...) and interactive methodologies (e.g., Liu and Davis 1995, Rustagi and Bare 1987, ...).

The above references clearly show how multi-criteria analysis is very well established in the forestry field. However in most of the applications reported, there is a certain disconnection between the solution provided by the models and the real and unknown optimum. In fact, the determination of the real optimum would comprehend the assessment of a utility function involving all the relevant criteria and it is well known that this task is full of conceptual and operational difficulties. These problems are specially important within a forestry context since not only the number of criteria involved but also the complex and continuous character of the feasible sets are usually important. Consequently, there are very few applications reported where this kind of multi-attribute utility approach is applied and in these cases, this approach is only used for very specific problems.

This paper presents a multi-criteria framework based upon compromise programming (CP) and the analytical hierarchy process (AHP), which by circumventing the difficulties associated to the assesment of multiattributes utility functions, provides solutions which can be interpreted to some extent in utility terms. The approach is applied to a timber management scheduling case although the basic analytical interpretation seems applicable to general forest management problems.

2 The Basic Model: "Dehesa de la Garganta" Forest

The "Dehesa de la Garganta" forest in the Segovia Mountains (Spain) is divided into three timber stands: the first two, with an area of 379.3 ha, are planned for conservational purposes and the third one, with an area of 2111.56 ha, is covered with *Pinus sylvestris* for production purposes. The main characteristics of this stand, representing the management unit, are shown in Table 1.

Table. 1 Forest "Dehesa de la Garganta"

	AGE CLASS			
Years	AREA (ha.)			
	Site Class I	Site Class II	Site Class III	Total Area
0-20	0	0	0	0
21-40	67.76	0	15.63	83.39
41-60	208.53	207.64	183.96	600.13
61-80	120.42	216.94	405.49	742.84
81-100	0	97. 88	326.27	424.15
101-120	14.68	102.62	135.92	253.23
121-140	0	0	0	0
141-160	0	0	782	7.82
	411.39	625.08	1075.09	2111.56

Source: Prieto & López Quero (1993)

To formulate the harvesting model the following criteria are considered: (a) the net present value of the forest over the planning horizon, (b) the equality of harvest volume in each cutting period, (c) the area control criterion which look for an ending regulated or even-aged forest, (d) the ending inventory criterion which ensures a solution for which the ending forest inventory is larger than or equal to the initial inventory and (e) the equality of cash flow in each cutting period.

Initially the model was formulated by considering the net present value criterion as the objective function and the four other criteria as rigid constraints. By taking into account that the planning horizon is 100 years, the time unit 10 years and the cutting age comprised between 70-120 years, a linear programming (LP) structure within a typical Model I formulation (Johnson & Scheurman 1977) was obtained.

LINEAR PROGRAMMING MODEL

OBJECTIVE FUNCTION:

MAX NET PRESENT VALUE: $$\sum_h \sum_i \sum_j Nv_{hij} \cdot X_{hij} \quad (1)$$

CONSTRAINT SET

Area accounting

$$\sum_j X_{hij} = A_{hi} \qquad \forall\, h, i \quad (2)$$

Volume control

$$\sum_h \sum_i V_{his} \cdot X_{hij} = H_s \quad (3)$$

$$-H_s + H_{s+1} = 0 \qquad \forall\, s \quad (4)$$

Area control

$$f_K = A \qquad \forall\, K \quad (5)$$

Ending forest inventory

$$\sum_i V^f_{hiT} \cdot X_{hiT} = I^f_h \qquad \forall\, h \quad (6)$$

$$I^f_h = I^i_h \qquad \forall\, h \quad (7)$$

Cash flow control

$$\sum_h \sum_i C_{his} \cdot X_{hij} = L_s \quad (8)$$

$$-L_s + L_{s+1} = 0 \qquad \forall\, s \quad (9)$$

Where:

T = planning horizon.

j = indicator of prescription; i.e. the schedule of every kind of action to be taken in each management unit.

X_{hij} = hectares harvested from hth class site, ith initial class age at jth prescription.

Nv_{hij} = net present value per hectare harvested from hth class site and ith initial class age at jth prescription

A_{hi} = hectares corresponding to hth class site and ith initial class age

V_{his} = volume per hectare harvested from hth class site, ith initial class age at sth cutting period.

H_s = volume harvested at sth cutting period.

f_K = hectares belonging to the kth final class age at the ending period.

A = forest area divided by the planning horizon and multiplied by the time span which defines the age class.
V^{f}_{hiT} = volume of ending forest inventory on hth site and ith class age.
I^{f}_{h} = volume of ending forest inventory on hth site.
I_{h}^{i} = volume of initial forest inventory on hth site.
C_{his} = cash flow per hectare harvested from hth class site, ith initial class age at sth cutting period
L_s = cash flow obtained at the sth cutting period

Although linear programming (LP) model (1) - (7) corresponds to a well known Model I structure, the following comments are aimed at clarifying its meaning as well as its size.

There are three class sites (h= 1,2,3) and six initial class ages (i= 1,2, ...6). As the planning horizon is T=100 years and the time unit 10 years, there will be 10 cutting periods (s= 1, 2, ..., 10). Moreover, as the time span which defines the age class is 20 years, there will be five final age classes (K= 1, 2, ..., 5);

The volume control restraints (3) and (4) impose a strict equality in the timber volume harvested at each of the ten cutting periods considered. Since the time span which defines the age class is 20 years, there will be five area control restraints. Moreover as the planning horizon is 100 years, area **A** is (2111.56 x 20)/(100) = 422.31 ha. As for the ending forest inventory, and given the pre-eminence of the initial situation, a restraint securing that the ending forest inventory is larger than or equal to the initial inventory is imposed. Finally, the cash flow constraints (8) and (9) impose a strict equality in the cash flows obtained at each of the ten cutting periods considered.

The LP model has a total of 130 decision variables. The area accounting block represents 14 constraints, the volume control block represents 10 accounting rows plus 9 volume constraints, the area control block represents 5 constraints (i.e., one constraint for each final class age), the ending forest inventory represents 3 accounting rows (one for each class site) plus 3 constraints (again one for each class site) and the cash flow control block represents 10 accounting rows plus 9 volume constraints. Therefore, the result is a (LP) model with 130 decisional variables and 63 constraints. Readers interested in finding out more details on the overall structure of LP model (1)-(9) can see Díaz-Balteiro (1995).

The above LP problem has no solution: that is, as is usual in this kind of formulation, the feasible set would be empty. In other cases the feasible set may not be empty but its size is so small that the net present value provided by the model would be very low. In short, the LP model (1)-(9) is too rigid and should be flexibilized in one way or in another. In the following sections the rigid character of the above problem will be mitigated with the help of a compromise programming (CP) formulation.

3 Pay-off Matrix Determination

In order to provide a pay-off matrix consistent with a CP model within a timber management context, its elements do not represent absolute values but deviations with respect to the corresponding target values. These targets represent the best achievable values and will be obtained by separately optimizing the five criteria considered. By operating in this way the following optimum values are obtained:

Net Present Value (NPV) = 3354028 thousands of pesetas in the whole planning horizon. Volume Control (VC) = 89928 m^3/decade. Area Control (AC) = 423 ha. Ending Forest Inventory (EFI) = 442903 m^3. Cash Flow Control (CFC) = 2551262 thousand of pesetas/period.

The pay-off matrix shown in Table 2 is obtained by minimizing the unwanted deviations between the achievements of each objective and the above target values and then computing the value of the other four objectives at each one of the optimal solutions. In this way the pay-off matrix shown in Table 2 is obtained. As usual, the elements in the main diagonal represent the ideal or anchor values whereas the figures in bold characters correspond to the anti-ideals or nadir values. It should be noted that all the objectives can achieve their optimum values separately except the criterion cash flow control, which reaches an optimum (minimum) deviation of 81340 thousand of pesetas with respect to the ideal value of 2551262 thousands of pesetas.

Table 2. Pay-off matrix for the five criteria considered (deviations with respect to target values)

	NPV thousands of pesetas	VC m^3	AC ha	EFI m^3	CFC thousands of pesetas
NPV	0	551478	546617	585528	**740884**
VC	**712134**	0	66750	249072	291987
AC	**1523**	136	0	845	1084
EFI	54480	21360	98154	0	**187920**
CFC	**2175345**	355543	398207	517107	81340

From Table 2 the following highlights should be remarked. There is a strong degree of conflict among the five criteria considered. This conflict is specially marked when the net present value objective is optimized. In fact, the timber harvest schedule of maximum net present value corresponds to the worst values for volume control, area control and cash flow criteria. It is also interesting to point out that the best outcome in terms of cash flow control is only compatible with the worst outcomes for net present value and ending forest inventory criteria. It can be said that no solution generated by the single optimization of any criterion seems attractive enough to be implemented in practice. Therefore a best-compromise solution among the five criteria considered should be sought.

4 Weights Elicitation

As a first step in the formulation of our CP model, weights $w_1, ..., w_5$ to be attached to the five criteria considered, have to be elicited. these weights play the following double role: (i) normalising the five criteria and (ii) measuring the decision maker's preferences for the five criteria considered.

Function (i) of the weights can be performed by resorting to any normalisation methods proposed in the multi-criteria field. We will adopt, as is usual in multi-criteria analysis, weights inversely proportional to the ranges of each criterion; i.e. anchor minus anti-ideal values (see Romero 1991, pp. 35-43 for the general problem of normalising objectives). The second function (ii) actually consists in eliciting the relative preferences of the decision maker. This task will be approached with the help of an interactive method, the AHP (Analytical Hierarchy Process) proposed by Saaty (1980,1994), which permits the elicitation of the preferential weights through a dialogue with a group of forestry experts as will be shown below.

The hierarchy structure of our problem is as follows. In the hierarchy first level, two aggregate criteria appear: forestry and financial aspects, respectively. The forestry aspect is grouped into three single criteria: volume control, area control and ending forest inventory. The financial aspect is grouped into two single criteria: net present value and cash flow control. The hierarchy criteria were presented to a group of academic members from the staff of the Forestry School at Technical University of Madrid for a pairwise comparison. These scholars played the role of an expert committee. The weights shown in Table 3 were then obtained with the help of the EXPERT CHOICE software (Forman, 1988).

Table 3. Preferential weights elicited for the hierarchy of criteria considered

FORESTRY	0.750	0.250	0.750	0.875	0.875	0.500	0.833	0.833
FINANCIAL	0.250	0.750	0.250	0.125	0.125	0.500	0.167	0.167
VOLUME CONTROL	0.281	0.207	0.243	0.078	0.149	0.429	0.091	0.149
AREA CONTROL	0.135	0.735	0.088	0.435	0.785	0.143	0.091	0.066
ENDING FOREST INVENTORY	0.584	0.058	0.669	0.487	0.066	0.429	0.818	0.785
NET PRESENT VALUE	0.250	0.833	0.250	0.500	0.833	0.250	0.900	0.250
CASH FLOW CONTROL	0.750	0.167	0.750	0.500	0.167	0.750	0.100	0.750

The aggregate preferential weights for the two levels of the hierarchy were determined in two different manners. First the average values were calculated. Second, because of the small sample size (only eight questionnaires) the average values cannot be significant. To redeem this problem a hierarchical cluster analysis was implemented. In this way two clusters were obtained. The second cluster comprises questionnaries corresponding to columns 2 and 5 of Table 3 while the first cluster comprises questionnaries corresponding to the other columns of Table 3. The three averages are shown in Table 4.

Table 4. Preferential weights (average, cluster1 and cluster2)

	CLUSTER 1	CLUSTER 2	AVERAGE
FORESTRY	0.757	0.563	0.708
FINANCIAL	0.243	0.438	0.292
VOLUME CONTROL	0.212	0.178	0.203
AREA CONTROL	0.160	0.760	0.310
ENDING FOREST INVENTORY	0.629	0.062	0.487
NET PRESENT VALUE	0.400	0.832	0.508
CASH FLOW CONTROL	0.600	0.167	0.492

In order to elicit the final preferential weights, a multiplicative aggregation between the weights corresponding to the two hierarchical levels shown in Table 3 was implemented. By resorting to simple multiplications the final preferential weights shown in Table 5 were obtained. These final weights divided by the corresponding ranges (ideal minus anti-ideal values) will be introduced into the CP formulations.

Table 5. Final preferential weights for the five criteria considered (average, cluster 1 and cluster 2)

	CLUSTER 1	CLUSTER 2	AVERAGE
NET PRESENT VALUE	0.0972	0.3646	0.1483
VOLUME CONTROL	0.1603	0.1001	0.1437
AREA CONTROL	0.1209	0.4275	0.2195
ENDING FOREST INVENTORY	0.4758	0.0349	0.3448
CASH FLOW CONTROL	0.1458	0.0729	0.1437

5 Efficient and Best-compromise Timber Harvest Schedules

In order to determine timber harvest schedules, our first step will consist in obtaining an approximation of the efficient set in the space of the five criteria considered. By resorting to the weighting method (Cohon 1978, chap. 6) more than 50 extreme efficient points were generated (Diaz-Balteiro 1995).

Presenting a decision maker with more than fifty possible timber harvest schedules would be an overhead of information that he is most unlikely to assimilate. In order to mitigate this problem, a filtering technique (Steuer and Harris 1980) was used. In this way, efficient solutions that do not sufficiently differ from another efficient solutions already retained by the filter will be discarded. By applying this filtering operation, the cluster of seven schedules shown in Table 6 is obtained. This cluster represents the seven most diverse efficient schedules measured in terms of deviations with respect to the anchor values previously defined.

Table 6. Cluster of efficient timber harvest schedules

EFFICIENT SCHEDULES	NPV	VC	AC	EFI	CFC
	thousand of pesetas	m^3	ha	m^3	thousand of pesetas
1	335378	15874	443	0	271070
2	0	713124	1523	54480	2175345
3	433360	429186	0	0	1145207
4	740994	291488	1084	187920	81340
5	643249	0	0	50720	627655
6	293207	0	943	135630	430397
7	521895	0	386	8398	320139

It is interesting to notice that efficient schedules 2 and 4 correspond to the first column of the pay-off matrix (net present value optimization) and to the fifth column of this matrix (cash flow control optimization). the other five efficient points represent different timber harvest schedules. These results can provide to the decision maker interesting insights about the gains and losses in terms of the five criteria associated to each efficient solution or schedule.

The second step in our analysis will be to determine the efficient schedule nearest to the ideal point for metrics L_1 and L_∞; that is, the formulation of a CP model (see e.g. Zeleny 1974, 1982; Yu 1985 for technical aspects related to CP). By using the anchor values represented in the main diagonal of the pay-off matrix shown in Table 2 and the weights derived in the preceding section the best-compromise timber harvest schedules shown in Table 7 were obtained.

It should be noted that as we are working with five criteria, we have a five-dimensional space of reference. Therefore the L_1 and L_∞ solutions do not necessarily bound a portion of the efficient set (*compromise set*) where the best-compromise solutions corresponding to other methods fall (Freimer & Yu 1976). The L_1 solution corresponds to a point of maximun efficiency where a separable and additive utility function linear in the attributes achieves its maximum (Ballestero and Romero 1993), whereas the L_∞ solution corresponds to a perfectly equilibrated point where the weighted gaps between the ideal values and the actual achievements are equal (Ballestero and Romero 1994).

Table 7. L_1 and L_∞ best-compromise timber harvest schedules

	CLUSTER 1			CLUSTER 2			AVERAGE		
	Weights	L_1	L_∞	Weights	L_1	L_∞	Weights	L_1	L_∞
NPV	0.0972	280094	400687	0.3646	180762	135109	0.1483	470136	231044
VC (m^3)	0.1603	229553	236770	0.1001	101752	472803	0.1437	0	229190
AC (ha)	0.1209	39	656	0.4275	0	237	0.2195	94	321
EFI (m^3)	0.4758	0	20589	0.0349	0	34566	0.3448	14991	25210
CFC	0.1458	780053	744133	0.0729	583034	1909028	0.1437	301193	674271

NPV and CFC are measured in thousands of pesetas

The L_1 and L_∞ solutions vary greatly between the two clusters and the average values. This result is easy to understand given the extremely different set of weights attached to each one of the three groups considered. In fact each group (clusters 1 and 2, and the average values) represent different subjective appreciations or orientations towards the importance of each criterion in forestry planning.

On the other hand, for a given group the differences between the L_1 and L_∞ solutions are in general very important. Hence it could be useful to contract these large differences by resorting to the method of the displaced ideal (Zeleny 1974). This task will require a certain interaction with the decision maker. Although this matter is interesting its development is beyond the scope and the permissible length of this paper.

Acknowledgements

Suggestions and editorial comments raised by one of the reviewers are highly appreciated.

References

Arp, P.A. and D.R. Lavigne. 1982. Planning whit goal programming: A case study for multiple-use of forested land. *For. Chronicle* 58: 225-232.

Ballestero, E. and C. Romero. 1993. Economic optimization by compromise programming: the joint production model. *J. Multi-Criteria Des. Analysis* 2: 65-72.

Ballestero, E. and C. Romero. 1994. Utility optimization when the utility function is virtually unknown. *Theory and Decision* 37: 233-243.

Bare, B.B. and G.A. Mendoza. 1994. A fuzzy approach to natural resource management from a regional perspective. *Int. Trans. Opl. Res.* 1: 51-58.

Cohon, J.L. 1978. *Multiobjective programming and planning*. Academic Press, New York.

Diaz-Balteiro, L. 1995. *Modelos de programación matemática para la ordenación de montes: desarrollos teóricos y aplicaciones al sector forestal español.* Doctoral dis, Forestry School, Technical University of Madrid.

Field, D.B. 1973. Goal programming for forest management. *For. Sci.* 19: 125-135.

Forman, E.H., T.L. Saaty, M.L. Selly and R. Waldrow. 1985. EXPERT CHOICE, Decision Support Software Inc., Pittsburgh.

Freimer, M. and P. L. Yu 1976. Some new results on compromise solutions for group decision problems. *Manag. Sci.* 22: 688-693.

Johnson, K.N. and H.L. Scheurman. 1977. Techniques for prescribing optimal harvest and investment under different objectives- Discussion and synthesis. *Forest Science Monographies* 18.

Kao, C., and J.D. Brodie. 1979. Goal programming for reconciling economic, even flow, and regulation objectives in forest harvest scheduling. *Can. J. For. Res.* 9: 525-531.

Liu, G. and L.S. Davis. 1995. Interactive resolution of multi-objective forest planning problems with shadow price and parametric analysis. *For. Sci.* 41: 452-469.

Mendoza, G.A., B.B. Bare and Z. Zhou. 1993. A fuzzy multiple objective linear programming approach to forest planning under uncertainty. *Agric. Systems* 41: 257-274.

Prieto, A. and M. López-Quero. 1993. *Manual de ordenación de montes*. Editorial Paraninfo, Madrid.

Rustagi, K.P. and B.B. Bare. 1987. Resolving multiple goal conflicts with interactive goal programming. *Can. J. For. Res.* 17: 1401-1407.

Romero, C. 1991. *Handbook of critical issues in goal programming*. Pergamon, Oxford.

Saaty, T.L. 1980. *The analytic hierarchy process: planning, priority setting, and resource allocation*. McGraw-Hill, New York.

Saaty, T.L. 1994. How to make a decision: the analytic hierarchy process. *Interfaces* 24: 19-43.

Steuer, R.E. and F.W. Harris. 1980. Intra-set point generation and filtering in decision and c.iterion space. *Comput. Oper. Res.* 7: 41-53.

Yu, P.L. 1985. *Multiple criteria decision making. Concepts, techniques and extensions*. Plenum Press, New York.

Zeleny, M. 1974. A concept of compromise solutions and the method of the displaced ideal. *Comput. Oper. Res.* 1: 479-496.

Zeleny, M. 1982. *Multiple criteria decision making*. McGraw-Hill, New York.

Location of a Semiobnoxious Facility. A Biobjective Approach

Emilio Carrizosa[1], Eduardo Conde[1] and Dolores Romero-Morales[2]

[1]Facultad de Matemáticas, Universidad de Sevilla, Sevilla, Spain
[2]E.U. Empresariales de Jerez, Universidad de Cádiz, Cádiz, Spain

Abstract. In the location of a semiobnoxious facility one has to consider both transportation and environmental (or social) costs to be optimized. Such objectives are modeled as functions of the distances to a set of demand points, leading to a biobjective optimization problem.

Since the usual solution set is in general infinite and so of dubious interest to Decision Makers, we propose as solution a finite feasible set representing the best compromise solutions. Such a finite set is obtained, using the concept of α-dominance, by a standard Global Optimization technique.

Keywords: Biobjective Optimization, Semiobnoxious Facilities, Branch & Bound, α-dominance.

1 Introduction

The term semiobnoxious refers to facilities which provide some kind of service to the population but at the same time they have negative effects such as pollution or risk.

Hence, it is natural to accommodate in semiobnoxious facility location problems both the minimization transportation costs and environmental impact.

These objectives are measuring different things, therefore the aggregation to a single objective, as proposed for example in [4, 14], does not seem to be plausible from managerial perspective. Instead, tackling the problem as biobjective is more correct, [1].

Since the usual solution concept for vector optimization problems, namely (weak) efficiency, [15], leads to disconnected infinite sets which do not appear to be easy to generate, we propose as surrogate a finite feasible set whose image is close to the set of (weakly) efficient outcomes, and could then be used as input for a posteriori analysis, perhaps with the aid of discrete MCDM techniques.

Substituting the efficient set by a finite set known to cover in some sense the efficient outcomes has been proposed, e.g. in [8] for a different problem. However, the approach in [8] for our problem would lead us to solve a large series of global optimization problems.

Instead, we show how the Global-Optimization method proposed in [6] for unconstrained problems can be used with very mild changes to attain our end of obtaining a finite set representing properly the efficient outcomes.

The rest of this paper is structured as follows. In Section 2 we propose a semiobnoxious facility location model. In Section 3 we define the α-dominance concept and we proof the existence of a finite α-dominator for our problem. In Section 4 we describe a vectorial Branch and Bound algorithm which finds a finite α-dominator after a finite number of iterations. Finally, in Section 5 we proof the convergence of this algorithm.

2 The Model

In this paper we consider the following problem:

$$\min_{x \in S}\{\sum_{a\in A} g_-^a(\|x-a\|_-^a), \sum_{a\in A} g_+^a(\|x-a\|_+^a)\} \tag{1}$$

where

- $A \subset \mathbb{R}^2$ is a nonempty finite set, representing the coordinates of the population centers.
- $S \subset \mathbb{R}^2$ is a nonempty compact robust set, representing the set of feasible locations for the facility. S is assumed to be a d.c. set, [9], which means that S can be written as

$$S = \{x \in \mathbb{R}^2 \ : \ h_1(x) - h_2(x) \leq 0\} \tag{2}$$

  h_1, h_2 being convex functions.
- For each population center $a \in A$, $\|\cdot\|_-^a$, $\|\cdot\|_+^a$ are arbitrary gauges in $\mathbb{R}^2$, [5], measuring the distance to the facility. Typical examples could by l_p norms, e.g. [6, 7], or the skewed gauges, [12].
- $g_-^a : [0, +\infty) \longmapsto (0, +\infty)$ is a nonincreasing convex function with finite right derivative at zero, measuring the environmental impact on the population center a caused by a facility at a distance $\|x-a\|_-^a$ from a. A typical example is $g_-^a(t) = \omega_a e^{-t}$, $\omega_a > 0$, e.g. [3, 14].
- $g_+^a : [0, +\infty) \longmapsto (0, +\infty)$ is a nondecreasing concave function with finite right derivative at zero, measuring the transportation costs from the population center a to a facility at a distance $\|x-a\|_+^a$, [12].

2.1 D.C. Problem

Problem (1) can be stated as

$$\min_{x \in S} f(x) = (f_1(x), f_2(x)) \tag{3}$$

where $f_1(x) = \sum_{a \in A} g_-^a(\|x - a\|_-^a)$, $f_2(x) = \sum_{a \in A} g_+^a(\|x - a\|_+^a)$.

In [14] it is proved that both objectives of Problem (3) are d.c. functions, that is, they can be expressed as the difference of two convex functions.

PROPOSITION 2.1 *The functions $f_1(x)$, $f_2(x)$ are d.c. functions.*

Hence, Problem (3) is a biobjective d.c. problem, which can be tackled with standard Global-Optimization techniques.

In particular, in [6] Hansen and Thisse consider $S = \mathbb{R}^2$ and propose a branch-and-bound procedure, which is stopped when all the elements (squares) in the partition have small diameter, yielding a finite set as near-optimal solutions to (3).

3 α-Domination

In this section we define a solution concept through the idea of dominance in the criterion vector.

DEFINITION 3.1 *We say x α-dominates y ($y \preceq_\alpha x$), with $0 < \alpha \leq 1$, if the following inequalities hold:*

$$f_i(y) > \alpha f_i(x) \quad i = 1, 2. \tag{4}$$

REMARK 3.1 If $\alpha = 1$, α-domination corresponds to the usual strong domination.

REMARK 3.2 Since we are assuming each function g_-^a, g_+^a is strictly positive both f_1 and f_2 are also strictly positive. Hence, since $0 < \alpha \leq 1$, taking logarithms at each inequality (4) we obtain the equivalent inequalities

$$\log(f_i(y)) > \log \alpha + \log(f_i(x)) \quad i = 1, 2. \tag{5}$$

Thus we see that $y \preceq_\alpha x$ is just the standard ε-dominance used, among others, by [10, 13, 16], considering $\varepsilon_i = \log \alpha$, for each $i = 1, 2$.

DEFINITION 3.2 *Let $0 < \alpha \leq 1$. We say x α-dominates E, if $y \preceq_\alpha x$ for each point $y \in E$.*

We say B is an α-dominator for Problem (3) iff for each $y \in S$, there exists $x \in B$, such that $y \preceq_\alpha x$.

Our aim is to offer a method to obtain a finite α-dominator.

REMARK 3.3 The relation $\preceq_\alpha$ is reflexive for $0 < \alpha < 1$ and only transitive for $\alpha = 1$.

REMARK 3.4 The μ-dominance relation implies the α-dominance relation, with $0 < \alpha \leq \mu \leq 1$.

The following result shows the existence of a finite α-dominator for Problem (3).

PROPOSITION 3.1 *If $0 < \alpha < 1$, then there exists a finite α-dominator for Problem (3).*

PROOF. For each $x \in S$ we denote by $D(x)$ the set of the feasible points α-dominated by x, i.e.,

$$\begin{aligned} D(x) &= \{y \in S \,:\, y \preceq_\alpha x\} \\ &= f_1^{-1}\left((\alpha f_1(x), +\infty)\right) \cap f_2^{-1}\left((\alpha f_2(x), +\infty)\right) \end{aligned} \tag{6}$$

and it follows that the set $D(x)$ is open. Moreover by Remark 3.3,

$$S = \bigcup_{x \in S} D(x),$$

and the result follows from the compactness of S. □

By Proposition 3.1, if $\alpha < 1$, it is always possible to find a finite α-dominator. The proof of that result is, however, nonconstructive. In the next section we present an algorithm which actually constructs such finite α-dominator.

4 Resolution

After formalizing in previous sections the intuitive idea of near optimality, we propose some modifications to the branch-and-bound algorithm in [6] to accommodate our aim of obtaining just a finite α-dominator for Problem (3).

Thus the algorithm proposed is the following:

ALGORITHM 4.1

Initialization.
Construct a square $M_0 \supset S$.
Initialize a list $\mathcal{L}$ with $\mathcal{L} = \{M_0\}$.

Set $B = \{v \in V(M_0) \; : \; h_1(v) - h_2(v) \leq 0\}$, where $V(M_0)$ is the vertex's set of M_0.

Iteration $k: 1, 2, \ldots$
Select M_k, a square of $\mathcal{L}$ with the largest diameter and remove it from $\mathcal{L}$.
Split M_k into four subsquares, M_k^l, $l = 1, \ldots, 4$.
For $l = 1, \ldots, 4$ add M_k^l to $\mathcal{L}$ if none of the following is true:

- M_k^l satisfies the conditions of Infeasibility Test.
(In other case, add to B each feasible known vertex of M_k^l and update this set suppressing 1-dominated points).
- M_k^l satisfies the conditions of Local α-Dominance Test.
- M_k^l satisfies the conditions of Global α-Dominance Test.

If $\mathcal{L} = \emptyset$, STOP: B is an α-dominator for Problem (3).
Else, go to the next iteration.

Now we discuss in detail the steps of the procedure.

4.1 Infeasibility Test

Contrary to the model discussed in [6], Problem (3) is assumed to be constrained. Then, it is necessary to check at each iteration the feasibility of the square under study.

If S has a manageable shape the strategies proposed in [11] could be used. In general, the following result, proposed in [9], uses standard bounding techniques for d.c. functions to offer an infeasibility test.

Hereafter we denote by ext(E) the extreme points of a convex set E.

PROPOSITION 4.1 *[9] Let $E \subset M_0$ be a closed convex set, and let ν be a subgradient of the function h^1 at the point $\bar{y} \in E$. Then, if*

$$\min_{y \in ext(E)} (h^1(\bar{y}) + \langle \nu, y - \bar{y} \rangle - h^2(y)) > 0,$$

the set E is infeasible.

4.2 Suppression by α-dominance

Since the aim of the algorithm proposed is to offer a finite α-dominator B for Problem (3), at each iteration we will check if the new squares obtained are α-dominated by some point known to be feasible.

In this sense we propose two α-dominance criteria, in which we compare a square E with one of its elements (local α-dominance) or elements in the list B but not necessarily close to E (global α-dominance).

4.2.1 Local α-dominance Test

In [6] it is proposed as stopping rule the size of squares is under a given value (tolerance). In this section it is proved how this tolerance level can be related to the vector-values from feasible points. In our case the tolerance level is directly related with the criterion space while in [6] it is related with the decision space.

We recall that the d.c. functions conserve some properties of convex functions such as being locally lipschitz, [9].

The aim of the following result is to find, a bound on the diameter of a set to conclude that this set is α-dominated by one of its feasible points.

PROPOSITION 4.2 *Let $E \subset M_0$ be a closed set, let L_i be a Lipschitz constant for f_i on E, $i = 1, 2$, and let $\bar{x} \in E \cap S$. If*

$$\delta(E) < \min \left\{ \frac{(1-\alpha) f_i(\bar{x})}{L_i} \ : \ i = 1, 2 \right\} \tag{7}$$

then $\bar{x}$ α-dominates E.

PROOF. By definition of Lipschitz constant, we have, for each $i = 1, 2$ and each $y \in E$, that

$$\begin{aligned} f_i(y) &\geq f_i(\bar{x}) - L_i \|y - \bar{x}\| \\ &\geq f_i(\bar{x}) - L_i \delta(E) \\ &> f_i(\bar{x}) - L_i \frac{(1-\alpha) f_i(\bar{x})}{L_i} \\ &= f_i(\bar{x}) - (1-\alpha) f_i(\bar{x}) \\ &= \alpha f_i(\bar{x}) \end{aligned}$$

Then the result follows.

□

Since the functions f_i are strictly positive and continuous on the compact set S there exists a constant β such that

$$0 < \beta < \min_{x \in S} f_i(x) \quad i = 1, 2. \tag{8}$$

COROLLARY 4.1 *Let E be a set whose diameter $\delta(E)$ satisfies*

$$\delta(E) < \frac{(1-\alpha)\beta}{L} \tag{9}$$

with β satisfying (8) and $L = \max_{i=1,2} L_i$. Then if $E \cap S \neq \emptyset$, each feasible point of E α-dominates E.

Corollary 4.1 implies that, when a square E is small enough (its diameter satisfies (9)) then any of its feasible points suitable represents the whole E.

This idea was previously used in the algorithm of [6]; however while we require α (thus the tolerance in the criterion space) as input, in [6] one needs to give the tolerance in decision space.

4.2.2 Global α-dominance Test

The dominance test proposed in [6] for a square E is based on comparing the outcomes of points in B with uniform lower bounds of the outcomes $\{(f_1(y), f_2(y)) : y \in E\}$.

In particular, if $l_1(E)$ and $l_2(E)$ represent such bounds, i.e.,

$$l_i(E) \leq \min_{y \in E} f_i(y) \quad i = 1, 2,$$

then according to [6], we could rule out E as soon as there exists some $x \in B$ such that

$$l_i(E) \geq f_i(x) \quad i = 1, 2. \tag{10}$$

Since our aim is to obtain an α-dominator, (10) can be strengthened to

$$l_i(E) \geq \alpha f_i(x) \quad i = 1, 2. \tag{11}$$

Hence a crucial step is the determination of $l_i(E)$.

In [6] such lower bounds are obtained using the monotonicity of f_i and properties of the projections of the norms in use, valid for l_p norms, see [6, 7, 11].

Since the practical performance of the procedure is strongly influenced by the accuracy of the bounds used, it would be of interest to calculate complementary or alternative bounds to those proposed in [6, 7, 11] (anyway, only valid for a very particular class of gauges).

In this sense, one can use the standard bounding strategy for d.c. functions, [9] (see [14] for a detailed description of such bounds, based on the right derivatives of g_-^a, g_+^a and the subdifferentials of $\|\cdot\|_-^a$, $\|\cdot\|_+^a$), or some sharper bounds based on Lagrangian relaxation, [2].

5 Convergence

The next result shows the finiteness of algorithm 4.1 ending with a finite α-dominator, with $0 < \alpha < 1$, for Problem (3).

THEOREM 5.1 *Let $\alpha \in (0,1)$. The algorithm 4.1 obtains in a finite number of iterations a finite α-dominator set $B \subset S$, where $0 < \alpha < 1$, for Problem (3).*

PROOF.

It is evident that, if the algorithm stops after a finite number of steps, the corresponding set B is a finite α-dominator. We then show that, indeed, the list $\mathcal{L}$ becomes empty after a finite number of iterations.

Let β be defined as in (8) and L be a common Lipschitz constant for f_1 and f_2.

Since at each step we select for inspection from $\mathcal{L}$ a square with highest diameter, it follows that, after a finite number of steps, all the squares in $\mathcal{L}$ have diameter less than $\frac{(1-\alpha)\beta}{L}$.

Then, let $\mathcal{L}$ consists of squares $E_1, \ldots, E_t$, with diameter less than $\frac{(1-\alpha)\beta}{L}$. We will show that all the points in each E_i will be removed after a finite number of steps.

Indeed, for any i, $1 \leq i \leq t$, it follows from Corollary 4.1 that, as soon as we find a feasible point $x_i \in E_i$, then x_i α-dominates E_i; moreover, by the selection rule used and the consistency of the bounds, we then would detect in a finite number of iterations that all the points in E_i are α-dominated by x_i (via local or global α-dominance tests).

The bounds proposed in the feasibility test are consistent, [9]. Hence, it will be detected in a finite number of steps that E_i is infeasible or that E_i intersects the interior of S.

In case E_i intersects S only in $bd(S)$, the boundary of S, the mere application of the feasibility test to subsquares obtained from E_i will not detect the real status of E_i.

However, since S is assumed to be robust, there exists another square E_j adjacent to E_i such that

$$E_j \cap S \supset E_i \cap S$$

and E_j intersects the interior of S.

Hence, after a finite number of steps it will be found some $x_j \in E_j \cap S$, known to α-dominate both E_i and E_j.

□

6 Acknowledgements

The research of E. Carrizosa and E. Conde has been partially supported by DGICYT grant PB93-0927. This support is gratefully acknowledged.

References

[1] BLOEMHOF-RUWAARD, J.M., VAN BEEK, P., HORDIJK, L., VAN WASSENHOVE, L.N., (1995), Interactions between Operational Research and Environmental Management, *European Journal of Operational Research* **85** 229-243.

[2] CARRIZOSA, E., CONDE, E., ROMERO-MORALES D., (1995), Location of a semiobnoxious facility. A Global-Optimization approach, *Prepublicación de la Facultad de Matemáticas de la Universidad de Sevilla, Sección de Estadística e Investigación Operativa*, **23**.

[3] CHEN, P.C., HANSEN, P., JAUMARD, B., TUY, H., (1992), Weber's Problem with Attraction and Repulsion, *Journal of Regional Science* **32** 467-486.

[4] DREZNER, Z., WESOLOWSKY, G.O., (1991), The Weber Problem on the Plane with some Negative Weights, *INFOR* **29** 87-99.

[5] DURIER, R., MICHELOT, C., (1985), Geometrical Properties of the Fermat-Weber Problem, *European Journal of Operational Research*, **20**, 332-343.

[6] HANSEN, P., THISSE, J.F., (1981), The Generalized Weber-Rawls Problem, in J.P. Brans (Ed.), *Operations Research*, North Holland, Amsterdam.

[7] HANSEN, P., PEETERS, D., RICHARD D., THISSE, J.F., (1985), The Minisum and Minimax Location Problems Revisited, *Operations Research* **33** 1251-1265.

[8] HELBIG, S., (1994), On a Constructive Approximation of the Efficient Outcomes in Bicriterion Vector Optimization, *Journal of Global Optimization* **5** 35-48.

[9] HORST, R., TUY, H., (1993), *Global Optimization (Deterministic Approaches)*, 2nd edition, Springer-Verlag, Berlin.

[10] LEMAIRE, B., (1992), Approximation in Multiobjective Optimization, *Journal of Global Optimization* **2** 117-132.

[11] PLASTRIA, F., (1992), GBSSS: The Generalized Big Square Small Square Method for Planar Single-Facility Location, *European Journal of Operational Research*, **62**, 163-174.

[12] PLASTRIA, F., (1995), Continuous Location Problems, in Z. Drezner (Ed.), 225-262, *Facility Location*, Springer-Verlag, Berlin.

[13] TAMMER, CHR., (1994), Stability results for approximately efficient solutions, *OR-Spektrum* **16** 47-52.

[14] TUY, H., AL-KHAYYAL, F.A., ZHOU, F., (1995), A D.C. Optimization Method for Single Facility Location Problems, *Journal of Global Optimization* **7** 209-227.

[15] WHITE, D.J., (1982), *Optimality and Efficiency*, John Wiley & Sons, Chichester.

[16] WHITE, D.J., (1986), Epsilon Efficiency, *Journal of Optimization Theory and Application* **49** 319-337.

A Hierarchical Location Model: Biobjective Programming vs FDH

M. Alminyana, F. Borrás and J.T. Pastor

Department of Statistics and Operations Research, Faculty of Sciences, University of Alicante, 03071-Alicante, SPAIN

Abstract. The pq-median problem, due to Serra and ReVelle, seeks to locate hierarchical facilities at two levels so as to obtain a coherent structure. Coherence requires that the entire area assigned to a facility at the inferior level must be assigned to one and the same facility at the next higher level of the hierarchy. Here we present a new heuristic wich combines the generation of points, by means of a "directed" branching procedure, with the selection of points, by means of the FDH-technique. We further compare our new heuristic with the two most relevant heuristics proposed by Serra and ReVelle.

Keywords. Location, pq-median problem, biobjective integer programming, heuristics, FDH

1. Introduction

It is widely accepted that many facility systems are hierarchical in nature, providing several levels of service. More specifically, a hierarchical system is one in which services are organized in different levels that are somehow related to one another in the complexity of function/service. An important consideration in the design of specific hierarchical location systems is that the services offered by the facilities are nested, i.e., a level k facility offers services of type k, k-1, ..., and 1.

In seeking a more responsive and effective relation among the different levels in a hierarchy, all the districts corresponding to a given hierarchical level should belong to one and the same district in the next level of the hierarchy. This property leads to a "coherent" districting structure. Serra and ReVelle (1993) are the first authors to address the issue of coherence in a hierarchical location model. While location and districting models have traditionally been studied separately, their integration promises improvements in the optimal allocation of resources in space.

The pq-median problem locates two levels of hierarchical facilities: the first level corresponds to facilities that offer basic services (called type A) and cover a smaller amount of area (or population) relative to a second level (called type B). It is a multiobjective integer programming model that combines two p-median problems, one for each hierarchical level, and where the relation between each level is expressed by two factors: (1) the facilities are nested, and (2), all areas assigned to a type A facility must be assigned to one and the same type B facility (coherence condition). Other important condition is that the number of desired facilities is known: p type A facilities and q type B facilities, where p>q. The mathematical formulation of the pq-median model follows:

$$\text{Min } Z_A = \sum_{i\in I}\sum_{j\in I} w_i d_{ij} x_{ij} \qquad \text{Min } Z_B = \sum_{i\in I}\sum_{k\in I} w_i d_{ik} y_{ik}$$

s.t.

$$\sum_{j\in I} x_{ij} = 1 \qquad \forall i \in I$$

$$\sum_{k\in I} y_{ik} = 1 \qquad \forall i \in I$$

$$\sum_{k\in I} z_{jk} = u_j \qquad \forall j \in I$$

$$u_j + v_j \le 1 \qquad \forall j \in I$$

$$x_{ij} \le u_j + v_j \qquad \forall i \in I\ \forall j \in I$$

$$y_{ik} \le v_k \qquad \forall i \in I, \forall k \in I$$

$$x_{ij} + z_{jk} \le y_{ik} + 1 \qquad \forall i \in I, \forall j \in I, \forall k \in I$$

$$\sum_{j\in I} u_j = p$$

$$\sum_{k\in I} v_k = q$$

$$x_{ij}, y_{ik}, z_{jk}, u_j, v_k = (0,1) \qquad \forall i \in I, \forall j \in I, \forall k \in I$$

where I represents the set of demand areas, i; j stands for type-A services and k for type-B services; w_i is the population at demand area i, d_{ij} the distance between i and j, x_{ij} is 1, if demand node i is assigned to a type-A facility at j, and 0 otherwise; y_{ik} means the same but for type B servers, z_{jk} is 1, if type A facility at j is assigned to type B facility at k (and 0 otherwise), u_j is 1 if there is a type A facility at j and the same holds for v_k in relation to type B facilities. All the variables are binary variables. If we admit that the set of demand areas, the set of possible location for type-A servers and the set of possible location for type-B servers are all different we are dealing with a more general problem, and, therefore, with a more complex formulation (see Serra and ReVelle (1993)).

The coherent hierarchical pq-median problem is very computationally intensive to solve, since it has a very large number of variables and constraints. One methodological approach to solve the problem at hand is to locate each level independently of the other in a successive manner, starting from de top of the hierarchy and proceeding down to the bottom or starting from the bottom of the hierarchy and proceeding up to the top. Serra and ReVelle (1994) have proposed four heuristics for solving the pq-median problem. All of them need to solve many r-median problems (*r=p* or *r=q* or *r=p+q*); in SAPHIER1 and SAPHIER2 heuristics the authors resort to the heuristic of Teitz and Bart (1968), using the greedy heuristic of Cornuejols et al. (1977) for finding an initial solution. In this paper we solve each r-median problem in the same way.

The heuristic we present in this paper, which we denote as Directed Branching Heuristic (DBH), combines the generation of points, by means of a "directed" branching procedure, the selection of non-inferior points and the evaluation of their efficiencies. Specifically, we use the Free Disposal Hull method (FDH, see Deprins,

Simar and Tulkens (1984)) to characterize the *relative spatial efficiency* (RSE) of possible sitting decisions. The RSE of a sitting decision was first discussed by Fisher and Rushton (1979), who suggested the need for creating analytical techniques when the access is defined by means of several alternative criteria. This is exactly what Desai and Storbeck (1990) did when they proposed to resort to the DEA methodology, followed by Desai et al. (1995). This approach departs from the traditional multiobjective programming point of view in the sense that the most important information at the output stage resides not only in the non-inferior set of solutions but on differing degrees of inferiority given by the (radial) efficiency scores. Moreover, Athanassopoulos and Storbeck (1995) proposed to turn over to the non-convex version of DEA, known as FDH, which is more appropriate in case of discrete space alternatives.

Here we develop a FDH model of spatial efficiency for evaluating the different feasible solutions of the pq-median problem. These solutions are generated by means of the directed branching procedure. In order to evaluate the performance of heuristic DBH we compare it with the two most relevant heuristics designed by Serra and ReVelle and identified as SAPHIER1 and SAPHIER2. Different combinations of type A and type B facilities were tested over a 52-node network.

2. The Directed Branching Heuristic (DBH)

The DBH heuristic starts by defining an initial set, S, of 9 points (or feasible solutions of the pq-median problem). After that, a point of S is selected and the iterative directed branching procedure starts. This procedure generates at most 50 new points. A selection among them follows, keeping only the FDH-efficient points. The set S is updated in order to guarantee that it contains all the generated efficient points. A new iteration starts by selecting another point of S. In each iteration a new point of S is visited and set S is updated. The heuristic stops when all the points in S have been visited. The DBH-heuristic ends by analyzing the points of the updated set S by means of the FDH-methodology. In this way we are able to offer to the decision-maker not only several FDH-efficient points but also many other inefficient points, some of which are close to be efficient. We will now explain each of the steps of heuristic DBH, i.e., (1) how the initial set S is defined, (2) how the directed branching procedure works, and (3), how the final results are presented.

2.1 Definition of the 9 points of the initial set S

The first point of S is computed as follows:

1. Locate q type-B facilities using a q-median model.
2. Form districts by assigning each population area to its closest type-B facility.
3. Modify the distance matrix to reflect the districts just created, so that the distances between demand areas which are in different districts are set to a very large number.
4. Locate p+q type-A facilities with the new distance matrix using a p+q-median formulation where q facilities are already fixed (the type-B points).
5. The (pq-median) feasible solution just obtained is the first point of S. Store it together with the corresponding values of Z_A and Z_B.

The other 8 points are obtained in a sequential way by worsening, as least as possible, either the Z_A or Z_B value of the first point of S. The new points are obtained using the following two movements:

worB: Given a feasible solution of the pq-median problem, this movement worsens the value of objective B as least as possible, while maintaining the value of objective A.
worA: This is a movement that worsens the value of objective A while maintaining the value of objective B for a given feasible solution of the pq-median problem.

The diagram associated with the initial set S of 9 feasible solutions of the pq-median problem, obtained using the worA and worB movements, is given in Fig. 1.

2.2 The directed branching procedure

After the selection of a point of S, our procedure generates three levels of branching. At the first level, our procedure generates al most 7 new points, according to the strategy drawn in Fig. 1. Among them, only the last 2 points (or 'extreme' points) are relevant. These 2 points are numbered '5' and '8' and constitute the starting points of the branching procedure at the second level. Since in any level we apply always the same branching-strategy, 4 new 'extreme' points are generated at the second level. As could be expected, we select the 4 'extreme' points of the second level as starting points of the third level and generate, finally, at most 28 points at this level. We finish this three-level branching process by selecting the FDH-efficient points among the set of all the generated points. These are the points that must be incorporated to set S in order to restart the iterative procedure by selecting a non-visited point of S. The procedure stops when all the points of S have been visited.

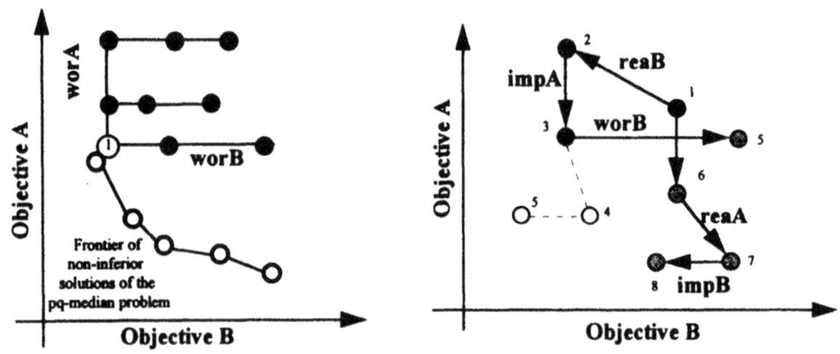

Fig. 1. The initial Set S and the branching strategy for DBH heuristic

Let us explain more in detail the branching strategy. Observe that we generate two branches by means of four different movements that have the following meaning:
impB: This is a movement that tries to improve the value of objective B while maintaining the value of objective A for a given feasible solution of the pq-median problem.
impA: This is a movement that tries to improve the value of objective A while maintaining the value of objective B for a given feasible solution. The step-by-step description is omitted because it is similar to the former one.
reaA: This is a movement that tries to improve the value of objective A while worsening the value of objective B for a given feasible solution.

reaB: This is a movement that tries to improve the value of objective B while worsening the value of objective A for a given feasible solution.

Observe that our branching strategy generates, for each starting point, at most 7 new points. If it happens that anyone of the seven movements has no effect, the corresponding branch is automatically cut and we add a further movement which is opposite to 'impB': worsen, as least as possible, the value of objective B while maintaining the type-A facilities and the objective value Z_A. This corresponds exactly to movement 'worB' defined previously. (See Figure 1.)

2.3 The presentation of the final results

After all the points of S have been visited, we finally evaluate the RSE of each of the points belonging to the updated set S by means of the FDH methodology (see Deprins et al. (1984)). In order to compute the radial inefficiency (given by the efficiency score θ) as well as the non-radial inefficiency (given by the value of the slack variables) of a given point (Z_A^0, Z_B^0) we must solve the following integer linear programming problem:

$$\begin{aligned}
&\text{Min } \theta_0 - \varepsilon(s_A + s_B) \\
&\text{s.t.} \\
&\sum_{j=1}^{k} Z_A^j \lambda_j + s_A = \theta_0 Z_A^0 \\
&\sum_{j=1}^{k} Z_B^j \lambda_j + s_B = \theta_0 Z_B^0 \\
&\sum_{j=1}^{k} \lambda_j = 1 \\
&\lambda_j \in (0,1) \quad j = 1,2,\ldots,k \quad s_A, s_B \geq 0,\ \theta \text{ free}
\end{aligned}$$

where the pair (Z_A^j, Z_B^j) are the values of the objective functions Z_A and Z_B for the jth point of the updated set S. Fortunately we do not have to solve the above problem as an integer programming problem because it is enough to perform vector comparisons, as explained in Tulkens (1993). For considering the effect of both types of inefficiencies we resort to the following 'global efficiency measure' (GEM, see Cooper and Pastor (1995)):

$$\Gamma_0 = \theta_0^* \left(1 - \frac{1}{2} \left[\frac{s_A^*}{\theta_0^* Z_A^0} + \frac{s_B^*}{\theta_0^* Z_B^0} \right] \right)$$

In this way we get the 'Pareto-efficient' points and we get, for the inefficient ones, a measure of their inefficiency. In fact a point is efficient if, and only if, $\Gamma_0 = 1$. Finally, the decision maker can select a solution considering not only efficient points but also inefficient ones which are close to be efficient.

3. Results and computational experience

The pq-median problem was solved on a test network using heuristics SAPHIER1, SAPHIER2 and DBH. The network had 52 nodes. Several different numbers of type A and type B facilities have been considered. Results and run times are reported. The routines and subroutines used for all heuristics have been coded in Fortran language, compiled with MICROSOFT FORTRAN COMPILER version 5.1 and solved with a Pentium-60.

Table 1 presents the 52-node network results obtained using the three heuristics. The table is set up so that all methods can be compared for each combination of p and q. The first column indicates the number of type A and type B facilities located. For each solution method, the computed values for both objectives are presented. With regard to SAPHIER1 and SAPHIER2 and for each set of facilities 11 weights were used to generate the trade-off between objectives. With regard to heuristic DBH we present only the efficient points selected by means of the FDH methodology. The first objective, Z_A, is the average distance per person (in miles) to facilities offering type A services. The second objective, Z_B, is the average distance per person to facilities offering type B services. In Fig. 2 we display only two cases. In any case, heuristic DBH outperforms the pair SAPHIER1-SAPHIER2.

Table 2 shows the computational time for the three heuristics. As the number of facilities increases, the heuristics need more time to find a solution. (This is because more exchanges are needed.)

Table 3 shows the complete results obtained by heuristic DBH for the case p=3, q=2. The final set S has 35 feasible solutions among which only 7 are efficient. Nevertheless Table 3 offers a classification of the whole set of points by means of the GEM measure. For instance, we can identify 6 non-efficient points with a GEM score over 0.99. That means that this points are close to be efficient and, therefore, they may be considered also as good solutions. Fig. 3 displays the relative position of the 35 solutions of Table 3.

4. Conclusions

In this paper we have developed a new heuristic for the pq-median problem using a directed branching procedure and the notion of FDH efficiency. From the point of view of the generation of non-inferior points our heuristic outperforms the pair SAPHIER1-SAPHIER2 and needs less computational time. From the point of view of the decision-maker we offer him a broader set of solutions among which to choose. This set of solutions includes non-inferior as well as inferior points. The latter ones are rated according to their global inefficiency.

Our paper suggests that multiobjective programming problems may be solved focusing not only on non-inferior points but also on inferior ones. The resort to a global efficiency measure similar to the one proposed here allows the rating of the different points according to their (global) efficiency and, consequently, offers a guide for selecting the most interesting solutions.

Table 1. Results for the pq-median model, 52-node network

N° of facility located	SAPHIER1		SAPHIER2		DBH	
p-q	Z_A	Z_B	Z_A	Z_B	Z_A	Z_B
3-2	34.51	44.50	27.36	44.50	27.36	44.50
	26.94	44.79	26.12	45.18	26.82	44.76
	25.32	45.98	25.42	65.33	26.12	45.18
	24.74	47.09			25.29	45.42
	24.59	48.33			24.72	46.59
					24.66	47.93
					24.59	48.33
4-2	25.48	44.50	24.21	44.50	24.21	44.50
	24.00	44.86	22.98	45.18	23.68	44.76
	23.29	45.28	21.88	50.66	22.48	45.16
	22.50	45.67	21.76	56.67	22.12	46.59
	22.52	45.62			21.78	46.74
	21.67	49.73			21.77	47.61
					21.75	47.72
					21.69	48.03
					21.67	48.46
5-2	22.41	44.50	21.79	44.50	21.79	44.50
	21.47	44.76	20.79	44.76	20.79	44.76
	19.88	45.61	20.44	45.18	20.44	45.18
	19.66	45.90	19.79	46.59	19.59	45.37
	19.59	45.37	19.43	60.93	19.52	47.28
	19.61	50.09			19.18	47.38
					19.07	47.72
6-2	19.18	44.96	19.86	44.50	19.86	44.50
	18.89	45.21	18.80	44.76	18.80	44.76
	18.47	45.56	18.25	46.59	17.97	45.34
	17.97	45.34	17.33	69.66	17.78	46.74
	17.37	47.72			17.37	47.72
	17.33	48.19			17.33	48.19

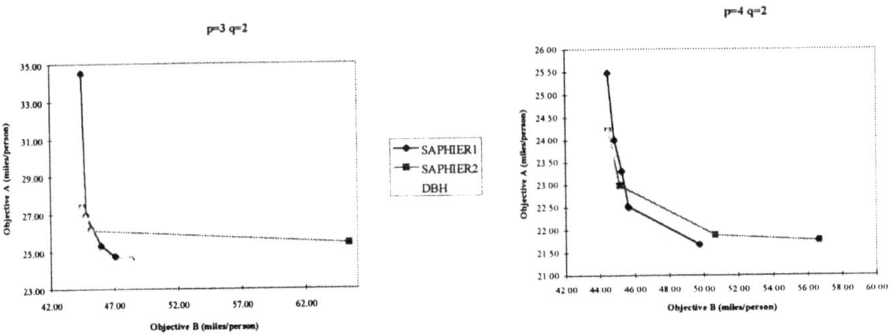

Fig. 2. Graphical display of the solutions of the three heuristics in two cases

Table 2. CPU time for Saphier1, Saphier2 and DBH. Number of visited points for DBH.

N° of facility located p-q	SAPHIER1 (CPU sec)	SAPHIER2 (CPU sec)	DBH (CPU sec)	Number of visited points
3-2	21	552	150	35
4-2	28	912	465	65
5-2	53	1515	438	46
6-2	55	2054	629	45

Table 3: Results of the FDH-analysis for the 3 type A, 2 type B facility case

Inputs		Efficiency	Slacks		GEM
Z_B	Z_A	θ	S_B	S_B	Γ
44.497	27.357	1.000	0.000	0.000	1.0000
54.038	27.357	0.899	0.253	0.000	0.8966
54.357	27.357	0.899	0.540	0.000	0.8940
44.497	27.363	1.000	0.000	0.006	0.9999
51.748	27.363	0.903	0.160	0.000	0.9018
54.234	27.363	0.899	0.419	0.000	0.8949
44.497	27.563	1.000	0.000	0.205	0.9963
54.357	27.563	0.892	0.176	0.000	0.8907
55.550	27.563	0.892	1.241	0.000	0.8811
47.806	24.999	0.989	0.682	0.000	0.9816
45.419	25.291	1.000	0.000	0.000	1.0000
44.759	26.825	1.000	0.000	0.000	1.0000
48.326	24.594	1.000	0.000	0.000	1.0000
47.380	24.975	0.990	0.308	0.000	0.9865
45.181	26.121	1.000	0.000	0.000	1.0000
47.683	25.353	0.977	0.000	0.051	0.9760
46.159	25.793	0.984	0.000	0.089	0.9823
45.454	26.323	0.994	0.000	0.044	0.9931
47.928	24.657	1.000	0.000	0.000	1.0000
46.587	24.719	1.000	0.000	0.000	1.0000
47.380	25.091	0.985	0.091	0.000	0.9842
45.284	26.640	0.998	0.000	0.458	0.9891
45.284	26.434	0.998	0.000	0.253	0.9929
48.594	24.931	0.989	0.132	0.000	0.9876
45.419	25.407	1.000	0.000	0.116	0.9977
47.090	24.744	0.999	0.456	0.000	0.9942
46.227	25.607	0.988	0.237	0.000	0.9851
46.707	25.531	0.991	0.847	0.000	0.9815
45.454	26.108	0.999	0.000	0.798	0.9840
47.683	25.358	0.977	0.000	0.057	0.9759
49.212	24.657	0.997	0.760	0.000	0.9897
47.276	25.320	0.985	0.000	0.233	0.9808
47.162	26.036	0.971	0.392	0.000	0.9672
47.153	26.146	0.967	0.192	0.000	0.9653
48.271	24.994	0.989	1.152	0.000	0.9771

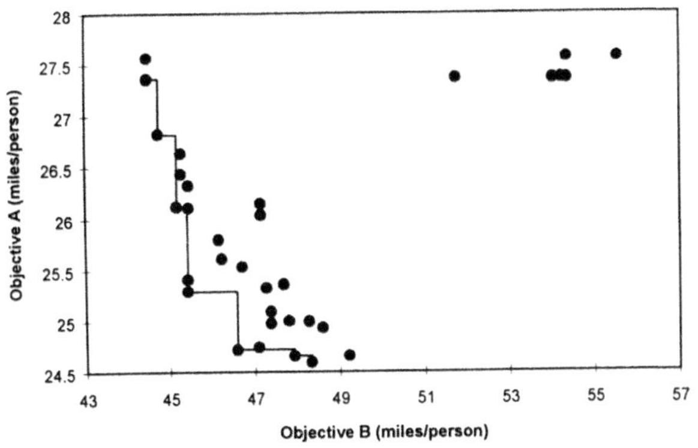

Fig. 3. Final Set of solutions of heuristic DBH with its FDH-frontier for the 3 type-A, 2 type-B facility case

5. References

Athanassopoulos, A. D. and Storbeck, J. E. (1995) Non-parametric models for spatial efficiency. *The Journal of Productivity Analysis*, **6**, 225-245.

Cooper, W. W. and Pastor, J. T. (1995) Global Efficiency Measurement in DEA. *Working Paper*, Universidad de Alicante, Spain.

Cornuejols, G.; Fisher, M.L. and Nemhauser, G.L. (1977) Location of bank accounts to optimize floats: an analytic study of exact & approximate algorithms. *Management Science*, **23**, 789-799.

Deprins, D.; Simar, L. and Tulkens, T. (1984) Measuring Labor-Efficiency in Post Offices, in Marchand, M; Pestiau, P. and Tulkens, T., eds., *The Performance of Public Enterprises: Concepts and Measurement*. North-Holland, Amsterdam.

Desai, A. and Storbeck, J. (1990) A data envelopment analysis for spatial efficiency. *Computers, Environment and Urban Systems*, **14**, 145-156.

Desai, A.; Haynes, K. and Storbeck, J. (1995) A spatial efficiency framework for the support of locational decisions. In a Charnes, A.; Cooper, W. W., Lewin, A. and L. Steiford (eds.) , *Data Envelopment Analysis: Theory, Methodology and Applications*, Kluwer Academic Publishers.

Fisher, M. and Rushton, G. (1979) Spatial efficiency of services locations and the regional development process. *Papers of the Regional Science Association*, **42**, 83-97.

Serra, D. and ReVelle, C. (1993) The pq-median problem, location and districting of hierarchical facilities. *Location Science*, **1**(4), 299-312.

Serra, D. and ReVelle, C. (1994) The pq-median problem: Location and districting of hierarchical facilities.-II. Heuristic solution methods. *Location Science*, **2**(2), 63-82.

Teitz, M. B. and Bart, P. (1968) Heuristics methods for estimating the generalized vertex median of a weighted graph. *Operations Research*, **16**, 955-965.

Tulkens, H. (1993) On FDH Efficiency Analysis: Some Methodological Issues and Applications to Retail Banking, Course, and Urban Transit. *Journal of Productivity Analysis*, **4** (1/2), 183-210.

Efficient Frontier Derived from Opinions and the Utility Optimization Problem

Enrique Ballestero[1]

[1] Universidad Politécnica de Madrid. E.T.S. Ingenieros Agrónomos, Ciudad Universitaria, 28040 Madrid, SPAIN

Abstract. When assessments and expectations on certain variables are performed by information from surveys, the consensus level of the corresponding opinions can be considered as a reliability index. We obtain mean-variance and goal-variance efficient frontiers by selecting activity mixes of maximum consensus. The utility optimum on the efficient frontier is approximated by resorting to a bounding theorem proven in the paper.

Keywords. Efficient frontier; Utility theory.

1. Introduction

Through the concept of efficiency, the analyst attempts to establish some comparison of certain outputs and its corresponding inputs (cost, risk, etc.). Up to now, efficiency analysis has raised two major problems. The first one is how to aggregate the various outputs (or inputs) by a system of weights or internal accounting prices. This weighting question can be solved subjectively (through preference weights) or objectively. An example of objective weighting is DEA (Charnes, Cooper and Rhodes, 1978).

The second main problem is to elicit efficient frontiers. When both aggregate input and output depend on the vector of activity levels $(y_1, y_2, \ldots, y_i \ldots, y_n)$ the efficient frontier is commonly obtained by minimizing the input subject to a parametric constraint related to the output level. Thus, in portfolio selection analysis, the variance of the portfolio returns $V(y_1, y_2, \ldots, y_i \ldots, y_n)$ is considered as an "input" to minimize under a parametric ouput restraint [such as expected returns $= E(y_1, y_2, \ldots, y_i \ldots, y_n) = E_0$] as well as $(y_1 + y_2 + \ldots + y_i + \ldots + y_n) = 1$ and $y_i \geq 0$ (for all i) where the y_i - vector is the portfolio composition (Markowitz, 1956). Nowadays this technique is useful even though it has been commented in the literature. See, e.g., Borch (1969), Feldstein (1969), Levy (1974).

Once the efficient frontier determined, another problem raises, the choice of the optimum utility vector $(\bar{y}_1, \bar{y}_2, \ldots, \bar{y}_i \ldots, \bar{y}_n)$ on the frontier. To avoid paradoxes related to the mean-variance efficient frontier, Kroll et al (1984) have resorted to the direct maximization of the utility function, however, this procedure is too cumbersome to be appllied. Even using the efficient frontier, the optimization process is a difficult step when the decision-maker's utility function is virtually unknown as often occurs. (See, e.g., Kallberg and Ziemba, 1983). An approximation solution has been given elsewhere. (Ballestero and Romero, 1991 and 1994).

This paper deals with one or more outputs which are estimated according to assessments and personal views obtained from surveys. As the consensus level amongts the different opinions in the survey is relevant to elicit reliable results, we introduce an index of disconsensus such as the variance of the opinions (if measurable) as the "input" to minimize for determining the efficient frontier. Moreover, we propose an optimization model (that we call Goal-Variance) which is usuful to derive the efficient frontier when there are two or more constraints with different parameters.

2. Mean-Variance and Goal-Variance Efficient Frontier from Opinions

We consider the following cases:

First case. [Mean-Variance (M-V) Efficient Frontier]. There is one output, and therefore, the survey only focuses on a single class of opinion. An example is opinions to predict the rerturn levels from the investments 1, 2, ..., i ... n for a given period of time. Let ξ_i be the return from the ith investment according to the different opinions. We denote by M_i the mean value of opinions ξ_i. The variances and covariances of opinions are denoted as V_i and V_{ih} respectively. Now, consider a portfolio (y_1, y_2, ..., y_i ..., y_n). Its return (estimated as an average opinion) will be the mean value:

$$M = \sum_{i=1}^{n} M_i \, y_i \tag{1}$$

On the other hand we have:

Consensus maximization = Min V =

$$= Min \; [\sum_{i=1}^{n} V_i \, y_i^2 + \sum_{i,h} V_{ih} \, y_i \, y_h] \tag{2}$$

subject to $M = M_0$ and $\sum_{i=1}^{n} y_i = 1$ with $y_i \geq 0 \; \forall \; i$ that is, the standard M-V parametric quadratic programming for efficient frontiers.

A M-V extension to several outputs is possible if the corresponding variables can be aggregated. Suppose two outputs, the first one being predictions about returns to receive at the end of the next α - year period, the second one analogous predictions related to the next β - year period. Thus, if ξ_{i1} and ξ_{i2} are the opinion variables predicting the ith-return for the α - year and β - year period respectively, we introduce the following aggregate variable:

$$\eta_i = (1 + r)^{-\alpha} \xi_{i1} + (1 + r)^{-\beta} \xi_{i2} \tag{3}$$

(where r is the discount rate) for reflecting opinions on the ith return.

Second case. (Goal-Variance Efficient Frontier) The survey focuses on two or more outputs (i.e., types of opinion) and there is not an aggregation criterion like (3). Suppose, e.g., the assessment of hospital units. A survey to obtain opinions from the patiens treated at these hospital units is carryed out. In this survey, the following outputs are investigated:

(i) Level of health care facilities at each hospital unit. These opinions are measured on a scale from 0 to 10.

(ii) Comfortability at each hospital unit (measured on a similar scale).

The problem is to obtain an assessment index for each hospital unit.

In general, we consider n units or activities (i = 1, 2, ..., n) and m outputs or types of opinion (j = 1, 2, ..., m). As noted, the purpose is to obtain an assessment index (between 0 and 1) for each unit or activity. The sum of these indexes should be equal to 1. Because of this condition, the assessment is competitive, i.e., high indexes for some units involve low indexes for another units. We introduce the goal-variance efficient frontier (G -V) model as follows.

First step. We determine the greatest M_j^* mean value of opinions attainable for each jth output.

Second step. The consensus maximization is given by the parametric quadratic programming (in matrix notation):

$$Min \sum_{j=1}^{m} w_j \; V_j \; (\xi_j' \; y)/(M_j^*)^2 \tag{4}$$

subject to

$$\bar{M} \; y + \eta = (1, 1, \dots 1)' \tag{5}$$

$$(1, 1, \dots 1) \; \eta = \lambda \tag{6}$$

$$(1, 1, \dots 1) \; y = 1 \tag{7}$$

$$y \geq 0 \quad \eta \geq 0$$

where:

ξ_j = vector of opinions for the jth output. The ith component of this vector (i = 1, 2, ... n) corresponds to opinions on the ith activity.

y = vector of activity levels. The ith component of this vector is interpreted as the assessment derived from the survey for the ith unit or activity.

V_j = variance of ξ_j' y. It is a well-known function of (y_1, y_2, ... y_i, ... y_n), its coefficients being variances V_i as well as covariances V_{ih}.

w_j = preference weight attached to the jth output in the objective function (4). Note this function is normalized as a sum of weighted ratios of the form variance/(mean)^2

M_{ji} = mean value of opinions for the jth output and the ith activity.

M_j^* = Max M_{ji} (over i) for each jth output.

$\bar{M}$ = matrix of the nomalized M_{ji}/M_j^* mean values.

η = vector of deviations from the normalized greatest M_j^*/M_j^* = 1 mean values.

λ = positive number used as parameter of the parametric quadratic programming.

Because of the formal similarity among constraint (5) and goal equations of goal programming, we have just introduce the term Goal-Variance for the above model.

Example 1. Consider 4 hospital units (activities $i = 1, 2, 3, 4$) as well as 2 outputs or types of opinion related to health care facilities ($j = 1$) and comfortability ($j = 2$). For simplicity, suppose all covariances equal to zero. Moreover, from the survey we obtain the following mean values and variances:

Health care facilities

$$\text{Mean} = M_1 = 2.1y_1 + 6.0y_2 + 1.4y_3 + 4.8y_4$$

$$\text{Maximum } M_{1i} \text{ (used as normalizer)} = M^*_1 = 6.0$$

$$\text{Variance} = V_1 = 0.90\, y^2_1 + 8.50y_2^2 + 0.46y_3^2 + 6.02y_4^2$$

Comfortability

$$\text{Mean} = M_2 = 3.4y_1 + 4.0y_2 + 5.3y_3 + 2.8y_4$$
$$\text{Maximum } M_{2i} \text{ (used as normalizer)} = M_2^* = 5.3$$
$$\text{Variance} = V_2 = 2.6y_1^2 + 4.1y_2^2 + 7.2y_3^2 + 1.8y_4^2$$

Weights in the objective function

$w_1 = 0.75$ (Health care facilities) and $w_2 = 0.25$ (comfortability)

Hence, the model (4) - (7) becomes:

$$\text{Min } 0.75\,[0.90y_1^2 + 8.50y_2^2 + 0.46y_3^2 + 6.02y_4^2]/(6.0)^2 + 0.25\,[2.6y_1^2 + 4.1y_2^2 + 7.2y_3^2 + 1.8y_4^2]/(5.3)^2$$

$$[2.1y_1 + 6.0y_2 + 1.4y_3 + 4.8y_4]/6.0 + \eta_1 = 1$$

$$[3.4y_1 + 4.0y_2 + 5.3y_3 + 2.8y_4]/5.3 + \eta_2 = 1$$

$$\eta_1 + \eta_2 = \lambda$$

$$y_1 + y_2 + y_3 + y_4 = 1$$

$$y_i \geq 0 \;\forall_i \quad \eta_1 \geq 0 \quad \eta_2 \geq 0$$

By solving the above parametric quadratic programming we obtain the efficient frontier in the table below.

λ	y_1	y_2	y_3	y_4	η_1	η_2	OFV
0.3	0	0.8716814	0	0.1283186	0.02566372	0.2743363	0.1646078
0.4	0	0.6616	0.109701	0.2286989	0.1298439	0.2701561	0.1017682
0.5	0	0.4660329	0.2845715	0.2493956	0.2680506	0.2319494	0.06114765
0.6	0.1030298	0.3249052	0.3384986	0.2335664	0.3731982	0.2268018	0.03914641
0.7	0.2761691	0.2208284	0.3101531	0.1928494	0.4558638	0.2441362	0.02595602
0.8	0.4493062	0.1167482	0.2818002	0.1521454	0.5385249	0.2614751	0.0249129
0.9	0.6224432	0.0126681	0.2534474	0.1114412	0.621186	0.2788139	0.02275223

OFV = Objective function values

3. Optimization on the Efficient Frontier

Consider an efficient frontier t $(x_1, x_2) = 0$ derived from opinions on a single output, where x_1 and x_2 represent mean and variance, respectively. Moreover, consider a utility function $u = u(x_1, x_2)$. Make the change:

$$\theta_1 = \frac{x_1 - x_{1*}}{x_1^* - x_{1*}} \quad and \quad \theta_2 = \frac{x_{2*} - x_2}{x_{2*} - x_2^*} \tag{8}$$

where:

x_1^* and x_{1*} are the ideal and the anti-ideal values for mean, respectively.

x_2^* and x_{2*} are the ideal and the anti-ideal values for variance, respectively. As variance is a measure of disconsensus, "less is better" for x_2. Then, we have $x_1^* < x_{2*}$.

Indices Θ_1 and Θ_2 are meaningful as measuring the mean and the consensus level, respectively. Note that $0 \leq \Theta_1 \leq 1$ and $0 \leq \Theta_2 \leq 1$. After the change (8) the above functions are normalized and become $T(\Theta_1, \Theta_2) = 0$ and $U = U(\Theta_1, \Theta_2)$. On the normalized frontier T we obviously have $\Theta_1 = 1$ for $\Theta_2 = 0$ and $\Theta_2 = 1$ for $\Theta_1 = 0$. Our information on the utility function U is very incomplete. In this context, we assume: *First*, U holds the law of diminishing marginal rate of substitution (MRS). Moreover its MRS has a constant value r_0 along the path $\Theta_1 = \Theta_2$. *Secondly*, the Lagrangean optimum for the linear utility $V = (r_0 \Theta_1 + \Theta_2)$ does exist on the efficient frontier, that is, on the space $\Theta_1 \geq 0$ and $\Theta_2 \geq 0$. Indeed these assumptions (the only ones we need in this paper) seem to be plausible, given our context of virtual ignorance. *Justification* (i) To assume the constant value of the MRS only along the path $\Theta_1 = \Theta_2$ does not substantially alter the utility map structure. Large families of functions satisfy the first assumption, and a good approximation to any empirical utility map can likely be achieved by selecting an appropriate function of those families. (ii) Even a constant value $r_0 \approx 1$ could be justified in accordance with the MRS law for average decision-makers (See Ballestero and Romero, 1994) (iii) Cobb-Douglas' utility functions require a much more restrictive additional hypothesis (the constant value of the MRS along *every* path $\Theta_2 = \lambda \Theta_1$) than our first assumption. Moreover, in Cobb-Douglas' utility the MRS must be proportional to (Θ_2/Θ_1). (iv) The second assumption, although slightly restrictive, is commonly implicitly accepted in models found in the literature. Now, we can prove the following:

Theorem I. The Lagrangean utility optimum of U on the frontier T = 0 is bounded between the points B and L defined as follows:

B is the intersection of $\Theta_1 = \Theta_2$ with the frontier curve $T(\Theta_1, \Theta_2) = 0$. Therefore, this point is the normalized $L\infty$ bound of the compromise set (without attaching preference weights).

L is the Lagrangean utility optimum for the linear utility $V = (r_0 \Theta_1 + \Theta_2)$, that is, the solution to Max $(r_0 \Theta_1 + \Theta_2)$ subject to the frontier $T(\Theta_1, \Theta_2) = 0$, and $\Theta_1 \geq 0$, $\Theta_2 \geq 0$. If $r_0 \neq 1$ the bound L differs from the L_1 bound of the compromise set (when L_1 is obtained without attaching preference weights).

(See proof in Appendix I)

4. Concluding Remark

Model (4)-(7) seems to be a suitable procedure to determine the efficient frontier when there are two or more outputs (i.e., types of opinion). Indeed, the parametric programming is reduced to a single parameter through equations (5) and (6). On the other hand, Theorem I allows for bounding the utility optimum with a limited information on the decision-maker's preferences. This information is reduced to parameter r_o which can be estimated by interactive techniques as a meaningful utility index.

APPENDIX I

Proving Theorem I. We use the notation T_1, T_2 and U_1, U_2 for partial derivatives of T and U respectively. According to the second assumption, only the two following cases can occur: (i) Point L lies between C (0, 1) and the Lagrangean utility optimum $M\ (\bar{\theta}_1, \bar{\theta}_2)$ of U on the frontier; (ii) Point L lies between M and A (1, 0). *Case (i).* In absolute values, the slope T_1/T_2 of the tangent at point C is T_1 (0,1)/T_2 (0,1) $< r_0$ (that is, the slope of the tangent at point L) as the slope at point C is the lowest one (in absolute value) on the frontier. Therefore we have T_1 (0,1) - r_0 T_2 (0,1) $<$ 0. Then the rule of signs involves $T_1\,(\bar{\theta}_1,\,\bar{\theta}_2) - r_0\,T_2\,(\bar{\theta}_1,\,\bar{\theta}_2) > 0$. Hence, taking into account the Lagrangean optimum condition:

$$U_1\,(\bar{\theta}_1,\,\bar{\theta}_2)/U_2\,(\bar{\theta}_1,\,\bar{\theta}_2) = T_1\,(\bar{\theta}_1,\bar{\theta}_2)/T_2\,(\bar{\theta}_1,\,\bar{\theta}_2) \tag{9}$$

we obtain $U_1\,(\bar{\theta}_1,\,\bar{\theta}_2)/U_2\,(\bar{\theta}_1,\,\bar{\theta}_2) > r_0$. From this relationship, according to the MRS law and the first assumption we have $\bar{\theta}_1 - \bar{\theta}_2 < 0$. By introducing the point A (1,0) into (Θ_1 - Θ_2) the result is 1 $>$ 0. Hence, the rule of signs implies that B lies between M and A. Therefore, M lies between L and B. *Case (ii)*. An analogous reasoning leads to the same result. Thus, Theorem I is demonstrated.

References

Ballestero, E. and Romero, C., 1991 "A theorem connecting utility function optimization and compromise programming", *Operations Research Letters*, 10, 421-427.

Ballestero, E. and Romero, C., 1994 "Utility optimization when the utility function is virtually unknown", *Theory and Decisions*, 37, 233-243.

Borch, K., 1969 "A note of uncertainty and indifference curves", *Review of Economic Studies*, 36, 1-4.

Charnes, A., Cooper, W.W. and Rhodes, E., 1978 "Measuring the efficiency of decision making units", *European Journal of Operational Research*, 2/6, 429-444.

Coleman, J.S., 1990 in R. Swedberg (Ed.) *Economics and Sociology* (Princeton University Press, New Jersey), 47-60.

Feldstein, M.S., 1969 "Mean variance analysis in the theory of liquidity preference and portfolio selection", *Review of Economic Studies*, 36, 5-12.

Kroll, Y., Levy, H. and Markowitz, H.M., 1984 "Mean-Variance versus direct utility maximization", *The journal of finance*, 39, 1, 47-60.

Levy, H., 1974 "The rationale of the mean-standard deviation analysis: Comment", *The American Economic Review*, 64, 3, 434-441.

Markowitz, H.M., 1956, "The optimization of a quadratic function subject to linear constraints", *Naval Research Logistics Quarterly*, 3, 111-133.

Finding Efficient Points in Multiobjective Quadratic Programming with Strictly Convex Objective Functions

Beato-Moreno A.[1], Luque-Calvo, P.[1], Osuna-Gómez, R.[1], Rufián-Lizana, A.[1]

[1] Departamento de Estadística e Investigación Operativa. Facultad de Matemáticas. Universidad de Sevilla, España.

Abstract. In this work we consider the unconstrained multiobjective quadratic problem with strictly convex objective functions, (PCM - D). Firstly we expose a technique to determine the equations of the efficient points supposing that there are only two objective functions. This method is based on results on quadratic forms withdrawals by Gantmacher. Secondly, the technique explained is extended to calculate the equations of the efficient points of the general problem (PCM - D).

1 The problem

In this work we study the multiobjective programming problem:

$$\textbf{(PM)} \qquad \begin{array}{ll} \min & (f_1(x), \ldots, f_m(x)) \\ s.t: & x \in S. \end{array}$$

where $f_i(x) : \mathbb{R}^n \longrightarrow \mathbb{R}$ is the i-th objective function, $i = 1, \ldots, m$ and $S \subseteq \mathbb{R}^n$ is the set of feasible points.

One of the problems that more attention has received in the literature is the problem (PM) where all the objective functions are linear and the set S comes defined by linear constraints, the multiobjective linear problem. The reason of such attention, besides their its applicability, is the consideration as one of the multiobjective problems more simplified where it is possible to obtain easytly the solutions.

The multiobjective linear problem is generalized when instead of linear functions we consider quadratic functions, that is,

$$\textbf{(PCM)} \qquad \begin{array}{ll} \min & (f_1(x), \ldots, f_m(x)) \\ s.t: & x \in S. \end{array}$$

where $f_i(x) = 0.5\, x^t A_i x + b_i^t x$, $A_i \in \mathcal{M}_{n \times n}$, $b_i \in \mathbb{R}^n$, $i = 1, \ldots, m$.

The objective of this work is to help to solve the unconstrained multiobjective quadratic problem, in the one which $S = \mathbb{R}^n$. Moreover, we will impose that the objective functions will be strictly convex, or rather, A_i must be a positive definite matrix, $\forall i = 1, \ldots, m$. We will call to this problem $(PCM - D)$.

The problems of the type $(PCM - D)$ appear in several fields of the matemathics. For example, the problems of localization with ellipsoidal norms are problems of multiobjective quadratic programming with strictly convex objective functions. The problems of estimation of parameters in statistical inference in those which is used the method of minimum squares are in its origin problems of multiobjective quadratic programming with strictly convex objective functions.

2 Caracterization of efficient points

In the multiobjective problem, the concept of solution is defined in wheel to the efficiency concept.

DEFINITION 2.1 *Given $x_0 \in S$, we will say that it is an efficient point for the problem (PM) if it does not exist another point $x \in S$ such that $f_i(x) \leq f_i(x_0)$, $\forall i = 1, \ldots, m$ with at least a strict inequality.*

DEFINITION 2.2 *We will call $E(PCM - D)$ to the set of efficient points for the problem $(PCM - D)$.*

DEFINITION 2.3 *Given the problem (PM) and $\lambda = (\lambda_1, \ldots, \lambda_m) \in \mathbb{R}^m$, $\lambda_i \geq 0$, $\forall i = 1, \ldots, m$, $\lambda \neq 0$, we will call weighted problem of (PM) associated with λ to*

$$(\mathbf{P}_\lambda) \qquad \begin{array}{l} \min \quad \lambda_1 f_1(x) + \ldots + \lambda_m f_m(x). \\ x \in \mathbb{R}^n \end{array}$$

Respect to the problem $(PCM - D)$ the caracterization of the efficient points is derived directly from the results 4.4 and 4.5 of [2].

THEOREM 2.1 *Given the problem $(PCM - D)$, $x_0 \in S$ is an efficient point if and only if it is unique solution of a weighted problem P_λ.*

3 Biobjective quadratic problem

To know the equation that verify the efficient points of a multiobjective quadratic problem with positive definite matrices and a number anyone m of objective functions we will begin studying that equations that verify the efficient points in the case of two objective functions, that is, m=2.

We will call $(PCM-D2)$ to the problem

$$(\mathbf{PCM-D2}) \qquad \min_{x \in \mathbb{R}^n} \quad (f_1(x), f_2(x))$$

where $f_i(x) = 0.5\, x^t A_i x + b_i^t x$, $i = 1, 2$, and each matrix A_i is positive definite, $i = 1, 2$. Consequently, it can be expressed$f_1(x) = 0.5\,(x-a_1)^t A_1(x-a_1)+c_1$ and $f_2(x) = 0.5\,(x-a_2)^t A_2(x-a_2)+c_2$, where a_1 minimizes $f_1(x)$ and a_2 minimizes $f_2(x)$. Also, we will suppose that $c_1 = c_2 = 0$, something which does not affect to the determination of the set $E(PCM-D2)$.

LEMMA 3.1 *The set $E(PCM-D2)$ is an arc that joins a_1 with a_2 and that is found within of the ellipsoids $E_1 = \{x \in \mathbb{R}^n \;/\; f_1(x) \leq f_1(a_2)\}$ and $E_2 = \{x \in \mathbb{R}^n \;/\; f_2(x) \leq f_2(a_1)\}$*

PROOF. It is a curve because the equations that they should verify the efficient points depends on a parameter $\lambda \geq 0$ and for each parameter exists only a solution.

This curve contains to a_1 because a_1 is the unique point that minimizes $f_1(x)$ in all $\mathbb{R}^n$ and therefore it is efficient. What is same occurs for a_2 and $f_2(x)$.

The points a_1 and a_2 are related by efficient point arcs because the functions are strongly concaves (see [5]).

Finally if $x \in \mathbb{R}^n \setminus E_1$, $f_1(x) > f_1(a_2)$. Also it is verified that $f_2(x) > f_2(a_2)$. Then x is not efficient. An analogue reasoning is continued if $x \in \mathbb{R}^n \setminus E_2$. □

Given two quadratic functions $f_1(x) = 0.5\,(x-a_1)^t A_1(x-a_1)$ and $f_2(x) = 0.5\,(x-a_2)^t A_2(x-a_2)$ with positive definite matrices, we can write $f_2(x) = 0.5\,(x-a_1)^t A_2(x-a_1)+(a_2-a_1)A_2(x-a_1)+0.5\,(a_1-a_2)^t A_2(a_1-a_2)$.

If we obtain a base about which the quadratic forms $x^t A_1 x$ and $x^t A_2 x$ will be diagonal at the same time, then, with respect to this base are $f_1(x) = 0.5\,(x-a_1)^t D_1(x-a_1)$ and $f_2(x) = 0.5\,(x-a_1)^t D_2(x-a_1) + b^t(x-a_1) + cte$.

For this we will use the definition of regular pencil of two quadratic and extracted results about this of the work of Gantmacher, [3]. To next, we expose some results on $E(PCM-D2)$.

THEOREM 3.1 *There exist a reference system for which $E(PCM-D2)$ verifies the following equations, for $\lambda \geq 0$:*

$$x_i^0 = \frac{1}{\lambda + \lambda_i}\, b_i, \qquad i = 1, \ldots, n,$$

with $\lambda_i \in \mathbb{R}$, $\lambda_i \neq 0$ *fixed and known,* $i = 1, \ldots, n$.

Moreover the efficient points verify: $0 \leq (x_i^0)^2 \leq (b_i/\lambda_i)^2$, $i = 1, \ldots, n$.

PROOF. In §6 of Chapter X of [3] is concluded that given two quadratic functions whose matrices are positive definites, there exists a reference system respect to which $f_1(y) = 0.5 \sum_{k=1}^n y_k^2$ and $f_2(y) = 0.5 \sum_{k=1}^n \lambda_k y_k^2 - b^t y$, $\lambda_k > 0$, $k = 1, \ldots, n$.

Therefore for to determine the geometry of the efficient points we can to refer us to the problem

$$\min_{x \in \mathbb{R}^n} \quad f(x) = (0.5 \textstyle\sum_{i=1}^n x_i^2,\ 0.5 \sum_{k=1}^n \lambda_k x_k^2 - b^t x).$$

The condition so that a point x^0 will be efficient is that there exist $\mu_1, \mu_2 \geq 0$ such that $\mu_1(Id\ x^0) + \mu_2(Dx^0 - b) = 0$, where $D = diag\ (\lambda_1, \ldots, \lambda_n)$, $\lambda_i > 0$, $i = 1, \ldots, n$ are the eigenvalues of the regular pencil $x^t A_1 x - \lambda x^t A_2 x$.

If $\mu_2 = 0$, it is evident that $x^0 = 0$. If $\mu_2 > 0$ we divide by μ_2 and we obtain that $(\mu_1/\mu_2) Id\ x^0 + Dx^0 - b = 0$. Let $\lambda = (\mu_1/\mu_2)$. Then

$$(\lambda Id + D)x^0 = b \quad \Longrightarrow x^0 = (\lambda Id + D)^{-1} b \quad \Longrightarrow x^0 = Eb,$$

where $E = \text{diag}\,(1/(\lambda+\lambda_1), \ldots, 1/(\lambda+\lambda_n))$, and $\lambda_1, \ldots, \lambda_n$ are the eigenvalues of the pencil $x^t A_1 x - \lambda x^t A_2 x$ that they are positive by the theorem 10 in §7 of chapter X of [3]. Also, as $\lambda \geq 0$, it is deduced that the denominator that appears in each equation is not zero for any value of λ.

Of the foregoing we conclude that for each component is verified, for $\lambda \geq 0$:

$$x_i^0 = \frac{1}{\lambda + \lambda_i}\, b_i \qquad i = 1, \ldots, n.$$

Of these equations are detached certain conditions on x_i^0:

- If $b_i = 0$, then $x_i^0 = 0$, $\forall \lambda \geq 0$. Then, evidently, $0 \leq (x_i^0)^2 \leq (b_i/\lambda)^2$.

- If $b_i > 0$, we obtain that $(\lambda + \lambda_i)x_i^0 = b_i$, of where is deduced that $(\lambda + \lambda_i) < 0$ and $x_i^0 < 0$, or $(\lambda + \lambda_i) > 0$ and $x_i^0 > 0$. The first is impossible since $\lambda \geq 0$ and $\lambda_i > 0$. Consequently, $x_i^0 > 0$ and

$$(\lambda + \lambda_i) = \frac{b_i}{x_i^0} \Longrightarrow \lambda = \frac{b_i}{x_i^0} - \lambda_i.$$

Imposing that $\lambda \geq 0$, we arrive to the fact that

$$\frac{b_i}{x_i^0} - \lambda_i \geq 0 \Longrightarrow \frac{b_i}{x_i^0} \geq \lambda_i \Longrightarrow b_i \geq \lambda_i x_i^0 \Longrightarrow x_i^0 \leq \frac{b_i}{\lambda_i}.$$

therefore is fulfilled that $0 \leq (x_i^0)^2 \leq (b_i/\lambda_i)^2$.

- If $b_i < 0$, by an analogous reasoning we arrive to $0 \leq (x_i^0)^2 \leq (b_i/\lambda_i)^2$.

With this finish we also prove that the set of efficient points is bounded.□

THEOREM 3.2 *The set $E(PCM - D2)$ form a finite arc of hyperbola whose extreme are the points that minimize each one of the objective functions, in the sense of the fact that the projection of $E(PCM - D2)$ to each par of certain system variables of reference comes given by the equation of an arc of hyperbola in $\mathbb{R}^2$.*

PROOF. As of the equations respect of the reference system to which alludes the theorem 3.1, in the case those which $x_i^0 \neq 0$ ($b_i^0 \neq 0$), we arrive to the following equations:

$$\lambda = (b_1/x_i^0) - \lambda_i,$$

where $\lambda \geq 0$. Consequently, the set of efficient points verifies, for $b_i \neq 0$:

$$(b_1/x_1^0) - \lambda_1 = \ldots = (b_n/x_n^0) - \lambda_n,$$

together with the condition of the fact that $0 \leq (x_i^0)^2 \leq (b_i/\lambda_i)^2$.

Alone lack to prove that it is the intersection of $n-1$ hyperbolics surfaces in $\mathbb{R}^n$ in the sense that is described in the stated. For each equation of the n-1 that they describe the efficient points emerges one of them.

We will catch for example the first equation. Supposing that $b_1 \neq 0$ and $b_2 \neq 0$, we obtain that $(b_1/x_1^0) - \lambda_1 = (b_2/x_2^0) - \lambda_2$, together with the condition $0 \leq (x_i^0)^2 \leq (b_i/\lambda_i)^2$, $i = 1, 2$. Developing:

$$(b_1 - \lambda_1 x_1^0)/x_1^0 = (b_2 - \lambda_2 x_2^0)/x_2^0 \Rightarrow (b_1 - \lambda_1 x_1^0)(x_2^0) = (b_2 - \lambda_2 x_2^0)(x_1^0) \Rightarrow$$
$$(b_1 x_2^0 - \lambda_1 x_1^0 x_2^0) = (b_2 x_1^0 - \lambda_2 x_2^0 x_1^0) \Longrightarrow (\lambda_2 - \lambda_1) x_1^0 x_2^0 + b_1 x_2^0 - b_2 x_1^0 = 0.$$

That it is the equation corresponding to the intersection of a quadratic form with the compact defined by the equations $0 \leq (x_i^0)^2 \leq (b_i/\lambda_i)^2$, $i = 1, 2$. The corresponding matrix to that quadratic form is:

$$\begin{pmatrix} 0 & (\lambda_2 - \lambda_1)/2 & 0 & \ldots & 0 \\ (\lambda_2 - \lambda_1)/2 & 0 & 0 & \ldots & 0 \\ \vdots & \vdots & \vdots & \ddots & \vdots \\ 0 & 0 & 0 & \ldots & 0 \end{pmatrix}$$

If $x_3, \ldots, x_n$ are fixed the quadratic part, in the variables $x_1$and x_2, is:

$$H = \begin{pmatrix} 0 & (\lambda_2 - \lambda_1)/2 \\ (\lambda_2 - \lambda_1)/2 & 0 \end{pmatrix},$$

of where $|H| = -((\lambda_2 - \lambda_1)/2)^2$. and therefore it is an hyperbola in $\mathbb{R}^2$, if $\lambda_1 \neq \lambda_2$. But correctly, it is the equation corresponding to the intersection of a quadratic form with a compact ($0 \leq (x_i^0)^2 \leq (b_i/\lambda_i)^2$, $i = 1, 2$).

If $\lambda_1 = \lambda_2$, is obtained a segment that part from $x^0 = 0$. Also when any $b_i = 0$, the crossing of variable two produces a segment.

In conclusion, each one of the $n-1$ previous equations is of type hyperbolic in the sense that we expose in the stated of the theorem. □

3.1 Algorithm to calculate the efficient points

In this paragraph we propose an algorithm that permit us to calculate in a effective way the set of efficient points by means of the determination of their implicites equations for the biobjective problem $(PCM - D2)$. We summarize the algorithm that we propose as follows:

1. Decompose the matrix A_i as $A_i = P_i^{-1} D_i P_i$, $i = 1, 2$, where D_i, $i = 1, 2$ are diagonal matrices whose diagonal elements are the eigenvalues of A_i, $i = 1, 2$. Build the matrices $P = P_2 P_1^{-1} D_1^{-1/2}$, $A = P^t D_2 P$ and the vector given by $b = D_1^{1/2} P_1 (a_2 - a_1)$.

2. Decompose $A = Q^{-1} D Q$, where D is a diagonal matrix whose diagonal elements are the eigenvalues of A, λ_i, $i = 1, \ldots, m$. It will be $c = DQb$.

3. The implicit equations of the efficient points come given by

$$\lambda_1 - \frac{(DQD_1^{1/2} P_1 (a_2 - a_1))_1}{(QD_1^{1/2} P_1 (x - a_1))_1} = \ldots = \lambda_n - \frac{(DQD_1^{1/2} P_1 (a_2 - a_1))_n}{(QD_1^{1/2} P_1 (x - a_1))_n}$$

   if $c_i \neq 0$, and $x_i = (a_1)_i$ if $c_i = 0$. Also we should to add the conditions

$$(QD_1^{1/2} P_1 (x - a_1))_i \cdot (DQD_1^{1/2} P_1 (a_2 - a_1))_i \geq 0.$$

$$0 \leq (QD_1^{1/2} P_1 (x - a_1)_i)^2 \leq \left(\frac{(DQD_1^{1/2} P_1 (a_2 - a_1))_i}{\lambda_i} \right)^2.$$

THEOREM 3.3 *The previous algorithm provides the implicit equations of the set $E(PCM - D2)$.*

PROOF. We know that the efficient points for the biobjective problem $(PCM - D2)$ are those which verify $\lambda A_1 (x - a_1) + A_2 (x - a_2) = 0$ for $\lambda \geq 0$, and the point a_1.

As A_1 is a real and symmetric matrix, there exists a diagonal matrix D_1 whose diagonal elements are the eigenvalues of A_1 and a orthogonal matrix P_1 compound by the respective eigenvectors so that $A_1 = P_1^{-1} D_1 P_1$. Analogously, there exist D_2 and P_2 of the same form such that $A_2 = P_2^{-1} D_2 P_2$. Therefore, the efficient points verify for $\lambda \geq 0$:

$$\lambda P_1^{-1} D_1 P_1 (x - a_1) + P_2^{-1} D_2 P_2 (x - a_2) = 0. \tag{1}$$

As D_1 is a positive definite matrix, there exists $D_1^{1/2}$ such that $D_1 = D_1^{1/2} D_1^{1/2}$, where

$$D_1^{1/2} = diag\,((\lambda_1^1)^{1/2}, \ldots, (\lambda_n^1)^{1/2})$$

and $\lambda_1^1, \ldots, \lambda_n^1 > 0$ are the eigenvectors of A_1.

It will be $y = D_1^{1/2} P_1(x - a_1)$, $b = D_1^{1/2} P_1(a_2 - a_1)$. Then, of (1) is deduced that

$$\begin{aligned} &\lambda P_1^{-1} D_1^{1/2} y \ + \ P_2^{-1} D_2 P_2 P_1^{-1} D_1^{-1/2}(y - b) = 0 \Longrightarrow \\ &\lambda Id \ y \ + \ D_1^{-1/2} P_1 P_2^{-1} D_2 P_2 P_1^{-1} D_1^{-1/2}(y - b) = 0. \end{aligned} \tag{2}$$

Let $P = P_2 P_1^{-1} D_1^{-1/2}$. Then (2) can be written as

$$\lambda y + P^t D_2 P(y - b) = 0 \quad \Longrightarrow \quad (\lambda Id + P^t D_2 P) y = P^t D_2 P b. \tag{3}$$

The matrix $P^t D_2 P$ is real and symmetric, since $(P^t D_2 P)^t = P^t D_2 P$, and their eigenvalues are positives, since it is easy to prove that it is positive definite. Then it is possible to obtain a decomposition of the form $P^t D_2 P = Q^t D Q$ where $Q^{-1} = Q^t$. Then of (3)

$$\begin{aligned} &(\lambda Id + Q^{-1} D Q) y = Q^{-1} D Q b \ \Rightarrow \ (\lambda Q^{-1} Q + Q^{-1} D Q) y = Q^{-1} D Q b \Rightarrow \\ &\Rightarrow \ Q^{-1}(\lambda Id + D) Q y = Q^{-1} D Q b \ \Rightarrow \ (\lambda Id + D) Q y = D Q b. \end{aligned} \tag{4}$$

Accomplishing the change $z = Qy$, $c = DQb$, (4) remain $(\lambda Id + D) z = c$.

If $c_i = 0$, then $z_i = 0, \quad \forall \lambda \geq 0$.

If $c_i \neq 0$, as the eigenvalues of D are positive, $\lambda + \lambda_i > 0, \quad \forall \lambda > 0$, where $\lambda_i, \quad i = 1, \ldots, m$ are the eigenvalues of the matrix D. Therefore

$$z = (\lambda Id + D)^{-1} c = Ec,$$

where $E = \ diag \ (1/(\lambda + \lambda_1), \ldots, 1/(\lambda + \lambda_n))$, of where

$$z_i = (1/(\lambda + \lambda_i)) c_i.$$

As $\lambda = (c_i / z_i^0) - \lambda_i \geq 0$ we deduced that

$$z_i \cdot c_i \geq 0, \qquad z_i^2 \leq (c_i / \lambda_i)^2$$

Rescinding the changes, $z = Qy$, $c = DQb$, $y = D_1^{1/2} P_1(x - a_1)$, $b = D_1^{1/2} P_1(a_2 - a_1)$, we obtain that

$$\begin{aligned} &(Q D_1^{1/2} P_1(x - a_1))_i = (1/\lambda + \lambda_i)(D Q D_1^{1/2} P_1(a_2 - a_1))_i \\ &(Q D_1^{1/2} P_1(x - a_1))_i \cdot (D Q D_1^{1/2} P_1(a_2 - a_1))_i \geq 0. \\ &0 \leq ((Q D_1^{1/2} P_1(x - a_1))_i)^2 \leq ((D Q D_1^{1/2} P_1(a_2 - a_1))_i / \lambda_i)^2. \end{aligned}$$

These equations are given in parametric form. For each value of the parameter $\lambda \geq 0$, results an efficient point. To obtain the implicit form, we must make a distinction. If $c_i = 0$, then the equation in implicit form corresponding is $z_i = 0$ ($x_i = (a_1)_i$), which is deduced of the second condition, for example. If $c_i \neq 0$, the implicit equations are

$$\lambda_1 - \frac{(D Q D_1^{1/2} P_1(a_1 - a_1 2))_1}{(Q D_1^{1/2} P_1(x - a_1))_1} = \ldots = \lambda_n - \frac{(D Q D_1^{1/2} P_1(a_2 - a_1))_n}{(Q D_1^{1/2} P_1(x - a_1))_n},$$

where in those equalities are eliminated the corresponding to components from $c_i = 0$, and we add the conditions

$$(QD_1^{1/2}P_1(x-a_1))_i \cdot (DQD_1^{1/2}P_1(a_2-a_1))_i \geq 0.$$

$$0 \leq (QD_1^{1/2}P_1(x-a_1)_i)^2 \leq \left(\frac{(DQD_1^{1/2}P_1(a_2-a_1))_i}{\lambda_i}\right)^2.$$

And we arrive to the conclusion that the proposed algorithm provides the set of efficient points for the problem $(PCM - D2)$. □

4 The general PCM-D problem

To know the behavior of the efficient points of a multiobjective quadratic problem with positive definite matrices, we propose an iterate method that relates the efficient points for m functions to the efficient points of the problem with $m-1$ functions. As consequence of this method we can conclude results on the form of conexion of the set of efficient points.

We consider the quadratic problem with positive definite matrices $(PCM - D)$. The efficient points for this problem are the solutions for the scalar problems

$$\min_{x \in \mathbb{R}^n} \quad \lambda_1 f_1(x) + \ldots + \lambda_m f_m(x)$$

varying $\lambda_1, \ldots, \lambda_m \geq 0$. We will suppose established that the efficient points for a quadratic problem with $m - 1$ functions and with positive definite matrices are constituted by surfaces that it is the union of arcs.

If $\lambda_m = 0$, the problems that appear are

$$\min_{x \in \mathbb{R}^n} \quad \lambda_1 f_1(x) + \ldots + \lambda_{m-1} f_{m-1}(x))$$

where $\lambda_1, \ldots, \lambda_{m-1} \geq 0$, and the solutions of this problems are the efficient points for the problem

$$\min_{x \in \mathbb{R}^n} \quad (f_1(x), \ldots, f_{m-1}(x))$$

that we suppose that form a surface constituted by arcs.

If $\lambda_m \neq 0$, fixing $\lambda_1, \ldots \lambda_{m-1}$, let be $f_{(\lambda_1,\ldots,\lambda_{m-1})}(x) = \lambda_1 f_1(x) + \ldots + \lambda_{m-1} f_{m-1}(x)$. The problems that appear are:

$$\min_{x \in \mathbb{R}^n} \quad (f_{(\lambda_1,\ldots,\lambda_{m-1})}(x) + \lambda_m f_m(x))$$

for $\lambda_m \geq 0$.

Varying λ_m in the previous problems we obtain the efficient set of the problem

$$\begin{array}{ll} \min & (f_{(\lambda'_1,\ldots,\lambda'_{m-1})}(x), f_m(x)) \\ x \in \mathbb{R}^n & \end{array}$$

except the point a_m. As the two objetive functions of this problems are quadratic, the set of efficient points form an arc that joint a_m with the point that minimize $f_{(\lambda_1,\ldots,\lambda_{m-1})}(x)$ wich is a point of the efficient set of the problem

$$\begin{array}{ll} \min & \lambda_1 f_1(x) + \ldots + \lambda_{m-1} f_{m-1}(x)) \\ x \in \mathbb{R}^n & \end{array}$$

Therefore we have arrived to a description of the efficient set in relation with the efficient points of certain multiobjective problem with a function less. We show this in the following result:

LEMMA 4.1 *The set of efficient points of the multiobjective quadratic problem with positive definite matrices are connected by finite arcs.*

References

[1] Beato-Moreno, A. *Eficiencia en Programación Cuadrática Multiobjetivo.* Ph. D. Thesis (Unpublished), 1995.

[2] Chankong, V. and Haimes, Y. Y. *Multiobjective Decision Making: Theory and Methodology.* Elsevier Science Publishing Co. Inc., North-Holland, 1983.

[3] Gantmacher, F.R. *The Theory of Matrices.* Vol I and II. Chelsea Publishing Company, New York, N.Y., 1959.

[4] Ruiz-Canales, P. - Rufián Lizana,A. *A Characterization of Weakly Efficient Points.* Mathematical Programming. **Vol 68** (1995), N. 2, pp. 205-212.

[5] Warburton, A.R. *Quasiconcave Vector Maximization: Connectedness of the Set Pareto-Optimal and Weak Pareto-Optimal Alternatives.* Journal of Optimization Theory and Applicattions. **Vol. 40** (1983), pp. 537-557.

Multiobjective Programming with a Concave Vector Value Function

Alfonso Mateos and Sixto Rios-Insua

Department of Artificial Intelligence, Technical University of Madrid
Campus de Montegancedo, Boadilla del Monte, 28660 Madrid, Spain

Abstract. We consider the multiobjective optimization problem under certainty with partial information on the decision maker's preferences structure by means of a vector value function with two components, which represents imprecision over his preferences. We introduce an approximation to the value efficient set which uses a vector value function with concave components on a convex feasible region. We propose an interactive algorithm to reduce the value efficient set and its approximation, based on information over the preferences revealed by the decision maker. Finally, a graphical example shows the method.

Keywords. Multiobjective programming, efficient set, vector value function

1 Introduction

Let us consider the multiobjective optimization problem of the form

$$\begin{array}{ll} \max & \mathbf{z}(x) = (z_1(x), ..., z_p(x)) \\ \text{s.t.} & x \in X \end{array} \tag{1}$$

where $z_1(\cdot), ..., z_p(\cdot)$ are p individual objective functions, X is the decision space and $Z = \mathbf{z}(X)$ is the feasible region in objective space. Assume partial information on the decision maker's (DM) preferences by a vector value function (Roberts, 1979) with two components, $\mathbf{v}(\cdot) = (v_1(\cdot), v_2(\cdot)) : Z \to \mathbb{R}^2$, which may be considered as a proxy for the underlying (scalar) value function $v(\cdot)$ and represents imprecision about his preferences. From a practical point of view, this vector function may be easier to assess, because each individual function $v_i(\cdot)$ will probably depend on a smaller set of objectives z_j than the global value function $v(\cdot)$.

Suppose further that the DM's preferences satisfy the conditions to agree with a value function as a linear function of the components in $\mathbf{v}(\cdot)$ (Chankong and Haimes, 1983), i.e., the DM's value function is $v^{\mathbf{k}}(\cdot) = k_1 v_1(\cdot) + k_2 v_2(\cdot)$, and there is uncertainty about the weighting factors $k_1, k_2 \in \mathbb{R}_+$. In this context, it is known that the isovalue or indifference curves are constant. Linear value functions are commonly assumed in many areas of economics, commerce, operations research,...

In the above framework, we can transform problem (1) into the vector optimization problem

$$\begin{array}{ll} \max & \mathbf{v}(\mathbf{z}) \\ \text{s.t.} & \mathbf{z} \in Z \end{array} \qquad (2)$$

and it arises the value efficient set

$$\mathcal{E}(Z, \mathbf{v}) = \{\mathbf{z} \in Z : \nexists\, \mathbf{z}' \in Z \text{ such that } \mathbf{v}(\mathbf{z}') \geq \mathbf{v}(\mathbf{z})\}$$

where the DM can confine his attention discarding all other solutions, because a nonefficient solution can never be optimal. The symbol $\geq$ represents the vector inequality, defined as

$$\mathbf{v}(\mathbf{z}') \geq \mathbf{v}(\mathbf{z}) \Leftrightarrow \begin{cases} v_i(\mathbf{z}') \geqq v_i(\mathbf{z}) & \text{for any } i = 1, 2, \text{ and} \\ v_j(\mathbf{z}') > v_j(\mathbf{z}) & \text{for at least one } j \in \{1, 2\} \end{cases}$$

while $\geqq$ is the inequality between real numbers.

From a mathematical point of view the resolution of problem (2) is analogous to the resolution of a biobjective problem, where we could consider $v_1(\cdot)$ and $v_2(\cdot)$ as individual objective functions and $V = \mathbf{v}(Z)$ would be the feasible region in the objective space.

If $v_1(\cdot)$ and $v_2(\cdot)$ were linear functions and Z a convex polyhedral set, there are several procedures to generate the whole or a part of the efficient set which are based upon the determination of efficient extreme points of the constraint polyhedron using a modified simplex algorithm, a weighted-sums objective function or a ϵ-constraint method (see, among others, Gal, 1972; Zeleny, 1974; Zionts and Wallenius, 1976, 1983; Steuer, 1986; Korhonen and Wallenius, 1990; Bryson, 1993; Armand, 1993).

In the more general case where Z is assumed to be a convex set, we may apply for two objectives the NISE method (Cohon, 1978; Cohon *et al.*, 1979) which provides a good approximation to the value efficient set, and it has been partially generalized by Balachandran and Gero (1985) in the case of three objectives. However, note that the generation of the whole efficient set is not usually the resolution of problem (2), because this set may have many points. Then, we propose an interactive method which generalizes the NISE method to concave functions. This new method uses information on preferences revealed by the DM to interactively reduce the value efficient set, where the most preferred solution will be included.

The rest of the paper is organized as follows. In the next section, we consider the multiobjective decision making problem where the two components of the vector value function are concave functions and the feasible region in the objective space is convex. We provide an approximation set which contains the value efficient set. In the third section, we generalize the above approximation by obtaining a general approximation which does not consider any assumption over the vector value function and the feasible region. Section four presents a practical way to obtain the DM's information on preferences and an interactive algorithm, which is considered in section five. A graphical example shows the interactive method in section six and, finally, some conclusions are provided.

2 An approximation to the value efficient set

We consider the multiobjective decision making problem (2) assuming that $v_1(\cdot)$ and $v_2(\cdot)$ are concave functions and Z is convex. These assumptions are not very restrictives because in many real problems the isovalue curves are convex. A way to obtain such property comes from (French, 1986):

Proposition 1 *If $v_1(\cdot)$ and $v_2(\cdot)$ are concave functions and $v^{\mathbf{k}}(\cdot) = k_1 v_1(\cdot) + k_2 v_2(\cdot)$ is the DM's value function, with $k_1, k_2 \in \mathbb{R}_+$, then the isovalue curves are convex.*

Now, we shall introduce a concept that will be the tool to characterize the approximation set to the value efficient set under the above conditions.

Definition 1 *Let $\mathbf{v}(\cdot) = (v_1(\cdot), v_2(\cdot))$ be a vector value function and $\mathbf{z}^{1*}$, $\mathbf{z}^{2*}$ the solutions where functions $v_1(\cdot)$ and $v_2(\cdot)$ achieve their optima, respectively. We define the bounded function associated to $\mathbf{v}(\cdot)$, as the straight line $r(\cdot)$ which passes through $\mathbf{v}(\mathbf{z}^{1*})$ and $\mathbf{v}(\mathbf{z}^{2*})$, i.e., $r(\mathbf{z}) \equiv v_2(\mathbf{z}) - m v_1(\mathbf{z}) - n$ with $\mathbf{z} \in Z$, $m = \frac{v_2(\mathbf{z}^{2*}) - v_2(\mathbf{z}^{1*})}{v_1(\mathbf{z}^{2*}) - v_1(\mathbf{z}^{1*})}$ and $n = \frac{v_1(\mathbf{z}^{2*}) v_2(\mathbf{z}^{1*}) - v_2(\mathbf{z}^{2*}) v_1(\mathbf{z}^{1*})}{v_1(\mathbf{z}^{2*}) - v_1(\mathbf{z}^{1*})}$.*

In case that the optima were not unique for each component, we should take those which provide the highest value to the other component in $\mathbf{v}(\cdot)$. Under the above assumptions, we introduce

Definition 2 *Given a vector value function $\mathbf{v}(\cdot) = (v_1(\cdot), v_2(\cdot))$, we define the approximation set to the value efficient set as $A(Z, \mathbf{v}) = \{\mathbf{z} \in Z : r(\mathbf{z}) \geqq 0\}$.*

From the above definitions, we can rewrite the approximation set

$$A(Z, \mathbf{v}) = \{\mathbf{z} \in Z : r(\mathbf{z}) \geqq r(\mathbf{z}^{1*}) = r(\mathbf{z}^{2*})\}$$

since $r(\mathbf{z}^{1*}) = r(\mathbf{z}^{2*}) = 0$. Moreover, $A(Z, \mathbf{v})$ contains the solutions of interest for the DM as we see in the next result.

Theorem 1 *Given a vector value function $\mathbf{v}(\cdot) = (v_1(\cdot), v_2(\cdot))$ with concave components and the convex feasible region $Z \subset \mathbb{R}^p$, then $\mathcal{E}(Z, \mathbf{v}) \subseteq A(Z, \mathbf{v})$.*

Proof. Assume that $\mathbf{z}^0 \in \mathcal{E}(Z, \mathbf{v})$ but $\mathbf{z}^0 \notin A(Z, \mathbf{v})$. First, we shall prove that whenever α fulfills

$$\frac{v_1(\mathbf{z}^0) - v_1(\mathbf{z}^{2*})}{v_1(\mathbf{z}^{1*}) - v_1(\mathbf{z}^{2*})} < \alpha \leqq \frac{v_2(\mathbf{z}^{2*}) - v_2(\mathbf{z}^0)}{v_2(\mathbf{z}^{2*}) - v_2(\mathbf{z}^{1*})} \tag{3}$$

solution $\alpha\mathbf{z}^{1*} + (1-\alpha)\mathbf{z}^{2*} \in Z$ dominates to $\mathbf{z}^0$. Due to the strict inequality in (3)

$$\begin{array}{c} v_1\left(z^0\right) < \alpha\left[v_1\left(\mathbf{z}^{1*}\right) - v_1\left(\mathbf{z}^{2*}\right)\right] + v_1\left(\mathbf{z}^{2*}\right) = \\ \alpha v_1\left(\mathbf{z}^{1*}\right) + (1-\alpha)\, v_1\left(\mathbf{z}^{2*}\right) \leqq v_1\left(\alpha\mathbf{z}^{1*} + (1-\alpha)\,\mathbf{z}^{2*}\right) \end{array} \tag{4}$$

and from the inequality in (3)

$$\begin{array}{c} v_2\left(z^0\right) \leqq \alpha\left[v_2\left(\mathbf{z}^{1*}\right) - v_2\left(\mathbf{z}^{2*}\right)\right] + v_2\left(\mathbf{z}^{2*}\right) = \\ \alpha v_2\left(\mathbf{z}^{1*}\right) + (1-\alpha)\, v_2\left(\mathbf{z}^{2*}\right) \leqq v_2\left(\alpha\mathbf{z}^{1*} + (1-\alpha)\,\mathbf{z}^{2*}\right) \end{array} \tag{5}$$

Thus, from (4) and (5), it is $\mathbf{v}\left(z^0\right) \leq \mathbf{v}\left(\alpha\mathbf{z}^{1*} + (1-\alpha)\,\mathbf{z}^{2*}\right)$, i.e., $\mathbf{z}^0$ is dominated by $\alpha\mathbf{z}^{1*} + (1-\alpha)\,\mathbf{z}^{2*}$, but this is a contradiction because $\mathbf{z}^0 \in \mathcal{E}(Z, \mathbf{v})$.

To complete the proof, let us show that it is possible to choose an α that fulfills (3). To do this, we only have to show that

$$\frac{v_1(\mathbf{z}^0) - v_1(\mathbf{z}^{2*})}{v_1(\mathbf{z}^{1*}) - v_1(\mathbf{z}^{2*})} < \frac{v_2(\mathbf{z}^{2*}) - v_2(\mathbf{z}^0)}{v_2(\mathbf{z}^{2*}) - v_2(\mathbf{z}^{1*})}.$$

In fact, because $\mathbf{z}^0 \notin A(Z, \mathbf{v})$, then

$$r(\mathbf{z}^0) < r(\mathbf{z}^{1*}) = r(\mathbf{z}^{2*}) = \alpha r\left(\mathbf{z}^{1*}\right) + (1-\alpha) r(\mathbf{z}^{2*}),$$

i.e.,

$$\begin{array}{c} v_2(\mathbf{z}^0) - m v_1(\mathbf{z}^0) - n < \\ < \alpha[v_2(\mathbf{z}^{1*}) - m v_1(\mathbf{z}^{1*}) - n] + (1-\alpha)[v_2(\mathbf{z}^{2*}) - m v_1(\mathbf{z}^{2*}) - n]. \end{array}$$

Hence

$$\begin{array}{c} v_2(\mathbf{z}^0) - m[v_1(\mathbf{z}^0) - v_1(\mathbf{z}^{2*})] < \\ < \alpha[v_2(\mathbf{z}^{1*}) - v_2(\mathbf{z}^{2*})] - \alpha[v_2(\mathbf{z}^{1*}) - v_2(\mathbf{z}^{2*})] + v_2(\mathbf{z}^{2*}) \end{array}$$

which implies that

$$-m[v_1(\mathbf{z}^0) - v_1(\mathbf{z}^{2*})] < v_2(\mathbf{z}^{2*}) - v_2(\mathbf{z}^0)$$

and this completes the proof. □

Note that this theorem is also true if the components of the vector value function were linear. Furthermore, to prove theorem 1, we have only used that $\alpha\mathbf{z}^{1*} + (1-\alpha)\,\mathbf{z}^{2*} \in Z$ if α holds (3). Then, we have

Corollary 1 *Given a vector value function* $\mathbf{v}(\cdot) = (v_1(\cdot), v_2(\cdot))$ *with concave components and the feasible region* $Z \subset I\!R^p$ *with* $\alpha\mathbf{z}^{1*} + (1-\alpha)\,\mathbf{z}^{2*} \in Z$ *for all* $\alpha \in [0,1]$, *then* $\mathcal{E}(Z, \mathbf{v}) \subseteq A(Z, \mathbf{v})$.

We observe that this corollary is a partial generalization of theorem 1. Moreover, let $\mathbf{v}(\mathbf{z}) = \mathbf{z}$ and Z as shown in figure 1. Hence, Z is nonconvex and nonpolyhedral but $\alpha \mathbf{z}^{1*} + (1-\alpha)\mathbf{z}^{2*} \in Z$ for all $\alpha \in [0, 1,]$. So, we cannot apply theorem 1 but corollary 1 is true and thus, we can construct the approximation set $A = A(Z, \mathbf{v})$ to the value efficient set $E = \mathcal{E}(Z, \mathbf{v})$ (see figure 1).

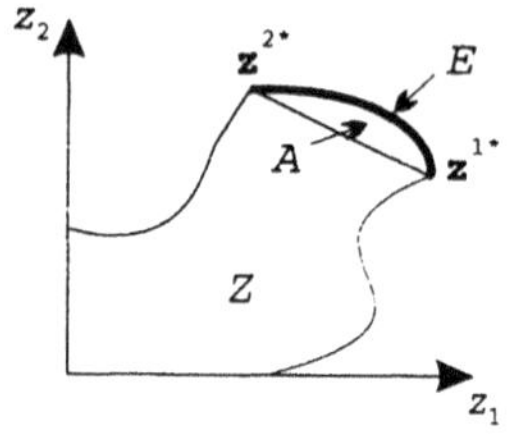

Figure 1. The value efficient set E and the approximation set A in a nonpolyhedral and nonconvex feasible region Z.

3 A generalization of the approximation

In this section, we present an extension to the above approximation for the general case (Mateos and Rios-Insua, 1997), i.e., no particular assumption is introduced for the vector value function $\mathbf{v}(\cdot) = (v_1(\cdot), v_2(\cdot))$ and for the feasible region Z. We shall consider such approximation only with the object of comparison to the former one and we shall not study it in detail.

Let n_1 and n_2 be the components of the nadir point, defined as $n_1 = v_1(\mathbf{z}^{2*})$ and $n_2 = v_2(\mathbf{z}^{1*})$. As before, in the case that the optima $\mathbf{z}^{1*}$ and $\mathbf{z}^{2*}$ were not unique, we shall take those which provide the highest value to the other component in $\mathbf{v}(\cdot)$.

Definition 3 *For a vector value function* $\mathbf{v}(\cdot) = (v_1(\cdot), v_2(\cdot))$, *we define the general approximation set to the value efficient set as*

$$A'(Z, \mathbf{v}) = \{\mathbf{z} \in Z : \ v_1(\mathbf{z}) \geqq n_1, \ v_2(\mathbf{z}) \geqq n_2\}$$

Again, this set contains the value efficient set.

Theorem 2 *Given a vector value function* $\mathbf{v}(\cdot) = (v_1(\cdot), v_2(\cdot))$, *then* $\mathcal{E}(Z, \mathbf{v}) \subseteq A'(Z, \mathbf{v})$.

We can see that $A(Z, \mathbf{v}) \subseteq A'(Z, \mathbf{v})$. Thus, $A = A(Z, \mathbf{v})$ provides a better approximation than $A' = A'(Z, \mathbf{v})$, however, the first one requires stronger assumptions. A graphical comparison between both approximations for a nonpolyhedric convex set Z and a linear value function $\mathbf{v}(\mathbf{z}) = \mathbf{z}$ is shown in figure 2.

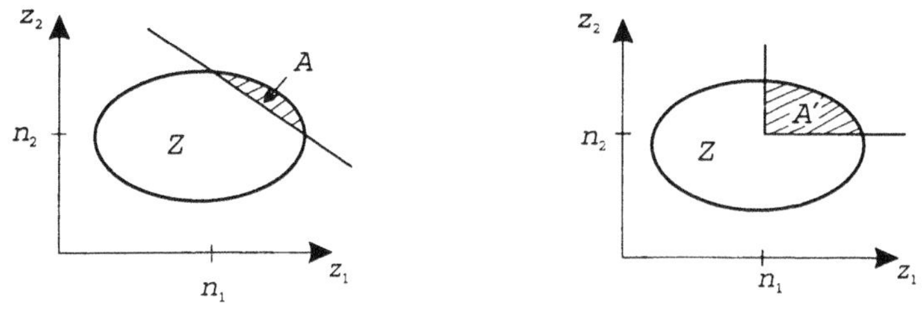

Figure 2. A graphical comparison between
the approximation sets A and A'.

4 Assessment of information on preferences

We consider a procedure to obtain information over the DM's preferences and to represent them in terms of constraints. To obtain such information, the DM must consider the comparison of pairs of solutions in V which are transformed from value efficient solutions in Z. We know that the value function is of the form $v^{\mathbf{k}}(\cdot) = k_1 v_1(\cdot) + k_2 v_2(\cdot)$, and there is uncertainty about $\mathbf{k} = (k_1, k_2)$, which firstly verifies $\mathbf{k} \in K_*^0$, where $K_*^0 = \mathbb{R}_+^2 \cap S_2$ (S_2 is the simplex on $\mathbb{R}^2$) is called *null information set* about the weighting factors, with generators $(1,0)$ and $(0,1)$. We shall denote such set by $K_*^0 = G\{(1,0),(0,1)\}$. However, suppose that the DM reveals that the solution $\mathbf{v}^{1*} = (v_1^{1*}, v_2^{1*}) = (v_1(\mathbf{z}^{1*}), v_2(\mathbf{z}^{1*}))$ is at least as preferred as the solution $\mathbf{v}^{2*} = (v_1^{2*}, v_2^{2*}) = (v_1(\mathbf{z}^{2*}), v_2(\mathbf{z}^{2*}))$, denoted $\mathbf{v}^{1*} \succeq \mathbf{v}^{2*}$. Hence, vector $\mathbf{k}$ must verify $\mathbf{k}\mathbf{v}^{1*} \geqq \mathbf{k}\mathbf{v}^{2*}$. Then, $\mathbf{k} \in K_*^1 = G\{(1,0),(a,b)\}$ where

$$a = \frac{v_2^{2*} - v_2^{1*}}{v_1^{1*} + v_2^{2*} - v_1^{2*} - v_2^{1*}} \quad \text{and} \quad b = \frac{v_1^{1*} - v_1^{2*}}{v_1^{1*} + v_2^{2*} - v_1^{2*} - v_2^{1*}}.$$

Otherwise, if $\mathbf{v}^{2*} \succeq \mathbf{v}^{1*}$, then $\mathbf{k} \in K_*^1 = G\{(a,b),(0,1)\}$. Under the assumption that $\mathbf{v}^{1*}$ is indifferent to $\mathbf{v}^{2*}$, we should have a scalar value function, because $k_1 = a$ and $k_2 = b$, and our decision problem becomes a scalar optimization problem. Thus, the new information lead us to consider a more reduced information set K_*^1. This idea has been considered in Zionts and Wallenius (1983) and Malakooti (1985), among others.

Associated to the new information set, we have a more precise vector value function defined from the generators of such information set.

Definition 4 *Given a vector value function $\mathbf{v}(\cdot) = (v_1(\cdot), v_2(\cdot))$ and a new information set $K_* = G\{\mathbf{k}^1, \mathbf{k}^2\}$, the vector value function associated with K_* is defined as $\mathbf{v}^K(\cdot) = (\mathbf{k}^1\mathbf{v}(\cdot), \mathbf{k}^2\mathbf{v}(\cdot))$, where each component is the scalar product.*

Observe that, for the null information set, we have $\mathbf{v}^{K^0}(\cdot) = \mathbf{v}(\cdot)$. If the current information about the preference structure is represented by $\mathbf{v}^K(\cdot)$, then, in analogous way as in theorem 1, we can prove

Theorem 3 *Given a vector value function* $\mathbf{v}(\cdot) = (v_1(\cdot), v_2(\cdot))$ *with concave components, a convex feasible region* $Z \subset I\!R^p$ *and an information set* K_*, *then* $\mathcal{E}(Z, \mathbf{v}^K) \subseteq A(Z, \mathbf{v}^K)$.

A similar result is valid for the general approximation.

Theorem 4 *Given a vector value function* $\mathbf{v}(\cdot) = (v_1(\cdot), v_2(\cdot))$ *and an information set* K_*, *then* $\mathcal{E}(Z, \mathbf{v}^K) \subseteq A'(Z, \mathbf{v}^K)$.

5 An interactive method

We present an interactive method which uses the above approximation to aid the DM to come up with his most preferred solution. Let us denote by $\mathbf{v}^{K^h}(\cdot) = (v_1^{K^h}(\cdot), v_2^{K^h}(\cdot))$ the vector value function associated to the information set K_*^h. The algorithm is:

Step 0. Let $h = 0$

Step 1. Calculate $\mathbf{z}^{1*}$ and $\mathbf{z}^{2*}$ such that $v_1^{K^h}(\mathbf{z}^{1*}) = \max\{v_1^{K^h}(\mathbf{z}) : \mathbf{z} \in Z\}$ and $v_2^{K^h}(\mathbf{z}^{2*}) = \max\left\{v_2^{K^h}(\mathbf{z}) : \mathbf{z} \in Z\right\}$. Denote $\mathbf{v}^{1*} = (v_1^{1*}, v_2^{1*}) = \mathbf{v}^{K^h}(\mathbf{z}^{1*})$ and $\mathbf{v}^{2*} = (v_1^{2*}, v_2^{2*}) = \mathbf{v}^{K^h}(\mathbf{z}^{2*})$.

Step 2. If $\mathbf{v}^{1*} = \mathbf{v}^{2*}$, stop, and the most preferred solution has been obtained. Otherwise, go to step 3.

Step 3. $\mathbf{v}^{1*}$ and $\mathbf{v}^{2*}$ are presented to the DM.

a) If $\mathbf{v}^{1*} \succeq \mathbf{v}^{2*}$, consider $K_*^{h+1} = G\{(1,0),(a,b)\}$, let $h = h+1$ and go to step 1.

b) If $\mathbf{v}^{2*} \succeq \mathbf{v}^{1*}$, consider $K_*^{h+1} = G\{(a,b),(0,1)\}$, let $h = h+1$ and go to step 1.

c) If $\mathbf{v}^{1*} \sim \mathbf{v}^{2*}$, then the DM's scalar value function is $v^{\mathbf{k}}(\cdot) = av_1(\cdot) + bv_2(\cdot)$. Calculate the optimal solutions and stop.

d) If the DM is unable to compare $\mathbf{v}^{1*}$ and $\mathbf{v}^{2*}$, $A(Z, \mathbf{v}^{K^h})$ is presented to him.

We note that the first step of the algorithm calculates the optima of each component of the vector value function associated to the current information set from which we determine the approximation set. The second step, verifies if it is feasible the ideal solution. The last one is an interactive step which provides, if possible, the new information set to start again the process.

6 Example

We present a graphical example for a linear problem to illustrate the steps of the method. Let us consider a biobjective problem with vector value function $\mathbf{v}(\mathbf{z}) = \mathbf{z}$ where $\mathbf{z} = (z_1, z_2)$. The feasible region Z is shown in figure 3, as well as the information set K_*^0 and the value efficient set $\mathcal{E}(Z, \mathbf{v}^{K^0})$ (dark line).

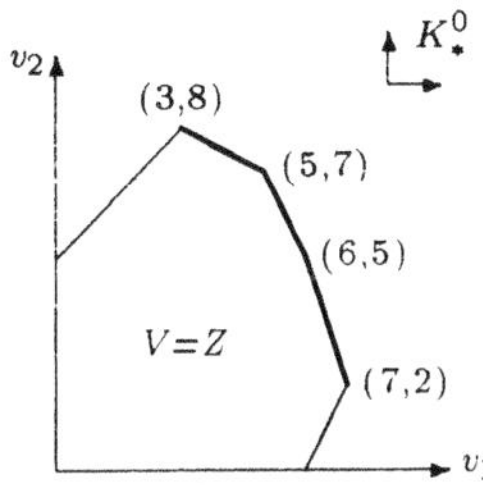

Figure 3

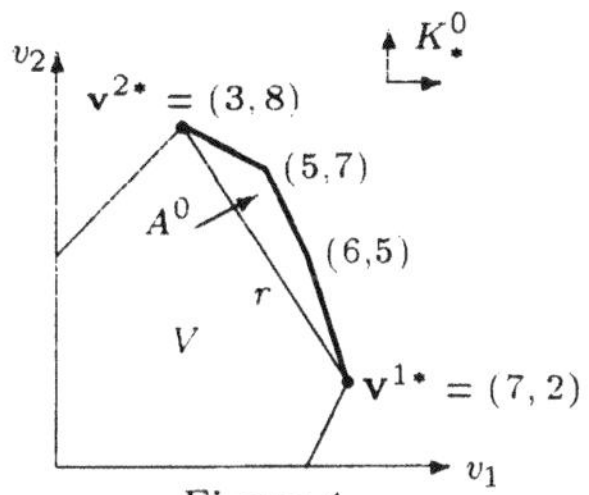

Figure 4

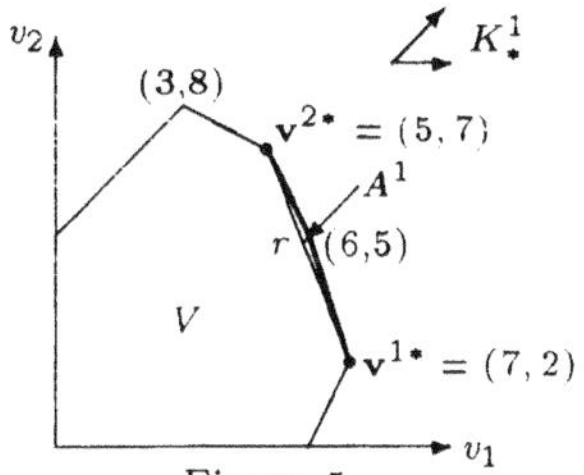

Figure 5

Figure 4 shows the approximation set $A^0 = A(Z, \mathbf{v}^{K^0})$ and the value efficient solutions $\mathbf{v}^{1*}$ and $\mathbf{v}^{2*}$. Assume the DM reveals always that $\mathbf{v}^{1*}$ is at least as preferred as $\mathbf{v}^{2*}$. Hence, figure 5 shows the new information set K_*^1, and then, the approximation set $A^1 = A(Z, \mathbf{v}^{K^1})$. Figures 6 and 7, show two more iterations and the most preferred solution given by $\mathbf{v}^{1*} = (7, 2)$.

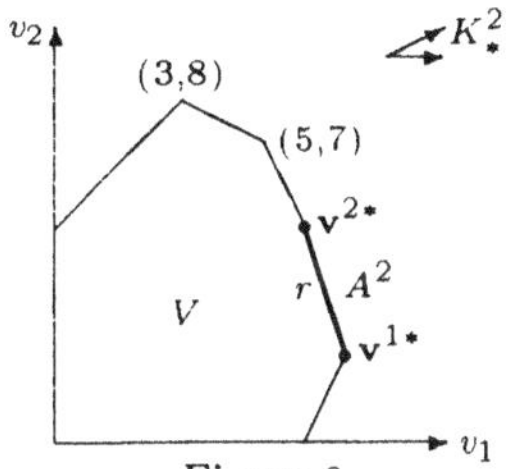

Figure 6

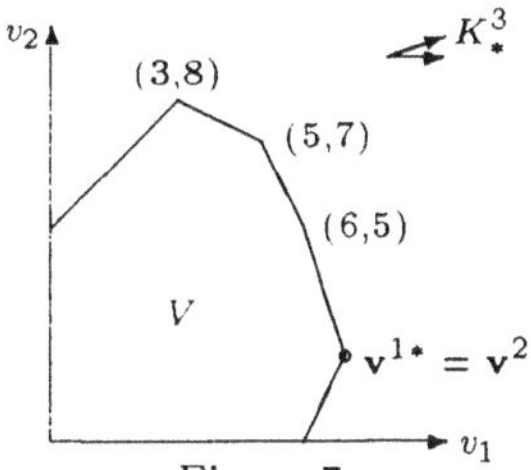

Figure 7

7 Conclusions

The generation of the efficient set in multiobjective decision making may not be considered as the resolution of the problem, because this set may have many points. It seems reasonable to look for procedures to aid a DM to reduce such set to come up with a unique solution. In this paper we have considered an interactive method based on an approximation set to the value efficient set, which intends to aid the DM. We assume partial information on the DM's preferences by a vector value function with two concave components and a convex feasible region in the objective space. This lead us to work with only some value efficient solutions and a good approximation in a small number of iterations. The method relaxes the linearity condition of the objective functions by allowing concave functions and thus, it intends to be an advance or a complement to some other methods existing in the literature. We have also suggested a more general approximation but it is only consider here with the object of comparison and must be more investigated.

Acknowledgments. This paper has been supported by the CICYT Project TIC95-0028. Authors are grateful to an anonymous referee for his valuable suggestions.

References

Armand, P. (1993), "Finding all Maximal Efficient Faces in Multiobjective Linear Programming", *Math. Prog.*, **61**, 357-375.

Balachandran, M. and J.S. Gero (1985), "The Noninferior Set Estimation (NISE) Method for Three Objective Problems", *Eng. Opt.*, **9**, 77-88.

Bryson, N. (1993), "Identifying the Efficient Extreme-Points of the Three-Objective Linear Programming Problem", *J. Opl. Res. Soc.* **44**, 1, 81-85.

Chankong, V. and Y.Y. Haimes (1983) *Multiobjective Decision Making: Theory and Methodology*, North Holland.

Cohon, J.L. (1978), *Multiobjective Programming and Planning*, Academic Press, New York.

Cohon, J.L., R.L. Church and D.P. Sheer (1979), "Generating Multiobjective Trade-offs: An Algorithm for Bicriterion Problems", *Water Resources Res.*, **15**, 1001-1010.

French, S. (1986), *Decision Theory: An Introduction to the Mathematics of Rationality*, Ellis Horwood, Chichester.

Gal. T. (1972), "A General Method for Determining the Set of all Efficient Solutions of a Linear Vector Maximum Problem", *Eur. J. Opl. Res.*, **1**, 307-322.

Korhonen, P. and J. Wallenius (1990), "A Multiple Objective Linear Programming Decision Support System", *Decision Support Systems* **6**, 243-251.

Malakooti, B. (1985). "Assessment through strength of preference", *Large Scale Systems*, **8**, 169-182.

Mateos, M. and S. Rios-Insua (1997), "An Approximation for the Value Efficient Set", in *Twenty Years of East-West Cooperation in MCDM*, T. Gal and T. Hanne (eds.), LNMES, Springer, Berlin. (to appear).

Roberts, F.S. (1979). *Measurement Theory*, Addison-Wesley, Reading, Mass.

Steuer, R. (1986), *Multiple Criteria Optimization: Theory, Computation, and Application*, Wiley.

Zeleny, M. (1974), *Linear Multiobjective Programming*, LNEMS 95, Springer.

Zionts, S. and J. Wallenius (1976), "An Interactive Programming Method for Solving the Multiple Criteria Problem", *Management Science* **22**, 652-663.

Zionts, S. and J. Wallenius (1983), "An Interactive Multiple Objective Linear Programming Method for a Class of Underlying Nonlinear Utility Functions", *Management Science* **29**, 519-529.

Modelling Best-Practice Frontiers When There Are Multiple Outputs

Dr. Chris Tofallis
Department of Economics, Statistics and Decision Sciences
The Business School
University of Hertfordshire
Mangrove Road
Hertford, Herts
SG13 8QF
U.K.

e-mail: c.tofallis@herts.ac.uk

Abstract. This paper presents a way of modelling relationships between *multiple dependent* and multiple independent variables. The method involves fitting coefficients to functions of these two sets of variables such that the resulting 'aggregate' functions have maximum correlation. The inclusion of constraints in the fitting procedure is also possible in order to satisfy any conditions felt to be appropriate, possibly for theoretical reasons. This approach may be described as *maximum correlation modelling* or *constrained canonical correlation analysis.*

The particular application here is to the construction of best-practice benchmark models of organisational units or branches carrying out similar activities which transform a set of inputs or resources into a set of outputs. The first stage involves identifying the non-dominated set (the efficient units) using data envelopment analysis; dominated units are then discarded from the subsequent stages. We then show how the best-practice model can be generated using a spreadsheet package which contains a constrained optimization facility. The technique is illustrated using data from university chemistry departments.

Keywords: data envelopment analysis, canonical correlation, modelling.

1 BACKGROUND

A production function gives the maximum possible output for given levels of the inputs, hence it implies a frontier with points lying on or below it, but not above it. Observations lying below the frontier are those which are technically inefficient. The efficiency of points below the frontier is measured by how close to the frontier they lie - various metrics exist for doing this. Frontiers also arise in connection with cost functions and profit functions. A cost function gives the minimum cost at which any given output level can be produced for given input prices. The profit function gives the maximum attainable profit for a given set of input and output prices. In all cases a point which lies away from the frontier is said to be 'dominated' (by those on the frontier - which are said to be 'non-dominated'). We shall be concerned with modelling these frontiers when presented with a set of observations (points).

In econometrics the application of least squares regression to such problems dates back to Aigner and Chu (1968). They ensured that no point appeared in front of the frontier by constraining all the residuals to have the same sign. This generated what are now referred to as 'deterministic frontiers'. The disadvantage with this approach is that the frontier is estimated by means of a regression on *all* the data points, both efficient and inefficient. As a result a cluster of inefficient data points can have a high leverage in pulling the frontier towards them, particularly as being inefficient they may have high residuals, something which the least squares method exaggerates by squaring. What is required is that the efficient frontier be estimated by reference to the efficient observations alone. (We shall be ensuring this occurs in the method that we shall present.)

In econometrics 'stochastic frontiers' are more popular than deterministic ones, although they too suffer from the same difficulty mentioned above. Stochastic frontiers permit points to lie beyond the frontier (possibly justified as being due to errors in the data). In an attempt to separate what is due to error from what is real it is assumed that the residuals are composed of a random error and a one-sided inefficiency term with some assumption being made about the distribution of points relative to the frontier in order that the two can be separated e.g. one might assume that inefficiency follows a half-normal or negative exponential distribution. We feel that such assumptions are difficult to justify. (A comprehensive survey paper for this subject is Knox Lovell, 1993.)

This paper will restrict itself to the simpler deterministic case and no distributional assumptions will need to be made. We shall in each instance have to specify the general functional form to be fitted to the frontier and so our method is parametric in character (we shall employ an optimization to estimate values for the parameters in the functional form). A non-parametric approach known as data envelopment analysis (DEA) uses straight line segments (or plane facets in higher dimensions) to connect adjacent points on the frontier, but does not generate a mathematical relationship to represent the frontier. (See Boussofiane et al 1991 for a review of DEA.) We shall employ DEA for the purpose of identifying the frontier points (the non-dominated set), all remaining observations (inefficient points) will be removed from subsequent stages of the analysis in which we shall attempt to model the frontier with a functional form.

Regression is a natural approach to use when modelling the relationship between a single dependent variable such as profit or cost and multiple independent variables. This paper seeks to go further by presenting a technique which is able to include the case of multiple *dependent* variables, this will allow us to construct production functions for multiple outputs and multiple inputs. Of course such a technique has a much wider scope for application in the area of mathematical model-building but we will restrict ourselves to this application in this paper.

We now review some previous pieces of work which individually contain some of the ideas we shall bring together here. In a stimulating paper bursting with new models Sengupta (1989) proposes the incorporation of a measure of correlation into the DEA model. To do this in the single output case he suggests literally adding in the correlation measure as another term in the DEA objective function. In attempting to deal with the case of multiple outputs he uses a minimax objective function (Chebyshev criterion) with the correlation added in. This implies a *common* set of weights (also called input and output multipliers) for all the organisational units being compared and so we lose the ability to distinguish frontier units from the dominated ones, which is a key strength of DEA. A simple DEA model can be described as attaching *individual* weights to inputs and outputs *for each unit* being compared, chosen so as to maximize the unit's efficiency score (sum of weighted outputs divided by sum of weighted inputs) subject to no score exceeding 100% when those weights are applied to all the other units. This involves solving an LP (linear programme) for each unit. However at the end of his paper, in what almost seems like an afterthought, Sengupta suggests a two-stage procedure: first use DEA to determine the set of efficient units and then 'develop a set of canonical variables by maximising the correlation' for the efficient set. We shall include these important ideas in the work presented here.

Canonical correlation analysis is a multivariate technique which, given two sets of variables (in the present instance the inputs and outputs), finds a linear combination of each set which maximises the correlation between them. The linear combinations are called canonical variates. Friedman and Sinuany-Stern (1994) applied such an approach to the complete set of organizational units in order to obtain a common set of weights which were then used to calculate efficiency scores (the ratio of the canonical variates). They did not separate the dominated from the non-dominated units. Their purpose was to obtain a ranking of all the units calculated according to a common standard. They point out that this approach avoids the use of zero weights for some factors by some units - something which is permitted in standard DEA models. They then employed statistical tests to compare their ranking with the DEA classification of units as being efficient or inefficient. A difficulty which they do not address is that there is nothing to prevent negative weights in canonical correlation analysis. This author has found these to occur with some data sets. The usual way of finding the canonical variates involves solving an eigenvalue problem (Manly, 1986) and so it is not possible to simply add non-negativity and other conditions as in a mathematical programming problem. This drawback (inability to impose necessary conditions on the parameters to be estimated) and the consequent difficulty in interpreting the results may partly explain the fact that the technique is one of the lesser used multivariate tools despite the fact that it was introduced as long ago as 1935 (by H.Hotelling).

Most of the work on production functions written by economists uses functional forms which have a single output (or its logarithm) as the dependent variable. If there are multiple outputs an obvious move is to aggregate them in some way. Following a method developed by Vinod (1968), Bardhan, Cooper and Kumbhakar (1994) used canonical correlation analysis for the purpose of finding aggregation weights for the outputs, it was then possible to regress the aggregate output on the inputs (or their logarithms) to fit a standard production function (the Cobb-Douglas) to the data. Unfortunately they found some negative canonical weights arising for their data set (which dealt with public secondary schools in Texas). Nevertheless they proceeded to carry out separate regressions for the DEA-efficient and DEA-inefficient schools and on comparing these felt that there was a significant difference in the production technology underlying the operation of these two sets of schools.

2 METHOD

Our aim is to find relationships which model the non-dominated set i.e. to find functional forms for the efficient frontier. Our strategy involves three stages:

(i) use DEA to identify which points are efficient, then remove all other points from the following stages. We shall use a DEA model which does not make any assumption regarding returns to scale (it is known as the Banker-Charnes-Cooper model after its originators).

(ii) construct linear combinations (X and Y) of the outputs (y_i) and of the inputs (x_j) respectively so as to maximize the correlation between the two, constraints are included to ensure non-negativity of the weights and to cover any other conditions which are deemed necessary (e.g. to incorporate any prior knowledge). This may be called *constrained* canonical correlation analysis (CCCA) and this appears to be the first time that this has ever been carried out. (We will show how this can easily be carried out using a well known spreadsheet package.)

If the value of the resulting correlation is felt to be sufficiently high, then proceed to stage three (one might also wish to carry out statistical tests of significance). If the correlation is too low then one may wish to enrich the model by including additional terms among the two sets of variables e.g. squared terms to deal with increasing or diminishing returns to scale, or cross-products to model interactions among the inputs (x_ix_j). Other transformations of the data may be appropriate, e.g. economists often take logs of the various inputs and outputs. CCCA can now be applied to the expanded and/or transformed set of variables.

One of the attractions of maximising the correlation is that the result is scale-invariant i.e. changing the units of measurement of one or more variables does not affect the correlation value that is achieved, nor will it affect the model generated e.g. if the values of a variable are multiplied by ten, then its coefficient will become one tenth of its original value.

(iii) A high correlation between the two combinations in the previous stage (i.e. between $X = \Sigma\, v_i\, f_i(x)$ and $Y = \Sigma\, u_j\, g_j(y)$) implies that a linear relationship will provide a good fit. Note that although our combinations may include non-linear functions of the original data, they are still linear in the weights (u,v). We now regress Y on X to find a model to represent the efficient points. Ordinary regression will not provide a true frontier since points will lie on either side of the line, however this may be appropriate if one wishes to allow for errors in the observations. A true frontier requires constrained regression to force all the residuals to have the same sign.

To perform CCCA (stage (ii)) version 5 of the Microsoft Excel spreadsheet was used. This has a facility called Solver which can carry out constrained optimization. One sets aside some cells in the spreadsheet to hold the weights and Solver is given these cell addresses as being the quantities to be varied to achieve the optimization. Elsewhere on the spreadsheet one multiplies the input and output values by the weights and calculates the linear combinations X and Y. As with all spreadsheets there is a built-in function for calculating correlations; choose a cell to hold this function: in Excel one types =Correl(X,Y). This cell is then used by Solver as the objective function to be maximised. Solver also needs to be given any constraints such as non-negativity conditions before it is allowed to proceed, the on-line help system shows you how to enter these.

3 EMPIRICAL APPLICATION

The (real) data deals with 52 British university chemistry departments (the 'units' to be compared) and has been extracted from Beasley 1990, it is publicly available on the Internet. We shall use two input measures: general expenditure (GEN, mostly salaries), and equipment expenditure (EQU) . Four output measures will be used: number of undergraduates (UG), number of taught postgraduates (PGT), number of research postgraduates (PGR), and research income (RSCH) attracted by the department. Beasley uses research income as a proxy measure for the quantity of research carried out by a department, in fact he also used it as an input measure. Using a particular variable both as an input and an output causes all units to appear efficient in DEA since they can place all their weight on this particular output/input 'pair' so that no other unit can exceed this ratio. He applies a whole series of subjective restrictions on the various factor weights and so avoids this difficulty. We prefer to avoid this degree of subjectivity. The present investigation also differs from Beasley (1990) in that we do not make use of four zero-one research rating variables each of which indicates the achievement of a particular research rating grade; we feel this would involve double counting for the research outcome.

It should be emphasised that we are merely using this data to help us illustrate the application of our method. Like Beasley we do not claim that this is an ideal set of variables for assessing the efficiency of the departments, nor that using data from a single year is adequate since, for instance, equipment expenditure may be lumpy - a major re-equipment occurring every few years with smaller amounts in between.

4 RESULTS

Of the 52 departments, 23 were found to be non-dominated when using the variables described above in a data envelopment analysis. The data for these DEA-efficient units appears in Table 1 where the monetary variables GEN, EQU, and RSCH are measured in thousands of pounds sterling.

Table 1 Data for the efficient departments.

University	INPUTS		OUTPUTS			
	GEN	EQU	RSCH	UG	PGT	PGR
1	446	21	183	62	0	37
2	670	53	288	137	0	43
3	2043	256	1050	253	27	118
4	686	46	436	137	18	27
5	1155	113	545	126	31	90
6	620	115	565	76	5	49
7	919	61	419	173	0	49
8	734	33	142	46	26	48
9	1760	742	1061	167	0	141
10	1238	67	490	157	4	60
11	1208	89	397	158	26	49
12	1758	196	1162	237	9	105
13	1409	122	527	263	0	94
14	3337	654	1780	707	0	211
15	1492	127	613	152	18	102
16	1620	420	1224	199	2	124
17	1324	144	565	189	13	104
18	947	146	724	236	7	54
19	370	32	102	58	1	10
20	360	73	122	89	0	7
21	849	32	258	158	3	53
22	619	21	136	73	0	13
23	2253	131	360	354	7	94

Next we seek to find a linear combination of the inputs and another of the outputs such that the correlation between the two is maximised subject to the condition that all the coefficients are positive. Note that we only do this for the efficient subset of the original data - we discard inefficient units. The resulting 'aggregate' or 'compound' functions were found to be:

$$X = GEN + 0.768\ EQU$$

(1)

$$Y = 0.02\ RSCH + UG + 1.58\ PGT + 4.97\ PGR$$

Multiplying all the inputs (or outputs) by a constant does not affect the correlation between the two functions. We have scaled the coefficients so that GEN and UG have unit coefficients. The correlation between the two functions X and Y came to 0.9735. This was felt to be sufficiently high that a more complicated model seemed unnecessary. So we proceed to regress the aggregate output on the aggregate input and find:

$$Y = 0.4353\ X - 9.04 \qquad (2)$$

The X-coefficient was statistically significant ($p < 0.001$) but the constant was not i.e. we cannot reject the hypothesis that it is zero. A negative constant here would imply a fixed 'start-up cost' of 9.04/0.4353 units of aggregate input before any output can occur. It also indicates that we do not have constant returns to scale since e.g. a doubling in X does not result in a doubling in Y according to (2).

Combining (1) and (2) provides our benchmark model for best-practice as:

$$0.02\ RSCH + UG + 1.58\ PGT + 4.97\ PGR = 0.435\ GEN + 0.334\ EQU - 9.04 \qquad \text{equation (3)}$$

with $r^2 = 0.95$. The coefficients on the left hand side of (3) may be interpreted as input consumption rates for efficient departments i.e. input consumed per unit output. It is reassuring to see that the three outputs measured in numbers of students (UG, PGT and PGR) have their coefficients ordered as we would expect: a research postgraduate would require more individual staff attention and equipment resources than a taught postgraduate, and the latter would consume more than an undergraduate because undergraduate classes would be larger and hence more economical. These results indicate that a taught postgraduate consumes about one and a half times the resources used by an undergraduate, and a research postgraduate five times as much. The coefficients on the right hand side can be interpreted as productivities (aggregate output per unit input). We see that a pound spent on staff (GEN) currently contributes 30% more to output than a pound spent on equipment (since 0.435/0.334 = 1.3).

5 CONCLUSION

The initial motivation for this research was to develop a method which permitted the construction of models which related the *multiple* outputs to the inputs in organisational units which were found to be efficient (non-dominated). Such models may be referred to as best-practice models in that they provide a benchmark for inefficient units to strive toward. Economists refer to them as production functions - though the economic literature has mainly concerned itself with the single output case. We have presented a technique which combines the strengths of the two general approaches: econometric production functions and data envelopment analysis. At the same time we have overcome some disadvantages of the former in that we have constructed frontiers which are not influenced by inefficient observations. We have also gone beyond DEA in that we have shown how a functional form (a single benchmark 'formula') can be obtained to relate all the variables for the efficient observations. In a future paper we will use such models to calculate efficiency scores for all the units being compared. Moreover it will be demonstrated that scores derived in this way have advantages over DEA efficiency scores in that they can be set up so as not to ignore slacks in particular variables. Under DEA an observation can have an efficiency score of 100% but still be inefficient as a result of slacks which the score does not take into account.

We wish to stress that the procedure presented here as constrained canonical correlation analysis potentially has much more widespread application. It is after all a generalised form of regression where multiple dependent variables are permitted. A more descriptive name for it might be '*maximum correlation modelling*'. Since it can be carried out on a spreadsheet there should be plenty of scope for immediate application.

Acknowledgement: I am grateful to R.I.Cooper of the University of Hertfordshire for introducing me to canonical correlation analysis.

6 REFERENCES

Aigner, DJ and SF Chu (1968). On Estimating the Industry Production Function. American Economic Review 58, 826-839.

Bardhan, IR, WW Cooper and SC Kumbhakar (1994). New Uses of DEA and Statistical Regressions for Efficiency Evaluation and Estimation. Submitted to Annals of Operations Research.

Beasley, JE (1990). Comparing University Departments. Omega: Int. J of Management Science 18, 171-183.

Boussofiane, A, RG Dyson and E Thanasssoulis (1991). Applied Data Envelopment Analysis. European Journal of Operational Research 52, 1-15.

Friedman, L and Z Sinuany-Stern (1994). Scaling Units via the Canonical Correlation Analysis and the DEA. Presented at EURO XIII conference, Glasgow.

Hotelling, H (1935). The Most Predictable Criterion. J. of Experimental Psychology 26, 139-142.

Knox Lovell, CA (1993). Production Frontiers and Productive Efficiency in 'The Measurement of Productive Efficeincy' ed. HO Fried, CA Knox Lovell and SS Schmidt, Oxford University Press.

Manly, BFJ (1986). Multivariate Statistical Methods: A Primer. Chapman and Hall, London.

Sengupta, JK (1989). Data envelopment with Maximum Correlation. Int. J. of Systems Sci. 20, 2085-2093.

Vinod, HD (1968). Econometrics of Joint Production. Econometrica 36, 322-336.

GPSR Compliance

The European Union's (EU) General Product Safety Regulation (GPSR) is a set of rules that requires consumer products to be safe and our obligations to ensure this.

If you have any concerns about our products, you can contact us on ProductSafety@springernature.com

In case Publisher is established outside the EU, the EU authorized representative is:

Springer Nature Customer Service Center GmbH
Europaplatz 3
69115 Heidelberg, Germany

Batch number: 10844669

Printed by Printforce, the Netherlands